EUCLIDEAN AND NON-EUCLIDEAN GEOMETRIES

EUCLIDEAN AND NON-EUCLIDEAN GEOMETRIES

Development and History

Third Edition

Marvin Jay Greenberg
University of California, Santa Cruz

W. H. Freeman and Company
New York

Cover image: This tessellation of the hyperbolic plane by alternately colored 30° –45° – 90° triangles, which appeared in a paper by H. S. M. Coxeter (Figure 7 in "Crystal Symmetry and Its Generalizations," *Transactions of the Royal Society of Canada* 3 ((51), 1957, pp. 1–11), inspired the Dutch artist M. C. Escher to create his (hyperbolic) "Circle Limit" patterns. In fact, by combining pairs of triangles that share long sides, one obtains a different tessellation by 45° –45° –90° triangles, which can also be alternately colored. It is easy to modify this latter tessellation to obtain *Circle Limit IV* (which appears on the front cover). © 1960 M. C. Escher Foundation, Baarn, Holland. All rights reserved.

Back cover image: This tessellation was generated by Douglas Dunham (Department of Computer Science, University of Minnesota, Duluth, MN 55812-2496).

Library of Congress Cataloging-in-Publication Data

Greenberg, Marvin J.
    Euclidean and non-Euclidean geometries : development and history /
Marvin Jay Greenberg. —3rd ed.
        p.   cm.
    Includes bibliographical references and indexes.
    ISBN-13: 978-0-7167-2446-9
    ISBN-10: 0-7167-2446-4
    1. Geometry. 2. Geometry, Non-Euclidean. 3. Geometry—
History. 4. Geometry, Non-Euclidean — History. I. Title.
QA445.G84 1994
516—dc20   93-7207
                                                                              CIP

Printed in the United States of America

Twelfth printing

W. H. Freeman and Company
41 Madison Avenue
New York, NY 10010
www.whfreeman.com

To Moses Ma, President of Velocity Development, without whose technical and emotional support this third edition could not have been written.

And to the memory of Stanford Professor Karel De Leeuw, inspiring teacher and outrageous friend.

The moral of this book is: Check your premises.

CONTENTS

PREFACE

This book presents the discovery of non-Euclidean geometry and the subsequent reformulation of the foundations of Euclidean geometry as a suspense story. The mystery of why Euclid's parallel postulate could not be proved remained unsolved for over two thousand years, until the discovery of non-Euclidean geometry and its Euclidean models revealed the impossibility of any such proof. This discovery shattered the traditional conception of geometry as the true description of physical space. Mainly through the influence of David Hilbert's *Grundlagen der Geometrie*, a new conception emerged in which the existence of many equally consistent geometries was acknowledged, each being a purely formal logical discipline that may or may not be useful for modeling physical reality. Albert Einstein stated that without this new conception of geometry, he would not have been able to develop the theory of relativity (see Einstein, 1921, Chapter I). The philosopher Hilary Putnam stated that "the overthrow of Euclidean geometry is the most important event in the history of science for the epistemologist" (1977, p. x). Chapter 8 of this book reveals the philosophical confusion that persists to this day.

This text is useful for several kinds of students. Prospective high school and college geometry teachers are presented with a rigorous treatment of the foundations of Euclidean geometry and an introduction to hyperbolic geometry (with emphasis on its Euclidean models). General education and liberal arts students are introduced to the history and philosophical implications of the discovery of non-Euclidean geometry (for example, the book was used very successfully as part of a course on scientific revolutions at Colgate University). Mathematics majors are given, in addition, detailed instruction in transformation geometry and hyperbolic trigonometry, challenging exercises, and a historical perspective that, sadly, is lacking in most mathematics texts.

I have used the development of non-Euclidean geometry to revive interest in the study of Euclidean geometry. I believe that this approach makes a traditional college course in Euclidean geometry more interesting: in order to identify the flaws in various attempted proofs of the Euclidean parallel postulate, we carefully examine the axiomatic foundations of Euclidean geometry; in order to prove the relative consistency of hyperbolic geometry, the properties of inversion in Euclidean circles are studied; in order to justify János Bolyai's construction of the limiting parallel rays, some ideas from projective geometry (cross-ratios, harmonic tetrads, perspectivities) are introduced.

I have used modified versions of Hilbert's axioms for Euclidean geometry, instead of the ruler-and-protractor postulates customary in current high school texts. The ruler-and-protractor statements are all included in Theorem 4.3 of Chapter 4, and from then on, measurement of segments and angles can be used in the customary manner. Thus, the change is less significant in practice than it is in principle. The principle here is that in a rigorous, historically motivated presentation of the foundations of geometry, it is important to separate the purely geometric ideas from the numerical ideas and to notice that the number system can be reconstructed from the geometry.

The number system so constructed could turn out to be different from the familiar real number system if we drop Dedekind's axiom of continuity; this opens the way to the new geometries discussed in Appendix B. In fact, continuity arguments are only used a few times in this book, and for all but one of those arguments, more elementary hypotheses (such as the elementary continuity principle or the circular continuity principle, or Archimedes' axiom) suffice. Dedekind's axiom is used here only to prove the existence of limiting parallel rays in hyperbolic geometry (Theorem 6.6 of Chapter 6); my recent research showed that even there, the elementary continuity principle and Aristotle's axiom suffice (but the proof is difficult). Of course, Dedekind's axiom is needed to obtain an axiom system that is categorical. But the remainder of the Hilbert-style axioms are closer to the spirit of Euclid's presentation of geometry, so that bright high school students and educated laymen will be able to understand this book.

A unique feature of this book is that some new results are developed *in the exercises* and then built upon in subsequent chapters. My experience teaching from earlier versions of this text convinced me that this method is very valuable for deepening students' understanding (students not only learn by doing, they enjoy developing new results on their own). *If students do not do a good number of exercises, they will have difficulty following subsequent chapters.* There are two sets of exercises for the first six chapters; the "major" exercises are the more challenging ones, which all students should attempt, but which mathematics majors are more likely to solve. This distinction is dropped in the last four chapters; most of the exercises for Chapters 7, 9, and 10 are "major," whereas the exercises for Chapter 8 are unusual for a mathematics text, consisting of historical and philosophical essay topics. Hints are given for most of the exercises. A solutions manual is available for instructors. The first six chapters also have projects at the end for further research in the library.

The main improvements in this third edition are as follows. Chapter 1 now contains the section warning about the danger in diagrams, but it also contains a new section on the power of diagrams for geometric insight, as illustrated by two dissection proofs of the Pythagorean theorem. In Chapter 2 (Logic and Incidence Geometry), I have added a brief section on projective and affine planes. Projective geometry, aside from its intrinsic interest, is essential for understanding certain properties of hyperbolic geometry, as can be seen from the new Major Exercise 13 in Chapter 6 plus two sections and the K-Exercises in Chapter 7. In Chapter 3, the section on axioms of continuity has been rewritten; some of the major exercises in previous editions that students found difficult are now worked out in that section. I also added Aristotle's axiom to that section, which replaces Archimedes' axiom if one wants to allow infinitesimals in geometry. Chapter 4 contains many new exercises to deepen the students' understanding.

It is Chapter 5 (History of the Parallel Postulate) that has changed the most, thanks to new historical insights gleaned from the recent treatises by Jeremy Gray (1989), B. A. Rosenfeld (1988), and Roberto Torretti (1978) (see the Bibliography). Clairaut's axiom is revealed. Legendre's many attempts to prove Euclid's parallel postulate are studied. And the remarkable glimpse by Lambert and Taurinus of the possibility of a geometry on "a sphere of imaginary radius" is highlighted; it is justified in a new section in Chapter 7 (A Model of the Hyperbolic Plane from Physics) as well as in Chapter 10 (Weierstrass coordinates). The historical part of Chapter 6 has been improved. Chapters 5, 6, and 7 also have important new exercises. Chapter 8 (Philosophical Implications) has been marginally changed, and I would appreciate readers' comments, since "working mathematicians" aren't particularly interested in philosophy. The rest of the book has minor improvements, except that the discussion of curvature and geodesics (the right way to think about "straight lines") in Appendix A is much improved.

Terminology and notation throughout the book are reasonably standard. I have followed W. Prenowitz and M. Jordan (1965) in using the term "neutral geometry" for the part of Euclidean geometry that is independent of the parallel postulate (the traditional name "absolute geometry" misleadingly implies that all other geometries depend on it). I have introduced the names "asymptotic" and "divergent" for the two types of parallels in hyperbolic geometry; I consider these a definite improvement over the welter of names in the literature. The theorems, propositions, and figures are numbered by chapter; for example, Theorem 4.1 is the first theorem in Chapter 4. Such directives as "see Coxeter (1968)" refer to the Bibliography at the back of the book (the Bibliography is arranged topically rather than strictly alphabetically).

Here are some suggested curricula for different courses:

1. A one-term course for prospective geometry teachers and/or mathematics majors, *with students of average ability.* Cover Chapters 1 – 6 and the first four sections of Chapter 7, adding Chapter 8 if there is time. In assigning exercises, omit the Major Exercises (except possibly for Chapter 1); omit most of the Exercises on Betweenness from Chapter 3; omit Exercises 21 – 31 from Chapter 4; omit Exercises 13 – 26 from Chapter 5; and assign only the Review Exercise and Exercises K-1, K-2, K-3, K-5, K-11, K-12, K-17, and K-18 from Chapter 7.

2. A one-term course for prospective geometry teachers and/or mathematics majors, *with better than average students.* Add to the curriculum of (1) the remainder of Chapter 7 and many of the exercises omitted in (1).

3. A one-term course for general education and/or liberal arts students. The core of this course would be Chapters 1, 2, and 5, the first three sections of Chapters 6 and 7, and all of Chapter 8. In addition, the instructor should selectively discuss material from Chapters 3 – 6 (such as Hilbert's axioms, the Saccheri-Legendre theorem, and some of the theorems in hyperbolic geometry), but should not impose too many proofs on these students. The essay topics of Chapter 8 are particularly appropriate for such a course.

4. A two-term course for mathematics majors. Cover as much of the book as time permits.

Thus this book is a resource for a wide variety of students, from the naive to the sophisticated, from the nonmathematical-but-educated to the mathematical wizards.

The late Errett Bishop once taught a liberal arts course in logic during which he realized the questionable nature of classical logic and wrote a book about doing mathematical analysis constructively. My own book has evolved from a liberal arts course in geometry I taught at the University of California at Santa Cruz in the early 1970s, when that campus was infused with joyful idealism and experimentation. Those were the days, my friend! (Unfortunately, our campus is losing that spirit — except for a few bright lights such as my friends the visionary Ralph Abraham, producer of a gorgeous series of books on visual mathematics and the multidisciplinary survey *Chaos, Eros and Gaia;* and the innovative chemist Frank Andrews, teacher of creative problem solving and author of *The Art and Practice of Loving.*) I am very pleased by the warm reception accorded earlier editions of this book for its unusual combination of rigor and history. It indicates that there is a real need to "humanize" mathematics texts and courses. For example, when I taught calculus to a large class recently, I was astonished at how much livelier the students (mainly nonmathematicians) became after they researched and then wrote essays about the history of calculus (many were fascinated by the strange personality of Isaac Newton), about the relevance of calculus to their own fields, and about their fear of this awesome subject. Also, such essays provide good practice in improving writing skills, which many students need. Instructors can assign essays from the Projects at the end of Chapters 1–6 and the topics in Chapter 8.

The history of the discovery of non-Euclidean geometry provides a valuable and accessible case study in the enormous difficulty we humans have in letting go of entrenched assumptions and opening ourselves to a new paradigm. It is delightfully instructive to observe the errors made by very capable people as they struggled with strange new possibilities they or their culture could not accept — Saccheri, working out the new geometry but rejecting it because it was "repugnant"; Legendre, giving one clever but false proof after another of Euclid's parallel postulate; Lambert, speculating about a possible geometry on a "sphere of imaginary radius"; Farkas Bolyai, publishing a false proof of Euclid's parallel postulate after his son had already published a non-Euclidean geometry; Gauss, afraid to publish his discoveries and not recognizing that his surfaces of constant negative curvature provided the tool for a proof that non-Euclidean geometry is consistent; or Charles Dodgson (alias Lewis Carroll), defending Euclid against his "modern rivals." It is inspiring to witness the courage it took János Bolyai and Lobachevsky to put forth the new idea before the surrounding culture could grasp it, and sad to see how little they were appreciated during their lifetimes.

Werner Erhard, who founded the *est* training taken by about a million people, understood the nontechnical message of this book. He read the Bolyai correspondence in Chapters 5–6 to thousands of people at an *est* gathering in San Francisco. I am happy to express my appreciation to him and to my students at Santa Cruz, whose enthusiasm for "having their minds blown" by this course has boosted my morale (especially Robert Curtis, who subsequently published an article in the *Journal of Geometry* on constructions in hyperbolic geometry). Suggestions from readers over the years have been helpful in improving the book, and I do welcome them. My thanks also to all the friendly people at W. H. Freeman and Company who helped produce this book, such as the late John Staples, without whose openness to innovation this book might not have appeared.

Marvin Jay Greenberg
San Francisco, California
June 1993

EUCLIDEAN AND NON-EUCLIDEAN GEOMETRIES

INTRODUCTION

Let no one ignorant of geometry enter this door.

ENTRANCE TO PLATO'S ACADEMY

1, 5 centuries ago 1850s

Most people are unaware that around a century and a half ago a revolution took place in the field of geometry that was as scientifically profound as the Copernican revolution in astronomy and, in its impact, as philosophically important as the Darwinian theory of evolution. "The effect of the discovery of hyperbolic geometry on our ideas of truth and reality has been so profound," writes the great Canadian geometer H. S. M. Coxeter, "that we can hardly imagine how shocking the possibility of a geometry different from Euclid's must have seemed in 1820." Today, however, we have all heard of the space-time geometry in Einstein's theory of relativity. "In fact, the geometry of the space-time continuum is so closely related to the non-Euclidean geometries that some knowledge of [these geometries] is an essential prerequisite for a proper understanding of relativistic cosmology."

Euclidean geometry is the kind of geometry you learned in high school, the geometry most of us use to visualize the physical universe. It comes from the text by the Greek mathematician Euclid, the *Elements,* written around 300 B.C. Our picture of the physical universe based on this geometry was painted largely by Isaac Newton in the late seventeenth century.

Geometries that differ from Euclid's own arose out of a deeper study of *parallelism.* Consider this diagram of two rays perpendicular to

Euclid (300 B.C) ⇒ Isaac Newton 1700s (late)

segment PQ:

In Euclidean geometry the perpendicular distance between the rays remains equal to the distance from P to Q as we move to the right. However, in the early nineteenth century two alternative geometries were proposed. In hyperbolic geometry (from the Greek *hyperballein*, "to exceed") the distance between the rays increases. In elliptic geometry (from the Greek *elleipein*, "to fall short") the distance decreases and the rays eventually meet. These non-Euclidean geometries were later incorporated in a much more general geometry developed by C. F. Gauss and G. F. B. Riemann (it is this more general geometry that is used in Einstein's general theory of relativity).[1]

We will concentrate on Euclidean and hyperbolic geometries in this book. Hyperbolic geometry requires a change in only one of Euclid's axioms, and can be as easily grasped as high school geometry. Elliptic geometry, on the other hand, involves the new topological notion of "nonorientability," since all the points of the elliptic plane not on a given line lie on the same side of that line. This geometry cannot easily be approached in the spirit of Euclid. I have therefore made only brief comments about elliptic geometry in the body of the text, with further indications in Appendix A. (Do not be misled by this, however; elliptic geometry is no less important than hyperbolic.) Riemannian geometry requires a thorough understanding of the differential and integral calculus, and is therefore beyond the scope of this book (it is discussed briefly in Appendix A).

Chapter 1 begins with a brief history of geometry in ancient times, and emphasizes the development of the axiomatic method by the Greeks. It presents Euclid's five postulates and includes one of Legendre's attempted proofs of the fifth postulate. In order to detect the

---

[1] Einstein's special theory of relativity, which is needed to study subatomic particles, is based on a simpler geometry of space-time due to H. Minkowski. The names "hyperbolic geometry" and "elliptic geometry" were coined by F. Klein; some authors misleadingly call these geometries "Lobachevskian" and "Riemannian," respectively.

flaw in Legendre's argument (and in other arguments), it will be necessary to carefully reexamine the foundations of geometry. However, before we can do any geometry at all, we must be clear about some fundamental principles of logic. These are reviewed informally in Chapter 2. In this chapter we consider what constitutes a rigorous proof, giving special attention to the method of indirect proof, or *reductio ad absurdum*. Chapter 2 introduces the very important notion of a *model* for an axiom system, illustrated by finite models for the axioms of incidence as well as real projective and affine models.

Chapter 3 begins with a discussion of some flaws in Euclid's presentation of geometry. These are then repaired in a thorough presentation of David Hilbert's axioms (slightly modified) and their elementary consequences. You may become restless over the task of proving results that appear self-evident. Nevertheless, this work is essential if you are to steer safely through non-Euclidean space.

Our study of the consequences of Hilbert's axioms, with the exception of the parallel postulate, is continued in Chapter 4; this study is called *neutral geometry*. We will prove some familiar Euclidean theorems (such as the exterior angle theorem) by methods different from those used by Euclid, a change necessitated by gaps in Euclid's proofs. We will also prove some theorems that Euclid would not recognize (such as the Saccheri-Legendre theorem).

Supported by the solid foundation of the preceding chapters, we will be prepared to analyze in Chapter 5 several important attempts to prove the parallel postulate (in the exercises you will have the opportunity to find flaws in still other attempts). Following that, your Euclidean conditioning should be shaken enough so that in Chapter 6 we may explore "a strange new universe," one in which triangles have the "wrong" angle sums, rectangles do not exist, and parallel lines may diverge or converge asymptotically. In doing so, we will see unfolding the historical drama of the almost simultaneous discovery of hyperbolic geometry by Gauss, J. Bolyai, and Lobachevsky in the early nineteenth century.

This geometry, however unfamiliar, is just as consistent as Euclid's. This is demonstrated in Chapter 7 by studying three Euclidean models that also aid in visualizing hyperbolic geometry. The Poincaré models have the advantage that angles are measured in the Euclidean way; the Beltrami-Klein model has the advantage that lines are represented by segments of Euclidean lines. In Chapter 7 we will also discuss topics in Euclidean geometry not usually covered in high school.

Chapter 8 takes up in a general way some of the philosophical implications of non-Euclidean geometries. The presentation is deliberately controversial, and the essay topics are intended to stimulate further thought and reading.

Chapter 9 introduces the new insights gained for geometry by the transformation approach (Felix Klein's *Erlanger Programme*). We classify all the motions of Euclidean and hyperbolic planes, use them to solve geometric problems, describe them analytically in the Cartesian and Poincaré models, characterize groups of transformations that are compatible with our congruence axioms, and introduce the fascinating topic of symmetry, determining all finite symmetry groups (essentially known by Leonardo da Vinci).

Chapter 10 is mainly devoted to the trigonometry of the hyperbolic plane, touching also upon area theory and surfaces of constant negative curvature. Among other results, we prove the hyperbolic analogue of the Pythagorean theorem, and we derive formulas for the circumference and area of a circle, for the relationships between right triangles and Lambert quadrilaterals, and for the circumscribed cycle of a triangle. We define various coordinate systems used to do analytic geometry in the hyperbolic plane. Appendix A tells more about elliptic geometry, which is mentioned throughout the book. We then introduce differential geometry, sketching the magnificent insights of Gauss and Riemann.

It is very important that you do as many exercises as possible, since new results are developed in the exercises and then built on in subsequent chapters. By working all the exercises, you may come to enjoy geometry as much as I do.

Hyperbolic geometry used to be considered a historical curiosity. Some practical-minded students always ask me what it is good for. Following Euclid's example, I may give them a coin (not having a slave to hand it to them) and tell them that I earn a living from it. Sometimes I ask them what great music and art are good for, or I refer them to essay topics 5 and 8 in Chapter 8. If they persist, I refer them to Luneburg's research on binocular vision (see Chapter 8), to classical mechanics, and to current research in topology, ergodic theory, and automorphic function theory (see Suggested Further Reading). This book and the course using it provide practical-minded people an opportunity to stretch their minds. As the great French mathematician Jacques Hadamard said, "Practical application is found by not looking for it, and one can say that the whole progress of civilization rests on that principle." Only impractical dreamers spent two thousand years wondering about proving Euclid's parallel postulate, and if they hadn't done so, there would be no spaceships exploring the galaxy today.

EUCLID'S GEOMETRY

The postulate on parallels . . . was in antiquity the
final solution of a problem that must have preoccupied
Greek mathematics for a long period before Euclid.

HANS FREUDENTHAL

THE ORIGINS OF GEOMETRY

The word "geometry" comes from the Greek *geometrein* (*geo-*,
"earth," and *metrein*, "to measure"); geometry was originally the
science of measuring land. The Greek historian Herodotus (5th century B.C.) credits Egyptian surveyors with having originated the subject of geometry, but other ancient civilizations (Babylonian, Hindu,
Chinese) also possessed much geometric information.

Ancient geometry was actually a collection of rule-of-thumb procedures arrived at through experimentation, observation of analogies,
guessing, and occasional flashes of intuition. In short, it was an empirical subject in which approximate answers were usually sufficient for
practical purposes. The Babylonians of 2000 to 1600 B.C. considered
the circumference of a circle to be three times the diameter; i.e., they
took π to be equal to 3. This was the value given by the Roman
architect Vitruvius and it is found in the Chinese literature as well. It
was even considered sacred by the ancient Jews and sanctioned in
scripture (I Kings 7:23) — an attempt by Rabbi Nehemiah to change

the value of π to $\frac{22}{7}$ was rejected. The Egyptians of 1800 B.C., according to the Rhind papyrus, had the approximation $\pi \sim (\frac{16}{9})^2 \sim 3.1604$.[1]

Sometimes the Egyptians guessed correctly, other times not. They found the correct formula for the volume of a frustum of a square pyramid — a remarkable accomplishment. On the other hand, they thought that a formula for area that was correct for rectangles applied to *any* quadrilateral. Egyptian geometry was not a science in the Greek sense, only a grab bag of rules for calculation without any motivation or justification.

The Babylonians were much more advanced than the Egyptians in arithmetic and algebra. Moreover, they knew the Pythagorean theorem — in a right triangle the square of the length of the hypotenuse is equal to the sum of the squares of the lengths of the legs — long before Pythagoras was born. Recent research by Otto Neugebauer has revealed the heretofore unknown Babylonian algebraic influence on Greek mathematics.

However, the Greeks, beginning with Thales of Miletus, insisted that geometric statements be established by deductive reasoning rather than by trial and error. Thales was familiar with the computations, partly right and partly wrong, handed down from Egyptian and Babylonian mathematics. In determining which results were correct, he developed the first logical geometry (Thales is also famous for having predicted the eclipse of the sun in 585 B.C.). The orderly development of theorems by *proof* was characteristic of Greek mathematics and entirely new.

The systematization begun by Thales was continued over the next two centuries by Pythagoras and his disciples. Pythagoras was regarded by his contemporaries as a religious prophet. He preached the immortality of the soul and reincarnation. He organized a brotherhood of believers that had its own purification and initiation rites, followed a vegetarian diet, and shared all property communally. The Pythagoreans differed from other religious sects in their belief that elevation of

---

[1] In recent years π has been approximated to a very large number of decimal places by computers; to five places, π is approximately 3.14159. In 1789 Johann Lambert proved that π was not equal to any fraction (rational number), and in 1882 F. Lindemann proved that π is a *transcendental number*, in the sense that it does not satisfy any algebraic equation with rational coefficients, which implies that in the Euclidean plane, it is impossible to square a circle using only straightedge and compass.

the soul and union with God are achieved by the study of music and mathematics. In music, Pythagoras calculated the correct ratios of the harmonic intervals. In mathematics, he taught the mysterious and wonderful properties of numbers. Book VII of Euclid's *Elements* is the text of the theory of numbers taught in the Pythagorean school.

The Pythagoreans were greatly shocked when they discovered irrational lengths, such as $\sqrt{2}$ (see Chapter 2, pp. 43–44). At first they tried to keep this discovery secret. The historian Proclus wrote: "It is well known that the man who first made public the theory of irrationals perished in a shipwreck, in order that the inexpressible and unimaginable should ever remain veiled." Since the Pythagoreans did not consider $\sqrt{2}$ a number, they transmuted their algebra into geometric form in order to represent $\sqrt{2}$ and other irrational lengths by segments ($\sqrt{2}$ by a diagonal of the unit square).

The systematic foundation of plane geometry by the Pythagorean school was brought to a conclusion around 400 B.C. in the *Elements* by the mathematician Hippocrates (not to be confused with the physician of the same name). Although this treatise has been lost, we can safely say that it covered most of Books I–IV of Euclid's *Elements*, which appeared about a century later. The Pythagoreans were never able to develop a theory of proportions that was also valid for irrational lengths. This was later achieved by Eudoxus, whose theory was incorporated into Book V of Euclid's *Elements*.

The fourth century B.C. saw the flourishing of Plato's Academy of science and philosophy (founded about 387 B.C.). In the *Republic* Plato wrote, "The study of mathematics develops and sets into operation a mental organism more valuable than a thousand eyes, because through it alone can truth be apprehended." Plato taught that the universe of ideas is more important than the material world of the senses, the latter being only a shadow of the former. The material world is an unlit cave on whose walls we see only shadows of the real, sunlit world outside. The errors of the senses must be corrected by concentrated thought, which is best learned by studying mathematics. The Socratic method of dialog is essentially that of indirect proof, by which an assertion is shown to be invalid if it leads to a contradiction. Plato repeatedly cited the proof for the irrationality of the length of a diagonal of the unit square as an illustration of the method of indirect proof (the *reductio ad absurdum* — see Chapter 2, pp. 42–44). The point is that this irrationality of length could never have been discovered by physical measurements, which always include a small experimental margin of error.

Euclid was a disciple of the Platonic school. Around 300 B.C. he produced the definitive treatment of Greek geometry and number theory in his 13-volume *Elements*. In compiling this masterpiece Euclid built on the experience and achievements of his predecessors in preceding centuries: on the Pythagoreans for Books I–IV, VII, and IX, Archytas for Book VIII, Eudoxus for Books V, VI, and XII, and Theaetetus for Books X and XIII. So completely did Euclid's work supersede earlier attempts at presenting geometry that few traces remain of these efforts. It's a pity that Euclid's heirs have not been able to collect royalties on his work, for he is the most widely read author in the history of mankind. His approach to geometry has dominated the teaching of the subject for over two thousand years. Moreover, the axiomatic method used by Euclid is the prototype for all of what we now call "pure mathematics." It is pure in the sense of "pure thought": no physical experiments need be performed to verify that the statements are correct — only the reasoning in the demonstrations need be checked.

Euclid's *Elements* is pure also in that the work includes no practical applications. Of course, Euclid's geometry has had an enormous number of applications to practical problems in engineering, but they are not mentioned in the *Elements*. According to legend, a beginning student of geometry asked Euclid, "What shall I get by learning these things?" Euclid called his slave, saying, "Give him a coin, since he must make gain out of what he learns." To this day, this attitude toward application persists among many pure mathematicians — they study mathematics for its own sake, for its intrinsic beauty and elegance (see essay topics 5 and 8 in Chapter 8).

Surprisingly enough, as we will see later, pure mathematics often turns out to have applications never dreamt of by its creators — the "impractical" outlook of pure mathematicians is ultimately useful to society. Moreover, those parts of mathematics that have not been "applied" are also valuable to society, either as aesthetic works comparable to music and art or as contributions to the expansion of human consciousness and understanding.[2]

---

[2] For more detailed information on ancient mathematics, see Bartel van der Waerden (1961).

THE AXIOMATIC METHOD

Mathematicians can make use of trial and error, computation of special cases, inspired guessing, or any other way to discover theorems. The axiomatic method is a method of *proving* that results are correct. Some of the most important results in mathematics were originally given only incomplete proofs (we shall see that even Euclid was guilty of this). No matter — correct proofs would be supplied later (sometimes much later) and the mathematical world would be satisfied.

So proofs give us assurance that results are correct. In many cases they also give us more *general* results. For example, the Egyptians and Hindus knew by experiment that if a triangle has sides of lengths 3, 4, and 5, it is a right triangle. But the Greeks proved that if a triangle has sides of lengths a, b, and c and if $a^2 + b^2 = c^2$, then the triangle is a right triangle. It would take an infinite number of experiments to check this result (and, besides, experiments only measure things approximately). Finally, proofs give us tremendous insight into relationships among different things we are studying, forcing us to organize our ideas in a coherent way. You will appreciate this by the end of Chapter 6 (if not sooner).

What is the axiomatic method? If I wish to persuade you by *pure reasoning* to believe some statement S_1, I could show you how this statement follows logically from some other statement S_2 that you may already accept. However, if you don't believe S_2, I would have to show you how S_2 follows logically from some other statement S_3. I might have to repeat this procedure several times until I reach some statement that you already accept, one I do not need to justify. That statement plays the role of an *axiom* (or *postulate*). If I cannot reach a statement that you will accept as the basis of my argument, I will be caught in an "infinite regress," giving one demonstration after another without end.

So there are two requirements that must be met for us to agree that a proof is correct:

REQUIREMENT 1. Acceptance of certain statements called "axioms," or "postulates," without further justification.

REQUIREMENT 2. Agreement on how and when one statement "follows logically" from another, i.e., agreement on certain rules of reasoning.

Euclid's monumental achievement was to single out a few simple postulates, statements that were acceptable without further justification, and then to deduce from them 465 propositions, many complicated and not at all intuitively obvious, which contained all the geometric knowledge of his time. One reason the *Elements* is such a beautiful work is that so much has been deduced from so little.

UNDEFINED TERMS

We have been discussing what is required for us to agree that a proof is correct. Here is one requirement that we took for granted:

REQUIREMENT 0. Mutual understanding of the meaning of the words and symbols used in the discourse.

There should be no problem in reaching mutual understanding so long as we use terms familiar to both of us and use them consistently. If I use an unfamiliar term, you have the right to demand a *definition* of this term. Definitions cannot be given arbitrarily; they are subject to the rules of reasoning referred to (but not specified) in Requirement 2. If, for example, I define a right angle to be a 90° angle, and then define a 90° angle to be a right angle, I would violate the rule against *circular reasoning.*

Also, we cannot define every term that we use. In order to define one term we must use other terms, and to define these terms we must use still other terms, and so on. If we were not allowed to leave some terms *undefined,* we would get involved in infinite regress.

Euclid did attempt to define all geometric terms. He defined a "straight line" to be "that which lies evenly with the points on itself." This definition is not very useful; to understand it you must already have the image of a line. So it is better to take "line" as an undefined term. Similarly, Euclid defined a "point" as "that which has no part" — again, not very informative. So we will also accept "point" as an undefined term. Here are the five undefined geometric terms that are the basis for defining all other geometric terms in plane Euclidean geometry:

> *point*
> *line*
> *lie on* (as in "two points *lie on* a unique line")
> *between* (as in "point C is *between* points A and B")
> *congruent*

For solid geometry, we would have to introduce a further undefined geometric term, "plane," and extend the relation "lie on" to allow points and lines to lie on planes. *In this book (unless otherwise stated) we will restrict our attention to plane geometry*, i.e., to one single plane. So we define *the plane* to be the set of all points and lines, all of which are said to "lie on" it.

There are expressions that are often used synonymously with "lie on." Instead of saying "point P *lies on* line *l*," we sometimes say "*l passes through* P" or "P is *incident* with *l*," denoted P I *l*. If point P lies on both line *l* and line *m*, we say that "*l* and *m* have point P *in common*" or that "*l* and *m* intersect (or *meet*) in the point P."

The second undefined term, "line," is synonymous with "straight line." The adjective "straight" is confusing when it modifies the noun "line," so we won't use it. Nor will we talk about "curved lines." Although the word "line" will not be defined, its use will be restricted by the axioms for our geometry. For instance, one axiom states that two given points lie on only one line. Thus, in Figure 1.1, *l* and *m* could not both represent lines in our geometry, since they both pass through the points P and Q.

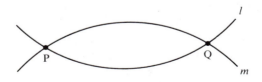

FIGURE 1.1

There are other mathematical terms that we will use that should be added to our list of undefined terms, since we won't define them; they have been omitted because they are not specifically geometric in nature, but are rather what Euclid called "common notions." Nevertheless, since there may be some confusion about these terms, a few remarks are in order.

↗undefined

The word "set" is fundamental in all of mathematics today; it is now used in elementary schools, so undoubtedly you are familiar with its use. Think of it as a "collection of objects." Two related notions are "belonging to" a set or "being an element (or member) of " a set, as in our convention that all points and lines *belong* to the plane. If every element of a set S is also an element of a set T, we say that S is "contained in" or "part of " or "a subset of " T. We will define "segment," "ray," "circle," and other geometric terms to be certain sets of points. A "line," however, is not a set of points in our treatment (for reasons of duality in Chapter 2). When we need to refer to the set of all points lying on a line l, we will denote that set by $\{l\}$.

In the language of sets we say that sets S and T are *equal* if every member of S is a member of T, and vice versa. For example, the set S of all authors of Euclid's *Elements* is (presumably) equal to the set whose only member is Euclid. Thus, "equal" means "identical."

Euclid used the word "equal" in a different sense, as in his assertion that "base angles of an isosceles triangle are *equal*." He meant that base angles of an isosceles triangle have an equal number of degrees, not that they are identical angles. So to avoid confusion we will not use the word "equal" in Euclid's sense. Instead, we will use the undefined term "congruent" and say that "base angles of an isosceles triangle are *congruent*." Similarly, we don't say that "if AB *equals* AC, then $\triangle$ABC is isosceles." (If AB *equals* AC, following our use of the word "equals," $\triangle$ABC is not a triangle at all, only a segment.) Instead, we would say that "if AB is *congruent* to AC, then $\triangle$ABC is *isosceles*." This use of the undefined term "congruent" is more general than the one to which you are accustomed; it applies not only to triangles but to angles and segments as well. To understand the use of this word, picture congruent objects as "having the same size and shape."

"equal"
ex: ∂
‖
CONGRUENT

Of course, we must specify (as Euclid did in his "common notions") that "a thing is congruent to itself," and that "things congruent to the same thing are congruent to each other." Statements like these will later be included among our axioms of congruence (Chapter 3).

The list of undefined geometric terms shown earlier in this section is due to David Hilbert (1862–1943). His treatise *The Foundations of Geometry* (1899) not only clarified Euclid's definitions but also filled in the gaps in some of Euclid's proofs. Hilbert recognized that Euclid's proof for the side-angle-side criterion of congruence in triangles was based on an unstated assumption (the principle of superposition), and that this criterion had to be treated as an axiom. He also built on the earlier work of Moritz Pasch, who in 1882 published the first rigorous treatise on geometry; Pasch made explicit Euclid's unstated assumptions about betweenness (the axioms on betweenness will be studied in Chapter 3). Some other mathematicians who worked to establish rigorous foundations for Euclidean geometry are: G. Peano, M. Pieri, G. Veronese, O. Veblen, G. de B. Robinson, E. V. Huntington, and H. G. Forder. These mathematicians used lists of undefined terms different from the one used by Hilbert. Pieri used only two undefined terms (as a result, however, his axioms were more complicated). The selection of undefined terms and axioms is *arbitrary;* Hilbert's selection is popular because it leads to an elegant development of geometry similar to Euclid's presentation.

EUCLID'S FIRST FOUR POSTULATES

Euclid based his geometry on five fundamental assumptions, called *axioms* or *postulates.*

EUCLID'S POSTULATE I.   For every point P and for every point Q not equal to P there exists a unique line *l* that passes through P and Q.

This postulate is sometimes expressed informally by saying "two points determine a unique line." We will denote the unique line that passes through P and Q by $\overleftrightarrow{PQ}$.

To state the second postulate, we must make our first definition.

DEFINITION.   Given two points A and B. The *segment* AB is the set whose members are the points A and B and all points that lie on the line $\overleftrightarrow{AB}$ and are between A and B (Figure 1.2). The two given points A and B are called the *endpoints* of the segment AB.[3]

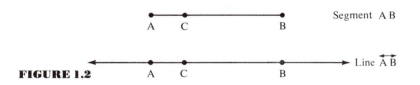

Segment A B

Line $\overleftrightarrow{AB}$

FIGURE 1.2

[3] Warning on notation: In many high school geometry texts the notation $\overline{AB}$ is used for "segment AB."

EUCLID'S POSTULATE II.   For every segment AB and for every segment CD there exists a unique point E such that B is between A and E
and segment CD is congruent to segment BE (Figure 1.3).

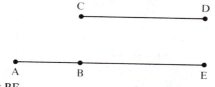

FIGURE 1.3   CD ≅ BE.

This postulate is sometimes expressed informally by saying that
"any segment AB can be extended by a segment BE congruent to a
given segment CD." Notice that in this postulate we have used the
undefined term "congruent" in the new way, and we use the usual
notation CD ≅ BE to express the fact that CD is congruent to BE.

In order to state the third postulate, we must introduce another
definition.

DEFINITION.   Given two points O and A. The set of all points P such
that segment OP is congruent to segment OA is called a *circle* with O as
center, and each of the segments OP is called a *radius* of the circle.

It follows from Euclid's previously mentioned common notion ("a
thing is congruent to itself") that OA ≅ OA, so A is also a point on the
circle just defined.

EUCLID'S POSTULATE III.   For every point O and every point A not
equal to O there exists a circle with center O and radius OA (Figure
1.4).

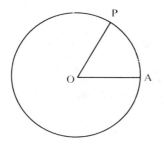

FIGURE 1.4   Circle with center O and radius OA.

Actually, because we are using the language of sets rather than that of Euclid, it is not really necessary to assume this postulate; it is a consequence of set theory that the set of all points P with OP ≅ OA exists. Euclid had in mind *drawing* the circle with center O and radius OA, and this postulate tells you that such a drawing is allowed, for example, with a compass. Similarly, in Postulate II you are allowed to extend segment AB by drawing segment BE with a straightedge. Our treatment "purifies" Euclid by eliminating references to drawing in our proofs.[4] But you should review the straightedge and compass constructions in Major Exercise 1.

DEFINITION. The *ray* $\overrightarrow{AB}$ is the following set of points lying on the line $\overleftrightarrow{AB}$: those points that belong to the segment AB and all points C on $\overleftrightarrow{AB}$ such that B is between A and C. The ray $\overrightarrow{AB}$ is said to *emanate from the vertex* A and to be *part* of line $\overleftrightarrow{AB}$. (See Figure 1.5.)

FIGURE 1.5 Ray $\overrightarrow{AB}$.

DEFINITION. Rays $\overrightarrow{AB}$ and $\overrightarrow{AC}$ are *opposite* if they are distinct, if they emanate from the same point A, and if they are part of the same line $\overleftrightarrow{AB} = \overleftrightarrow{AC}$ (Figure 1.6.).

---

[4] However, it is a fascinating mathematical problem to determine just what geometric constructions are possible using only a compass and straightedge. Not until the nineteenth century was it proved that such constructions as trisecting an arbitrary angle, squaring a circle, or doubling a cube were impossible using only a compass and straightedge. Pierre Wantzel proved this by translating the geometric problem into an algebraic problem; he showed that straightedge and compass constructions correspond to a solution of certain algebraic equations using only the operations of addition, subtraction, multiplication, division, and extraction of square roots. For the particular algebraic equations obtained from, say, the problem of trisecting an arbitrary angle, such a solution is impossible because cube roots are needed. Of course, it is possible to trisect angles using other instruments, such as a marked straightedge and compass (see Major Exercise 3 and Projects 1, 2, and 4), and J. Bolyai proved that in the hyperbolic plane, it is possible to "square" the circle.

FIGURE 1.6 Opposite rays.

DEFINITION. An "*angle* with *vertex* A" is a point A together with two distinct nonopposite rays $\overrightarrow{AB}$ and $\overrightarrow{AC}$ (called the *sides* of the angle) emanating from A.[5] (See Figure 1.7.)

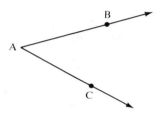

FIGURE 1.7 Angle with vertex A.

We use the notations ∢A, ∢BAC, or ∢CAB for this angle.

DEFINITION. If two angles ∢BAD and ∢CAD have a common side $\overrightarrow{AD}$ and the other two sides $\overrightarrow{AB}$ and $\overrightarrow{AC}$ form opposite rays, the angles are *supplements* of each other, or *supplementary angles* (Figure 1.8).

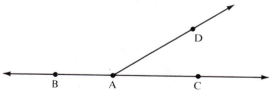

FIGURE 1.8 Supplementary angles.

DEFINITION. An angle ∢BAD is a *right angle* if it has a supplementary angle to which it is congruent (Figure 1.9).

---

[5] According to this definition, there is no such thing as a "straight angle." We eliminated this expression because most of the assertions we will make about angles do not apply to "straight angles." The definition excludes zero angles as well.

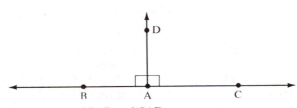

FIGURE 1.9   Right angles ∢BAD ≅ ∢CAD.

We have thus succeeded in defining a right angle without referring to "degrees," by using the undefined notion of congruence of angles. "Degrees" will not be introduced formally until Chapter 4, although we will occasionally refer to them in informal discussions.

We can now state Euclid's fourth postulate.

EUCLID'S POSTULATE IV.   All right angles are congruent to each other.

This postulate expresses a sort of homogeneity; even though two right angles may be "very far away" from each other, they nevertheless "have the same size." The postulate therefore provides a natural standard of measurement for angles.[6]

THE PARALLEL POSTULATE

Euclid's first four postulates have always been readily accepted by mathematicians. The fifth (parallel) postulate, however, was highly controversial. In fact, as we shall see later, consideration of alternatives to Euclid's parallel postulate resulted in the development of non-Euclidean geometries.

At this point we are not going to state the fifth postulate in its original form, as it appeared in the *Elements*. Instead, we will present a simpler postulate (which we will later show is logically equivalent to Euclid's original). This version is sometimes called *Playfair's postulate*

---

[6] On the contrary, there is no natural standard of measurement for *lengths* in Euclidean geometry. Units of length (one foot, one meter, etc.) must be chosen arbitrarily. The remarkable fact about hyperbolic geometry, on the other hand, is that it does admit a natural standard of length (see Chapter 6).

because it appeared in John Playfair's presentation of Euclidean geometry, published in 1795 — although it was referred to much earlier by Proclus (A.D. 410–485). We will call it *the Euclidean parallel postulate* because it distinguishes Euclidean geometry from other geometries based on parallel postulates. The most important definition in this book is the following:

DEFINITION.   Two lines *l* and *m* are *parallel* if they do not intersect, i.e., if no point lies on both of them. We denote this by *l* ‖ *m*.

Notice first that we assume the lines lie in the same plane (because of our convention that all points and lines lie in one plane, unless stated otherwise; in space there are noncoplanar lines which fail to intersect and they are called *skew lines,* not "parallel"). Notice secondly what the definition does *not* say: it does not say that the lines are equidistant, i.e., it does not say that the distance between the two lines is everywhere the same. Don't be misled by drawings of parallel lines in which the lines appear to be equidistant. We want to be rigorous here and so should not introduce into our proofs assumptions that have not been stated explicitly. At the same time, don't jump to the conclusion that parallel lines are *not* equidistant. We are not committing ourselves either way and shall reserve judgment until we study the matter further. At this point, the only thing we know for sure about parallel lines is that they do not meet.

THE EUCLIDEAN PARALLEL POSTULATE.   For every line *l* and for every point P that does not lie on *l* there exists a unique line *m* through P that is parallel to *l*. (See Figure 1.10.)

FIGURE 1.10   Lines *l* and *m* are parallel.

Why should this postulate be so controversial? It may seem "obvious" to you, perhaps because you have been conditioned to think in Euclidean terms. However, if we consider the axioms of geometry as abstractions from experience, we can see a difference between this postulate and the other four. The first two postulates are abstractions from our experiences drawing with a straightedge; the third postulate derives from our experiences drawing with a compass. The fourth postulate is perhaps less obvious as an abstraction; nevertheless it derives from our experiences measuring angles with a protractor (where the sum of supplementary angles is 180°, so that if supplementary angles are congruent to each other, they must each measure 90°).

The fifth postulate is different in that we cannot verify empirically whether two lines meet, since we can draw only segments, not lines. We can extend the segments further and further to see if they meet, but we cannot go on extending them forever. Our only recourse is to verify parallelism indirectly, by using criteria other than the definition.

What is another criterion for *l* to be parallel to *m*? Euclid suggested drawing a *transversal* (i.e., a line *t* that intersects both *l* and *m* in distinct points), and measuring the number of degrees in the interior angles α and β on one side of *t*. Euclid predicted that if the sum of angles α and β turns out to be less than 180°, the lines (if produced sufficiently far) would meet on the same side of *t* as angles α and β (see Figure 1.11). This, in fact, is the content of Euclid's fifth postulate.

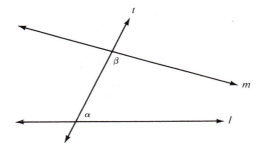

FIGURE 1.11

The trouble with this criterion for parallelism is that it turns out to be logically equivalent to the Euclidean parallel postulate that was just stated (see the section Equivalence of Parallel Postulates in Chapter 4.). So we cannot use this criterion to convince ourselves of the correctness of the parallel postulate — that would be circular reasoning. Euclid himself recognized the questionable nature of the parallel postulate, for he postponed using it for as long as he could (until the proof of his 29th proposition).

ATTEMPTS TO PROVE THE PARALLEL POSTULATE

Remember that an axiom was originally supposed to be so simple and intuitively obvious that no one could doubt its validity. From the very beginning, however, the parallel postulate was attacked as insufficiently plausible to qualify as an unproved assumption. For two thousand years mathematicians tried to derive it from the other four postulates or to replace it with another postulate, one more self-evident. All attempts to derive it from the first four postulates turned out to be unsuccessful because the so-called proofs always entailed a hidden assumption that was unjustifiable. The substitute postulates, purportedly more self-evident, turned out to be logically equivalent to the parallel postulate, so that nothing was gained by the substitution. We will examine these attempts in detail in Chapter 5, for they are very instructive. For the moment, let us consider one such effort.

The Frenchman Adrien Marie Legendre (1752 – 1833) was one of the best mathematicians of his time, contributing important discoveries to many different branches of mathematics. Yet he was so obsessed with proving the parallel postulate that over a period of 29 years he published one attempt after another in different editions of his *Éléments de Géométrie*.[7] Here is one attempt (see Figure 1.12):

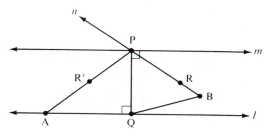

FIGURE 1.12

Given P not on line l. Drop perpendicular PQ from P to l at Q. Let m be the line through P perpendicular to $\overleftrightarrow{PQ}$. Then m is parallel to l,

---

[7] Davies' translation of the *Éléments* was the most popular geometry textbook in the United States during the nineteenth century. Legendre is best known for the method of least squares in statistics, the law of quadratic reciprocity in number theory, and the Legendre polynomials in differential equations. His attempts to prove the parallel postulate led to two important theorems in neutral geometry (see Chapter 4).

Adrien Marie Legendre

since l and m have the common perpendicular $\overleftrightarrow{PQ}$. Let n be any line through P distinct from m and $\overleftrightarrow{PQ}$. We must show than n meets l. Let $\overrightarrow{PR}$ be a ray of n between $\overrightarrow{PQ}$ and a ray of m emanating from P. There is a point R′ on the opposite side of $\overleftrightarrow{PQ}$ from R such that ∢QPR′ ≅ ∢QPR. Then Q lies in the interior of ∢RPR′. Since line l passes through the point Q interior to ∢RPR′, l must intersect one of the sides of this angle. If l meets side $\overrightarrow{PR}$, then certainly l meets n. Suppose l meets side $\overrightarrow{PR'}$ at a point A. Let B be the unique point on side $\overrightarrow{PR}$ such that PA ≅ PB. Then △PQA ≅ PQB (SAS); hence ∢PQB is a right angle, so that B lies on l (and n).

You may feel that this argument is plausible enough. Yet how could you tell if it is correct? You would have to justify each step, first defining each term carefully. For instance, you would have to define what was meant by two lines being "perpendicular" — otherwise, how could you justify the assertion that lines *l* and *m* are parallel simply because they have a common perpendicular? (You would first have to prove that as a separate theorem, if you could.) You would have to justify the side-angle-side (SAS) criterion of congruence in the last statement. You would have to define the "interior" of an angle, and prove that a line through the interior of an angle must intersect one of the sides. In proving all of these things, you would have to be sure to use only the first four postulates and not any statement equivalent to the fifth; otherwise the argument would be circular.

Thus there is a lot of work that must be done before we can detect the flaw. In the next few chapters we will do this preparatory work so that we can confidently decide whether or not Legendre's proposed proof is valid. (Legendre's argument contains several statements that cannot be proved from the first four postulates.) As a result of this work we will be better able to understand the foundations of Euclidean geometry. We will discover that a large part of this geometry is independent of the theory of parallels and is equally valid in hyperbolic geometry.

THE DANGER IN DIAGRAMS

Diagrams have always been helpful in understanding geometry — they are included in Euclid's *Elements* and they are included in this book. But there is a danger that a diagram may suggest a fallacious argument. A diagram may be slightly inaccurate or it may represent only a special case. If we are to recognize the flaws in arguments such as Legendre's, we must not be misled by diagrams that *look* plausible.

What follows is a well-known and rather involved argument that pretends to prove that all triangles are isosceles. Place yourself in the context of what you know from high school geometry. (After this chapter you will have to put that knowledge on hold.) Find the flaw in the argument.

Given △ABC. Construct the bisector of ∢A and the perpendicular bisector of side BC opposite to ∢A. Consider the various cases (Figure 1.13).

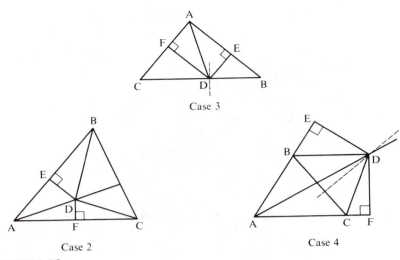

Case 3

Case 2

Case 4

FIGURE 1.13

Case 1. The bisector of ⊰A and the perpendicular bisector of segment BC are either parallel or identical. In either case, the bisector of ⊰A is perpendicular to BC and hence, by definition, is an altitude. Therefore, the triangle is isosceles. (The conclusion follows from the Euclidean theorem: if an angle bisector and altitude from the same vertex of a triangle coincide, the triangle is isosceles.)

Suppose now that the bisector of ⊰A and the perpendicular bisector of the side opposite are not parallel and do not coincide. Then they intersect in exactly one point, D, and there are three cases to consider:

Case 2. The point D is inside the triangle.

Case 3. The point D is on the triangle.

Case 4. The point D is outside the triangle.

For each case construct DE perpendicular to AB and DF perpendicular to AC, and for cases 2 and 4 join D to B and D to C. In each case, the following proof now holds (see Figure 1.13):

DE ≅ DF because all points on an angle bisector are equidistant from the sides of the angle; DA ≅ DA, and ∢DEA and ∢DFA are right angles; hence, △ADE is congruent to △ADF by the hypotenuse-leg theorem of Euclidean geometry. (We could also have used the SAA theorem with DA ≅ DA, and the bisected angle and right angles.) Therefore, we have AE ≅ AF. Now, DB ≅ DC because all points on the perpendicular bisector of a segment are equidistant from the ends of the segment. Also, DE ≅ DF, and ∢DEB and ∢DFC are right angles. Hence, △DEB is congruent to △DFC by the hypotenuse-leg theorem, and hence FC ≅ BE. It follows that AB ≅ AC — in cases 2 and 3 by addition and in case 4 by subtraction. The triangle is therefore isosceles.

THE POWER OF DIAGRAMS

Geometry, for human beings (perhaps not for computers), is a visual subject. Correct diagrams are extremely helpful in understanding proofs and in discovering new results. One of the best illustrations of this is Figure 1.14, which reveals immediately the validity of the

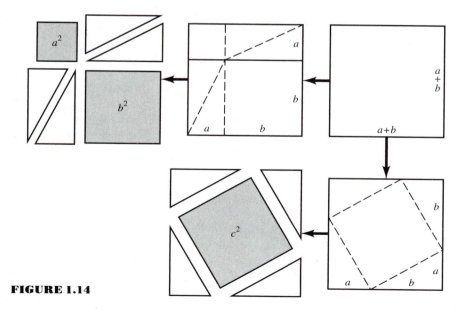

FIGURE 1.14

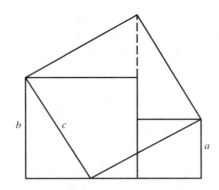

FIGURE 1.15

Pythagorean theorem in Euclidean geometry. (Euclid's proof was much more complicated.) Figure 1.15 is a simpler diagram suggesting a proof by dissection.

REVIEW EXERCISE

Which of the following statements are correct?

(1) The Euclidean parallel postulate states that for every line *l* and for every point P not lying on *l* there exists a unique line *m* through P that is parallel to *l*.

(2) An "angle" is defined as the space between two rays that emanate from a common point.

(3) Most of the results in Euclid's *Elements* were discovered by Euclid himself.

(4) By definition, a line *m* is "parallel" to a line *l* if for any two points P, Q on *m*, the perpendicular distance from P to *l* is the same as the perpendicular distance from Q to *l*.

(5) It was unnecessary for Euclid to assume the parallel postulate because the French mathematician Legendre proved it.

(6) A "transversal" to two lines is another line that intersects both of them in distinct points.

(7) By definition, a "right angle" is a 90° angle.

(8) "Axioms" or "postulates" are statements that are assumed, without further justification, whereas "theorems" or "propositions" are proved using the axioms.

(9) We call $\sqrt{2}$ an "irrational number" because it cannot be expressed as a quotient of two whole numbers.

(10) The ancient Greeks were the first to insist on proofs for mathematical statements to make sure they were correct.

EXERCISES

In Exercises 1 – 4 you are asked to define some familiar geometric terms. The exercises provide a review of these terms as well as practice in formulating definitions with precision. In making a definition, you may use the five undefined geometric terms and all other geometric terms that have been defined in the text so far or in any preceding exercises.

Making a definition sometimes requires a bit of thought. For example, how would you define *perpendicularity* for two lines *l* and *m*? A first attempt might be to say that "*l* and *m* intersect and at their point of intersection these lines form right angles." It would be legitimate to use the terms "intersect" and "right angle" because they have been previously defined. But what is meant by the statement that *lines* form right angles? Surely, we can all draw a picture to show what we mean, but the problem is to express the idea verbally, using only terms introduced previously. According to the definition on p. 17, an angle is formed by two nonopposite *rays* emanating from the same vertex. We may therefore define *l* and *m* as *perpendicular* if they intersect at a point A and if there is a ray $\overrightarrow{AB}$ that is part of *l* and a ray $\overrightarrow{AC}$ that is part of *m* such that ∢BAC is a right angle (Figure 1.16). We denote this by *l* ⊥ *m*.

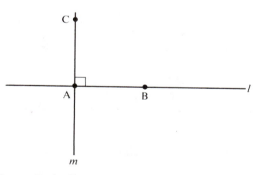

FIGURE 1.16   Perpendicular lines.

1. Define the following terms:
   (a) *Midpoint* M of a segment AB.
   (b) *Perpendicular bisector* of a segment AB (you may use the term "midpoint" since you have just defined it).
   (c) Ray $\overrightarrow{BD}$ *bisects* angle ∢ABC (given that point D is between A and C).
   (d) Points A, B, and C are *collinear*.
   (e) Lines *l*, *m*, and *n* are *concurrent* (see Figure 1.17).

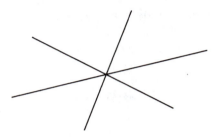

FIGURE 1.17   Concurrent lines.

2. Define the following terms:
   (a) The *triangle* △ABC formed by three noncollinear points A, B, and C.
   (b) The *vertices*, *sides*, and *angles* of △ABC. (The "sides" are segments, not lines.)
   (c) The sides *opposite to* and *adjacent to* a given vertex A of △ABC.
   (d) *Medians* of a triangle (see Figure 1.18).
   (e) *Altitudes* of a triangle (see Figure 1.19).
   (f) *Isosceles* triangle, its *base*, and its *base angles*.
   (g) *Equilateral* triangle.
   (h) *Right* triangle.

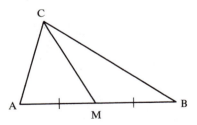

FIGURE 1.18   Median.

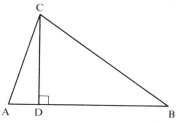

FIGURE 1.19 Altitude.

3. Given four points, A, B, C, and D, no three of which are collinear and such that any pair of the segments AB, BC, CD, and DA either have no point in common or have only an endpoint in common. We can then define the *quadrilateral* □ABCD to consist of the four segments mentioned, which are called its *sides*, the four points being called its *vertices;* see Figure 1.20. (Note that the order in which the letters are written is essential. For example, □ABCD may not denote a quadrilateral, because, for example, AB might cross CD. If □ABCD did denote a quadrilateral, it would not denote the same one as □ACDB. Which permutations of the four letters A, B, C, and D do denote the same quadrilateral as □ABCD?) Using this definition, define the following notions:

   (a) The *angles* of □ABCD.
   (b) *Adjacent* sides of □ABCD.
   (c) *Opposite* sides of □ABCD.
   (d) The *diagonals* of □ABCD.
   (e) A *parallelogram.* (Use the word "parallel.")

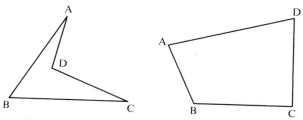

FIGURE 1.20 Quadrilaterals.

4. Define *vertical angles* (Figure 1.21). How would you attempt to prove that vertical angles are congruent to each other? (Just sketch a plan for a proof—don't carry it out in detail.)

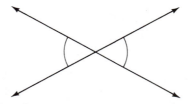

FIGURE 1.21   Vertical angles.

5. Use a common notion (p. 13) to prove the following result: If P and Q are any points on a circle with center O and radius OA, then OP ≅ OQ.

6. (a) Given two points A and B and a third point C between them. (Recall that "between" is an *undefined* term.) Can you think of any way to prove from the postulates that C lies on line $\overleftrightarrow{AB}$?

    (b) Assuming that you succeeded in proving C lies on $\overleftrightarrow{AB}$, can you prove from the definition of "ray" and the postulates that $\overrightarrow{AB} = \overrightarrow{AC}$?

7. If S and T are any sets, their *union* $(S \cup T)$ and *intersection* $(S \cap T)$ are defined as follows:

    (i) Something belongs to $S \cup T$ if and only if it belongs either to S or to T (or to both of them).

    (ii) Something belongs to $S \cap T$ if and only if it belongs both to S and to T.

Given two points A and B, consider the two rays $\overrightarrow{AB}$ and $\overrightarrow{BA}$. Draw diagrams to show that $\overrightarrow{AB} \cup \overrightarrow{BA} = \overleftrightarrow{AB}$ and $\overrightarrow{AB} \cap \overrightarrow{BA} = AB$. What additional axioms about the undefined term "between" must we assume in order to be able to *prove* these equalities?

8. To further illustrate the need for careful definition, consider the following possible definitions of *rectangle:*

    (i) A quadrilateral with four right angles.

    (ii) A quadrilateral with all angles congruent to one another.

    (iii) A parallelogram with at least one right angle.

In this book we will take (i) *as our definition.* Your experience with Euclidean geometry may lead you to believe that these three definitions are equivalent; sketch informally how you might prove that, and notice carefully which theorems you are tacitly assuming. In hyperbolic geometry these definitions give rise to three different sets of quadrilaterals (see Chapter 6). Given the definition of "rectangle," use it to define "square."

9. Can you think of any way to prove from the postulates that for every line *l*

    (a) There exists a point lying on *l*?

    (b) There exists a point not lying on *l*?

10. Can you think of any way to prove from the postulates that the plane is nonempty, i.e., that points and lines exist? (Discuss with your instructor what it means to say that mathematical objects, such as points and lines, "exist.")

11. Do you think that the Euclidean parallel postulate is "obvious"? Write a brief essay explaining your answer.

12. What is the flaw in the "proof" that all triangles are isosceles? (All the theorems from Euclidean geometry used in the argument are correct.)

13. If the number π is defined as the ratio of the circumference of any circle to its diameter, what theorem must first be proved to legitimize this definition? (For example, if I "define" a new number φ to be the ratio of the area of any circle to its diameter, that would not be legitimate. The required theorem is proved in Section 21.2 of Moise, 1990.)

14. Do you think the axiomatic method can be applied to subjects other than mathematics? Is the U.S. Constitution (including all its amendments) the list of axioms from which the federal courts logically deduce all rules of law? Do you think the "truths" asserted in the Declaration of Independence are "self-evident"?

15. Write a commentary on the application of the axiomatic method finished in 1675 by Benedict de Spinoza, entitled: *Ethics Demonstrated in Geometrical Order and Divided into Five Parts Which Treat (1) of God; (2) of the Nature and Origin of the Mind; (3) of the Nature and Origin of the Emotions; (4) of Human Bondage, or of the Strength of the Emotions; (5) of the Power of the Intellect, or of Human Liberty.* (Devote the main body of your review to Parts 4 and 5.)

MAJOR EXERCISES

1. In this exercise we will review several basic Euclidean constructions with a straightedge and compass. Such constructions fascinated mathematicians from ancient Greece until the nineteenth century, when all classical construction problems were finally solved.

    (a) Given a segment AB. Construct the perpendicular bisector of AB. (Hint: Make AB a diagonal of a rhombus, as in Figure 1.22.)

    (b) Given a line *l* and a point P lying on *l*. Construct the line through P perpendicular to *l*. (Hint: Make P the midpoint of a segment of *l*.)

    (c) Given a line *l* and a point P *not* lying on *l*. Construct the line through P perpendicular to *l*. (Hint: Construct isosceles triangle $\triangle ABP$ with base AB on *l* and use (a).)

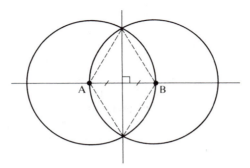

FIGURE 1.22

(d) Given a line *l* and a point P not lying on *l*. Construct a line through P parallel to *l*. (Hint: use (b) and (c).)

(e) Construct the bisecting ray of an angle. (Hint: Use the Euclidean theorem that the perpendicular bisector of the base on an isosceles triangle is also the angle bisector of the angle opposite the base.)

(f) Given △ABC and segment DE ≅ AB. Construct a point F on a given side of line $\overleftrightarrow{DE}$ such that △DEF ≅ △ABC.

(g) Given angle ∢ABC and ray $\overrightarrow{DE}$. Construct F on a given side of line $\overleftrightarrow{DE}$ such that ∢ABC ≅ ∢FDE.

2. Euclid assumed the compass to be *collapsible*. That is, given two points P and Q, the compass can draw a circle with center P passing through Q (Postulate III); however, the spike cannot be moved to another center O to draw a circle of the same radius. Once the spike is moved, the compass collapses. Check through your constructions in Exercise 1 to see if they are possible with a collapsible compass. (For purposes of this exercise, being "given" a line means being given two or more points on it.)

(a) Given three points P, Q, and R. Construct with a straightedge and collapsible compass a rectangle □PQST with PQ as a side and such that PT ≅ PR (see Figure 1.23).

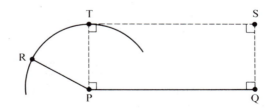

FIGURE 1.23

(b) Given a segment PQ and a ray $\overrightarrow{AB}$. Construct the point C on $\overrightarrow{AB}$ such that PQ ≅ AC. (Hint: Using (a), construct rectangle □PAST with

PT $\cong$ PQ, and then draw the circle centered at A and passing through S.)

Exercise (b) shows that you can transfer segments with a collapsible compass and a straightedge, so you can carry out all constructions as if your compass did not collapse.

3. The straightedge you used in the previous exercises was supposed to be *unruled* (if it did have marks on it, you weren't supposed to use them). Now, however, let us mark two points on the straightedge so as to mark off a certain distance *d*. Archimedes showed how we can then trisect an arbitrary angle:

For any angle, draw a circle γ of radius *d* centered at the vertex O of the angle. This circle cuts the sides of the angle at points A and B. Place the marked straightedge so that one mark gives a point C on line $\overleftrightarrow{OA}$ such that O is between C and A, the other mark gives a point D on circle γ, and the straightedge must simultaneously rest on the point B, so that B, C, and D are collinear (Figure 1.24). Prove that ∢COD so constructed is one-third of ∢AOB. (Hint: Use Euclidean theorems on exterior angles and isosceles triangles.)

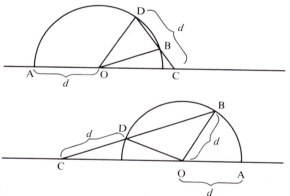

FIGURE 1.24

4. The number $\rho = (1 + \sqrt{5})/2$ was called the *golden ratio* by the Greeks, and a rectangle whose sides are in this ratio is called a *golden rectangle*.[8] Prove that a golden rectangle can be constructed with straightedge and compass as follows:

(a) Construct a square □ABCD.

---

[8] For applications of the golden ratio to Fibonacci numbers and phyllotaxis, see Coxeter (1969), Chapter 11.

(b) Construct midpoint M of AB.
(c) Construct point E such that B is between A and E and MC ≅ ME
    (Figure 1.25).

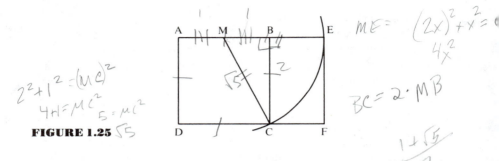

$ME = (2x)^2 + x^2 =$

$4x^2$

$2^2 + 1^2 = (MC)^2$
$4 + 1 = MC^2$
$5 = MC^2$

$BC = 2 \cdot MB$

$\dfrac{1 + \sqrt{5}}{2}$

FIGURE 1.25 $\sqrt{5}$

(d) Construct the foot F of the perpendicular from E to $\overleftrightarrow{DC}$.
(e) Then □AEFD is a golden rectangle (use the Pythagorean theorem
    for △MBC).
(f) Moreover, □BEFC is another golden rectangle (first show that
    $1/\rho = \rho - 1$).

The next two exercises require a knowledge of trigonometry.

5. The Egyptians thought that if a quadrilateral had sides of lengths a, b, c,
    and d, then its area S was given by the formula $(a + c)(b + d)/4$. Prove
    that actually

$$4S \leqq (a + c)(b + d)$$

with equality holding only for rectangles. (Hint: Twice the area of a
triangle is $ab \sin \theta$, where θ is the angle between the sides of lengths a, b
and $\sin \theta \leqq 1$, with equality holding only if θ is a right angle.)

6. Prove analogously that if a triangle has sides of lengths a, b, c, then its area
    S satisfies the inequality

$$4S\sqrt{3} \leqq a^2 + b^2 + c^2$$

with equality holding only for equilateral triangles. (Hint: If θ is the angle
between sides b and c, chosen so that it is at most 60°, then use the
formulas

$$2S = bc \sin \theta$$
$$2bc \cos \theta = b^2 + c^2 - a^2 \text{ (law of cosines)}$$
$$\cos (60° - \theta) = (\cos \theta + \sqrt{3} \sin \theta)/2$$

7. Let △ABC be such that AB is not congruent to AC. Let D be the point of
    intersection of the bisector of ∢A and the perpendicular bisector of side

BC. Let E, F, and G be the feet of the perpendiculars dropped from D to $\overleftrightarrow{AB}$, $\overleftrightarrow{AC}$, $\overleftrightarrow{BC}$, respectively. Prove that:

(a) D lies outside the triangle on the circle through ABC.

(b) One of E or F lies inside the triangle and the other outside.

(c) E, F, and G are collinear.

(Use anything you know, including coordinates if necessary.)

PROJECTS

1. Write a paper explaining in detail why it is impossible to trisect an arbitrary angle or square a circle using only a compass and unmarked straightedge; see Jones, Morris, and Pearson (1991); Eves (1963–1965); Kutuzov (1960); or Moise (1990). Explain how arbitrary angles can be trisected if in addition we are allowed to draw a parabola or a hyperbola or a conchoid or a limaçon (see Peressini and Sherbert, 1971).

2. Here are two other famous results in the theory of constructions:

   (a) The Danish mathematician G. Mohr and the Italian L. Mascheroni discovered independently that all Euclidean constructions of points can be made with a compass alone. A line, of course, cannot be drawn with a compass, but it can be determined with a compass by constructing two points lying on it. In this sense, Mohr and Mascheroni showed that the straightedge is unnecessary.

   (b) On the other hand, the German J. Steiner and the Frenchman J. V. Poncelet showed that all Euclidean constructions can be carried out with a straightedge alone if we are first given a single circle and its center.

   Report on these remarkable discoveries (see Eves, 1963–1965, and Kutuzov, 1960).

3. Given any $\triangle ABC$. Draw the two rays that trisect each of its angles, and let P, Q, and R be the three points of intersection of adjacent trisectors. Prove Morley's theorem[9] that $\triangle PQR$ is an equilateral triangle (see Figure 1.26 and Coxeter, 1969).

4. An n-sided polygon is called *regular* if all its sides (respectively, angles) are congruent to one another. Construct a regular pentagon and a regular hexagon with straightedge and compass. The regular septagon cannot be so constructed; in fact, Gauss proved the remarkable theorem that the regular n-gon is constructible if and only if all odd prime factors of n occur

---

[9] For a converse and generalization of Morley's theorem, see Kleven (1978).

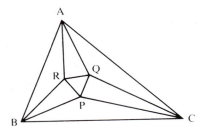

FIGURE 1.26   Morley's theorem.

to the first power and have the form $2^{2^m} + 1$ (e.g., 3, 5, 17, 257, 65,537). Report on this result, using Klein (1956). Primes of that form are called *Fermat primes*. The five listed are the only ones known at this time. Gauss did not actually construct the regular 257-gon or 65,537-gon; he only showed that the minimal polynomial equation satisfied by cos $(2\pi/n)$ for such n could be solved in the surd field (see Moise, 1990). Other devoted (obsessive?) mathematicians carried out the constructions. The constructor for $n = 65,537$ labored for 10 years and was rewarded with a Ph.D. degree; what is the reward for checking his work?

5. Write a short biography of Archimedes (Bell, 1961, is one good reference). Archimedes discovered some of the ideas of integral calculus 14 centuries before Newton and Leibniz.

"I THINK YOU SHOULD BE MORE EXPLICIT HERE IN STEP TWO."

CHAPTER 2

LOGIC AND INCIDENCE GEOMETRY

Reductio ad absurdum . . . is a far finer gambit than any chess gambit: a chess player may offer the sacrifice of a pawn or even a piece, but a mathematician offers the *game*.

G. H. HARDY

INFORMAL LOGIC

In the previous chapter we were introduced to the postulates and basic definitions of Euclid's geometry, slightly rephrased for greater precision. We would like to begin proving some theorems or propositions that are logical consequences of the postulates. However, the exercises of the previous chapter may have alerted you to expect some difficulties that we must first clear up. For example, there is nothing in the postulates that guarantees that a line has any points lying on it (or off it)! You may feel this is ridiculous — it wouldn't be a line if it didn't have any points lying on it. (What kind of a line is he feeding us anyway?) In a sense, your protest would be legitimate, for if my concept of a line were so different from yours, we would not understand each other, and Requirement 0 — that there be mutual understanding of words and symbols used — would be violated.

So let me be perfectly clear. We must play this game according to the rules, the rules mentioned in Requirement 2 but not spelled out. Unfortunately, to discuss them completely would require changing the content of this book from geometry to symbolic logic. Instead, I will only remind you of some basic rules of reasoning that you, as a rational being, already know.

LOGIC RULE 0.   No unstated assumptions may be used in a proof.

The reason for taking the trouble in Chapter 1 to list all our axioms was to be explicit about our basic assumptions, including even the most obvious. Although it is "obvious" that two points determine a unique line, Euclid stated this as his first postulate. So if in some proof we want to say that every line has points lying on it, we should list this statement as another postulate (or prove it, but we can't). In other words, all our cards must be out on the table. If you reread Exercises 6, 7, 9, and 10 in Chapter 1, you will find some "obvious" assumptions that we will have to make explicit. This will be done later.

Perhaps you have realized by now that there is a vital relation between axioms and undefined terms. As we have seen, we must have undefined terms in order to avoid infinite regress. But this does not mean we can use these terms in any way we choose. The axioms tell us exactly what properties of undefined terms we are allowed to use in our arguments. You may have some other properties in your mind when you think about these terms, but you're not allowed to use them in a proof (Rule 0). For example, when you think of the unique line determined by two points, you probably think of it as being "straight," or as "the shortest path between the two points." Euclid's postulates do not allow us to assume these properties. Besides, from one viewpoint, these properties could be considered contradictory. If you were traveling the surface of the earth, say, from San Francisco to Moscow, the shortest path would be an arc of a *great circle* (a straight path would

FIGURE 2.1   The shortest path between two points on a sphere is an arc of a *great circle* (a circle whose center is the center of the sphere and whose radius is the radius of the sphere, e.g., the equator).

bore through the earth). Indeed, pilots in a hurry fly their aircraft over great circles.

THEOREMS AND PROOFS

All mathematical theorems are conditional statements, statements of the form

If [hypothesis] *then* [conclusion].

In some cases a theorem may state only a conclusion; the axioms of the particular mathematical system are then implicit (assumed) as a hypothesis. If a theorem is not written in the conditional form, it can nevertheless be translated into that form. For example,

Base angles of an isosceles triangle are congruent.

can be interpreted as

If a triangle has two congruent sides, then the angles opposite those sides are congruent.

Put another way, a conditional statement says that one condition (the hypothesis) *implies* another (the conclusion). If we denote the hypothesis by H, the conclusion by C, and the word "implies" by an arrow $\Rightarrow$, then every theorem has the form $H \Rightarrow C$. (In the example above, H is "two sides of a triangle are congruent" and C is "the angles opposite those sides are congruent.")

Not every conditional statement is a theorem. For example, the statment

If $\triangle ABC$ is any triangle, then it is isosceles.

is not a theorem. Why not? You might say that this statement is "false" whereas theorems are "true." Let's avoid the loaded words "true" and "false," for they beg the question and lead us into more complicated issues.

In a given mathematical system the only statements we call *theorems*[1] are those statements for which a *proof* has been supplied. We can

---

[1] Or sometimes *propositions, corollaries,* or *lemmas.* "Theorem" and "proposition" are interchangeable; a "corollary" is an immediate consequence of a theorem, and a "lemma" is a "helping theorem." Logically, they all mean the same; the title is just an indicator of the author's emphasis.

TEST #I5

disprove the assertion that every triangle is isosceles by exhibiting a triangle that is not isosceles, such as a 3-4-5 right triangle.

The crux of the matter then is the notion of *proof*. By definition, a proof is a list of statements, together with a justification for each statement, ending up with the conclusion desired. Usually, each statement in a proof will be numbered in this book, and the justification for it will follow in parentheses.

LOGIC RULE 1.   The following are the six types of justifications allowed for statements in proofs:

(1) "By hypothesis. . . ."
(2) "By axiom. . . ."
(3) "By theorem . . ." (previously proved).
(4) "By definition. . . ."
(5) "By step . . ." (a previous step in the argument).
(6) "By rule . . . of logic."

TEST I3

Later in the book our proofs will be less formal, and justifications may be omitted when they are obvious (Be forewarned, however, that these omissions can lead to incorrect results.) *A justification may involve several of the above types.*

Having described proofs, it would be nice to be able to tell you how to find or construct them. Yet that is the mystery of doing mathematics. Certain techniques for proving theorems are learned by experience, by imitating what others have done. But there is no rote method for proving or disproving every statement in mathematics. (The nonexistence of such a rote method is, when stated precisely, a deep theorem in mathematical logic and is the reason why computers will never put mathematicians out of business—see DeLong, 1970, Chapter 5).

However, some suggestions may help you construct proofs. First, make sure you clearly understand the meaning of each term in the statement of the proposed theorem. If necessary, review their definitions. Second, keep reminding yourself of what it is you are trying to prove. If it involves parallel lines, for example, look up previous propositions that give you information about parallel lines. If you find another proposition that seems to apply to the problem at hand, check carefully to see whether it really does apply. Draw pictures to help you visualize the problem.

RAA PROOFS

The most common type of proof in this book is proof by *reductio ad absurdum,* abbreviated RAA. In this type of proof you want to prove a conditional statement, $H \Rightarrow C$, and you begin by assuming the contrary of the conclusion you seek. We call this contrary assumption the *RAA hypothesis,* to distinguish it from the hypothesis H. The RAA hypothesis is a temporary assumption from which we derive, by reasoning, an *absurd statement* ("absurd" in the sense that it denies something known to be valid). Such a statement might deny the hypothesis of the theorem or the RAA hypothesis; it might deny a previously proved theorem or an axiom. Once it is shown that the negation of C leads to an absurdity, it follows that C must be valid. This is called the *RAA conclusion.* To summarize:

LOGIC RULE 2. To prove a statement $H \Rightarrow C$, assume the negation of statement C (RAA hypothesis) and deduce an absurd statement, using the hypothesis H if needed in your deduction.

Let us illustrate this rule by proving the following proposition (Proposition 2.1): If l and m are distinct lines that are not parallel, then l and m have a unique point in common.

Proof:
(1) Because l and m are not parallel, they have a point in common (by definition of "parallel").
(2) Since we want to prove uniqueness for the point in common, we will assume the contrary, that l and m have two distinct points A and B in common (RAA hypothesis).
(3) Then there is more than one line on which A and B both lie (step 2 and the hypothesis of the theorem, $l \neq m$).
(4) A and B lie on a unique line (Euclid's Postulate I).
(5) Intersection of l and m is unique (3 contradicts 4, RAA conclusion). ∎

Notice that in steps 2 and 5, instead of writing "Logic Rule 2" as justification, we wrote the more suggestive "RAA hypothesis" and "RAA conclusion," respectively.

As another illustration, consider one of the earliest RAA proofs, discovered by the Pythagoreans (to their great dismay). In giving this proof, we will use some facts about Euclidean geometry and numbers that you know, and we will be informal.

Suppose $\triangle ABC$ is a right isosceles triangle with right angle at C. We can choose our unit of length so that the legs have length 1. The theorem then says that the length of the hypotenuse is irrational (Figure 2.2).

By the Pythagorean theorem, the length of the hypotenuse is $\sqrt{2}$, so we must prove that $\sqrt{2}$ is an irrational number, i.e., that it is not a rational number.

What is a rational number? It is a number that can be expressed as a quotient p/q of two integers p and q. For example, $\frac{1}{2}$, $\frac{2}{3}$, and $5 = \frac{5}{1}$ are rational numbers. We want to prove that $\sqrt{2}$ is not one of these numbers.

We begin by assuming the contrary, that $\sqrt{2}$ is a rational number (RAA hypothesis). In other words, $\sqrt{2} = p/q$ for certain unspecified whole numbers p and q. You know that every rational number can be written in lowest terms, i.e., such that the numerator and denominator have no common factor. For example, $\frac{4}{6}$ can be written as $\frac{2}{3}$, where the common factor 2 in the numerator and denominator has been canceled. Thus we can assume all common factors have been canceled, so that p and q have no common factor.

Next, we clear denominators:

$$\sqrt{2}q = p$$

and square both sides:

$$2q^2 = p^2.$$

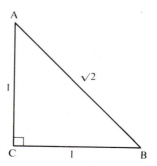

FIGURE 2.2

This equation says that p^2 is an even number (since p^2 is twice another whole number, namely, q^2). If p^2 is even, p must be even, for the square of an odd number is odd, as you know. Thus,

$$p = 2r$$

for some whole number r (that is what it means to be even). Substituting $2r$ for p in the previous equations gives

$$2q^2 = (2r)^2 = 4r^2.$$

We then cancel 2 from both sides to get

$$q^2 = 2r^2.$$

This equation says that q^2 is an even number; hence q must be even.

We have shown that numerator p and denominator q are both even, meaning that they have 2 as a common factor. Now this is absurd, because all common factors were canceled. Thus, $\sqrt{2}$ is irrational (RAA conclusion). ∎

NEGATION

In an RAA proof we begin by "assuming the contrary." Sometimes the contrary or negation of a statement is not obvious, so you should know the rules for negation.

First, some remarks on notation. If S is any statement, we will denote the negation or contrary of S by $\sim S$. For example, if S is the statement "p is even," then $\sim S$ is the statement "p is not even" or "p is odd."

The rule below applies to those cases where S is already a negative statement. The rule states that two negatives make a positive.

LOGIC RULE 3. The statement "$\sim(\sim S)$" means the same as "S."

We followed this rule when we negated the statement "$\sqrt{2}$ is irrational" by writing the contrary as "$\sqrt{2}$ is rational" instead of "$\sqrt{2}$ is not irrational."

Another rule we have already followed in our RAA method is the rule for negating an implication. We wish to prove $H \Rightarrow C$, and we assume, on the contrary, H does not imply C, i.e., that H holds and at the same time $\sim C$ holds. We write this symbolically as $H \,\&\, \sim C$, where & is the abbreviation for "and." A statement involving the connective "and" is called a *conjunction*. Thus:

LOGIC RULE 4. The statement "$\sim[H \Rightarrow C]$" means the same as "$H \,\&\, \sim C$."

Let us consider, for example, the conditional statement "if 3 is an odd number, then 3^2 is even." According to Rule 4, the negation of this is the declarative statement "3 is an odd number and 3^2 is odd."

How do we negate a conjunction? A conjunction $S_1 \,\&\, S_2$ means that statements S_1 and S_2 both hold. Negating this would mean asserting that one of them does not hold, i.e., asserting the negation of one or the other. Thus:

LOGIC RULE 5. The statement "$\sim[S_1 \,\&\, S_2]$" means the same as "$[\sim S_1$ or $\sim S_2]$."   *& vice versa*

A statement involving the connective "or" is called a *disjunction*. The mathematical "or" is not exclusive like "or" in everyday usage. Consider the conjunction "$1 = 2$ and $1 = 3$." If we wish to deny this, we must write (according to Rule 5) "$1 \neq 2$ or $1 \neq 3$." Of course, both inequalities are valid. So when a mathematician writes "S_1 or S_2" he means "either S_1 holds or S_2 holds *or they both hold*."

Finally let us be more precise about what is an absurd statement. It is the conjunction of a statement S with the negation of S, i.e., "$S \,\&\, \sim S$." A statement of this type is called a *contradiction*. A system of axioms from which no contradiction can be deduced is called *consistent*.

QUANTIFIERS

Most mathematical statements involve *variables*. For instance, the Pythagorean theorem states that for any right triangle, if a and b are the lengths of the legs and c the length of the hypotenuse, then

$c^2 = a^2 + b^2$. Here *a*, *b*, and *c* are variable numbers, and the triangle whose sides they measure is a variable triangle.

Variables can be quantified in two different ways. First, in a *universal* way, as in the expressions:

"For any *x*, ..."
"For every *x*, ..."
"For all *x*, ..."
"Given any *x*, ..."
"If *x* is any . . . ."

Second, in an *existential* way, as in the expressions:

"For some *x*, ..."
"There exists an *x* . . . ."
"There is an *x* . . . ."
"There are *x* . . . ."

Consider Euclid's first postulate, which states informally that two points P and Q determine a unique line *l*. Here P and Q may be any two points, so they are quantified universally, whereas *l* is quantified existentially, since it is asserted to exist, once P and Q are given.

It must be emphasized that a statement beginning with "For every . . ." does not imply the existence of anything. The statement "every unicorn has a horn on its head" does not imply that unicorns exist.

If a variable *x* is quantified universally, this is usually denoted as ∀*x*, (read as "for all *x*"). If *x* is quantified existentially, this is usually denoted as ∃*x* (read as "there exists an *x* . . ."). After a variable *x* is quantified, some statement is made about *x*, which we can write as $S(x)$ (read as "statement *S* about *x*"). Thus, a universally quantified statement about a variable *x* has the form ∀*x*$S(x)$.

We wish to have rules for negating quantified statements. How do we deny that statement $S(x)$ holds for all *x*? We can do so clearly by asserting that for some *x*, $S(x)$ does not hold.

LOGIC RULE 6. The statement "~[∀*x*$S(x)$]" means the same as "∃*x* ~ $S(x)$."

For example, to deny "All triangles are isosceles" is to assert "There is a triangle that is not isosceles."

Similarly, to deny that there exists an x having property $S(x)$ is to assert that all x fail to have property $S(x)$.

LOGIC RULE 7. The statement "$\sim[\exists x S(x)]$" means the same as "$\forall x \sim S(x)$."

For example, to deny "There is an equilateral right triangle" is to assert "Every right triangle is nonequilateral" or, equivalently, to assert "No right triangle is equilateral."

Since in practice quantified statements involve several variables, the above rules will have to be applied several times. Usually, common sense will quickly give you the negation. If not, follow the above rules.

Let's work out the denial of Euclid's first postulate. This postulate is a statement about all pairs of points P and Q; negating it would mean, according to Rule 6, asserting the existence of points P and Q that do not satisfy the postulate. Postulate I involves a conjunction, asserting that P and Q lie on some line *l and* that *l* is unique. In order to deny this conjunction, we follow Rule 5. The assertion becomes either "P and Q do not lie on any line" *or* "they lie on more than one line." Thus, the negation of Postulate I asserts: "There are two points P and Q that either do not lie on any line or lie on more than one line."

If we return to the example of the surface of the earth, thinking of a "line" as a great circle, we see that there do exist such points P and Q—namely, take P to be the north pole and Q the south pole. Infinitely many great circles pass through both poles. (See Figure 2.3.)

Mathematical statements are sometimes made informally, and you may sometimes have to rephrase them before you will be able to

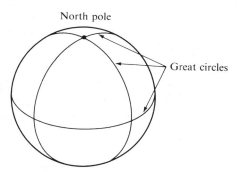

North pole

Great circles

FIGURE 2.3

negate them. For example, consider the following statement:

If a line intersects one of two parallel lines, it also intersects the other.

This appears to be a conditional statement, of the form "if . . . then . . ."; its negation, according to Rule 4, would appear to be:

A line intersects one of two parallel lines and does not intersect the other.

If this seems awkward, it is because the original statement contained *hidden quantifiers* that have been ignored. The original statement refers to *any* line that intersects one of two parallel lines, and these are *any* parallel lines. There are universal quantifiers implicit in the original statement. So we have to follow Rule 6 as well as Rule 4 in forming the correct negation, which is:

There exist two parallel lines and a line that intersects one of them and does not intersect the other.

IMPLICATION

Another rule, called the *rule of detachment*, or *modus ponens*, is the following:

LOGIC RULE 8. If $P \Rightarrow Q$ and P are steps in a proof, then Q is a justifiable step.

This rule is almost a definition of what we mean by implication. For example, we have an axiom stating that if ∢A and ∢B are right angles, then ∢A ≅ ∢B (Postulate IV). Now in the course of a proof we may come across two right angles. Rule 8 allows us to assert their congruence as a step in the proof.

You should beware of confusing a conditional statement $P \Rightarrow Q$ with its *converse* $Q \Rightarrow P$. For example, the converse of Postulate IV states that if ∢A ≅ ∢B then ∢A and ∢B are right angles, which is absurd.

However, it may sometimes happen that both a conditional statement and its converse are valid. In case $P \Rightarrow Q$ and $Q \Rightarrow P$ both hold, we write simply $P \Leftrightarrow Q$ (read as "*P if and only if Q*" or "*P is logically equivalent* to *Q*"). All definitions are of this form. For example, three points are collinear if and only if they lie on a line. Some theorems are also of this form, such as the theorem "a triangle is isosceles if and only if two of its angles are congruent to each other." The next rule gives a few more ways that "implication" is often used in proofs.

LOGIC RULE 9. (a) $[[P \Rightarrow Q] \ \& \ [Q \Rightarrow R]] \Rightarrow [P \Rightarrow R]$.
(b) $[P \ \& \ Q] \Rightarrow P, \ [P \ \& \ Q] \Rightarrow Q$.
(c) $[\sim Q \Rightarrow \sim P] \Leftrightarrow [P \Rightarrow Q]$.

Part (c) states that every implication $P \Rightarrow Q$ is logically equivalent to its *contrapositive* $\sim Q \Rightarrow \sim P$. All parts of Rule 9 are called *tautologies*, because they are valid just by their form, not because of what P, Q, and R mean; by contrast, the validity of a formula such as $P \Rightarrow Q$ does depend on the meaning, as we have just seen. There are infinitely many tautologies, and the next rule gives the most infamous.

LAW OF EXCLUDED MIDDLE AND PROOF BY CASES

LOGIC RULE 10. For every statement P, "*P* or $\sim P$" is a valid step in a proof (law of excluded middle).[2]

For example, given point P and line *l*, we may assert that either P lies on *l* or it does not. If this is a step in a proof, we will usually then break the rest of the proof into cases — giving an argument under the case assumption that P lies on *l* and giving another argument under the case assumption that P does not. Both arguments must be given, or

---

[2] The law of excluded middle characterizes classical two-valued logic: either a statement holds or it does not; there is no middle ground such as "we don't know." Constructivist mathematicians (such as Brouwer, Bishop, Beeson, and Stolzenberg) reject the unqualified use of this rule when applied to existence statements. They insist that in order to meaningfully prove that a mathematical object *exists*, one must supply an effective method for constructing it. It is uninformative merely to assume that the object does not exist (RAA hypothesis) and then derive a contradiction (so they also reject Logic Rule 6 when applied to infinite sets). The "constructive" aspect of Euclid's geometry traditionally refers to "straightedge and compass constructions" (see the Major Exercises of Chapter 1). We will pay close attention to this aspect throughout this book.

else the proof is incomplete. A proof of this type is given for Proposition 3.16 in Chapter 3, which asserts that there exists a line through P perpendicular to *l*.

Sometimes there are more than two cases. For example, it is a theorem that either an angle is acute or it is right or it is obtuse — three cases. We will have to give three arguments — one for each case assumption. You will give such arguments when you prove the SSS criterion for congruence of triangles in Exercise 32 of Chapter 3. This method of *proof by cases* was used (correctly) in the incorrect attempt in Chapter 1 to prove that all triangles are isosceles.

LOGIC RULE 11. Suppose the disjunction of statements S_1 or S_2 or . . . or S_n is already a valid step in a proof. Suppose that proofs of C are carried out from each of the *case assumptions* $S_1, S_2, . . . , S_n$. Then C can be concluded as a valid step in the proof (proof by cases).

And this concludes our discussion of logic. No claim is made that all the rules of logic have been listed, just that those listed should suffice for our purposes. For further discussion, see DeLong (1970) and his bibliography.

INCIDENCE GEOMETRY

Let us apply the logic we have developed to a very elementary part of geometry, *incidence geometry*. We assume only the undefined terms "point" and "line" and the undefined relation "incidence" between a point and a line, written "P lies on *l*" or P ɪ *l* or "*l* passes through P" as before. We don't discuss "betweenness" or "congruence" in this restricted geometry (but we are now beginning the new axiomatic development of geometry that fills the gaps in Euclid and applies to other geometries as well; that development will continue in future chapters, and the formal definitions given in Chapter 1 will be used). These undefined terms will be subjected to three axioms, the first of which is the same as Euclid's first postulate.

INCIDENCE AXIOM 1. For every point P and for every point Q not equal to P there exists a unique line *l* incident with P and Q.

INCIDENCE AXIOM 2. For every line *l* there exist at least two distinct points incident with *l*.

INCIDENCE AXIOM 3. There exist three distinct points with the property that no line is incident with all three of them.

These axioms fill the gap mentioned in Exercises 9 and 10, Chapter 1. We can now assert that every line has points lying on it — at least two, possibly more — and that the points do not all lie on one line. Moreover, we know that the geometry must have at least three points in it, by the third axiom and Rule 9(b) of logic. Namely, Incidence Axiom 3 is a conjunction of two statements:

1. *There exist distinct points A, B, and C.*
2. *For every line, at least one of these points does not lie on the line.*

Rule 9(b) tells us that a conjunction of two statements implies each statement separately, so we can conclude that three distinct points exist (Rule 8).

Incidence geometry has some defined terms, such as "collinear," "concurrent," and "parallel," defined exactly as they were in Chapter 1. Incidence Axiom 3 can be rewritten as "there exist three noncollinear points." Parallel lines are still lines that do not have a point in common.

What sort of results can we prove using this meager collection of axioms? None that are very exciting, but here are a few you can prove as exercises.

PROPOSITION 2.1. If *l* and *m* are distinct lines that are not parallel, then *l* and *m* have a unique point in common.

PROPOSITION 2.2. There exist three distinct lines that are not concurrent.

PROPOSITION 2.3. For every line there is at least one point not lying on it.

PROPOSITION 2.4. For every point there is at least one line not passing through it.

PROPOSITION 2.5. For every point P there exist at least two lines through P.

MODELS

In reading over the axioms of incidence in the previous section, you may have imagined dots and long dashes drawn on a sheet of paper. With this representation in mind, the axioms appear to be correct statements. We will take the point of view that these dots and dashes are a *model* for incidence geometry.

More generally, if we have any axiom system, we can interpret the undefined terms in some way, i.e., give the undefined terms a particular meaning. We call this an *interpretation* of the system. We can then ask whether the axioms, so interpreted, are correct statements. If they are, we call the interpretation a *model*. When we take this point of view, interpretations of the undefined terms "point," "line," and "incident" other than the usual dot-and-dash drawings become possible.

Example 1. Consider a set {A, B, C} of three letters, which we will call "points." "Lines" will be those subsets that contain exactly two letters — {A, B}, {A, C}, and {B, C}. A "point" will be interpreted as "incident" with a "line" if it is a member of that subset. Thus, under this interpretation, A lies on {A, B} and {A, C} but does not lie on {B, C}. In order to determine whether this interpretation is a *model*, we must check whether the interpretations of the axioms are correct statements. For Incidence Axiom 1, if P and Q are any two of the letters, A, B, and C, {P, Q} is the unique "line" on which they both lie. For Axiom 2, if {P, Q} is any "line," P and Q are two distinct "points" lying on it. For Axiom 3, we see that A, B, and C are three distinct "points" that are not collinear.

What is the use of models? The main property of any model of an axiom system is that all theorems of the system are correct statements in the model. This is because logical consequences of correct statements are themselves correct. (By definition of "model," axioms are correct statements when interpreted in models; theorems are logical consequences of axioms.) Thus, we immediately know that the five propositions in the previous section hold in the three-point geometry above (Example 1).

Suppose we have a statement in the formal system but don't yet know whether it is a theorem, i.e., we don't yet know whether it can be proved. We can look at our models and see whether the statement is correct in the models. If we can find *one* model where the interpreted statement fails to hold, we can be sure that no proof is possible. You are undoubtedly familiar with testing for the correctness of geometric statements by drawing pictures. Of course, the converse does not work; just because a drawing makes a statement *look* right does not mean you can *prove* it. This was illustrated on pp. 23–25.

The advantage of having several models is that a statement may hold in one model but not in another. Models are "laboratories" for experimenting with the formal system.

Let us experiment with the Euclidean parallel postulate. This is a statement in the formal system incidence geometry: "For every line *l* and every point P not lying on *l* there exists a unique line through P that is parallel to *l*." This statement appears to be correct according to our drawings (although we cannot verify the uniqueness of the parallelism, since we cannot extend our dashes indefinitely). But what about our three-point model? It is immediately apparent that *no parallel lines exist* in this model: {A, B} meets {B, C} in the point B and meets {A, C} in the point A; {B, C} meets {A, C} in the point C. (We say that this model has the *elliptic parallel property*.)

Thus, we can conclude that *no proof of the Euclidean parallel postulate from the axioms of incidence alone is possible; in fact, in incidence geometry it is impossible to prove that parallel lines exist.* Similarly, the statement "any two lines have a point in common" (the elliptic parallel property) cannot be proved from the axioms of incidence geometry, for if you

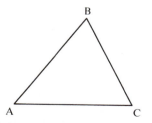

FIGURE 2.4 Elliptic parallel property (no parallel lines). A 3-point incidence geometry.

could prove it, it would hold in the usual drawn model (and in the models that will be described in Examples 3 and 4).

The technical description for this situation is that the statement "parallel lines exist" is "independent" of the axioms of incidence. We call a statement *independent* of given axioms if it is impossible to either prove or disprove the statement from the axioms. Independence is demonstrated by constructing two models for the axioms: one in which the statement holds and one in which it does not hold. This method will be used very decisively in Chapter 7 to settle once and for all the question of whether the parallel postulate can be proved.

An axiom system is called *complete* if there are no independent statements in the language of the system, i.e., every statement in the language of the system can either be proved or disproved from the axioms. Thus, the axioms for incidence geometry are incomplete. The axioms for Euclidean and hyperbolic geometries given later in the book can be proved to be complete (see Tarski's article in Henkin, Suppes, and Tarski, 1959).

Example 2. Suppose we interpret "points" as points on a sphere, "lines" as great circles on the sphere, and "incidence" in the usual sense, as a point lying on a great circle. In this interpretation there are again no parallel lines. However, this interpretation is *not* a model for incidence geometry, for, as was already mentioned, the interpretation of Incidence Axiom 1 fails to hold — there are an infinite number of great circles passing through the north and south poles on the sphere (see Figure 2.3).

Example 3. Let the "points" be the four letters A, B, C, and D. Let the "lines" be all six sets containing exactly two of these letters:

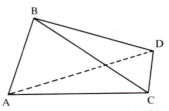

FIGURE 2.5 Euclidean parallel property. A 4-point incidence geometry.

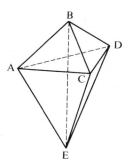

FIGURE 2.6 Hyperbolic parallel property. A 5-point incidence geometry.

{A, B}, {A,C}, {A, D}, {B, C}, {B, D}, and {C, D}. Let "incidence" be set membership, as in Example 1. As an exercise, you can verify that this is a model for incidence geometry and that in this model the Euclidean parallel postulate does hold (see Figure 2.5).

Example 4. Let the "points" be the five letters A, B, C, D, and E.[3] Let the "lines" be all 10 sets containing exactly two of these letters. Let "incidence" be set membership, as in Examples 1 and 3. You can verify that in this model the following statement about parallel lines, characteristic of hyperbolic geometry, holds: "For every line *l* and every point P not on *l* there exist at least two lines through P parallel to *l*." (See Figure 2.6).

Let us summarize the significance of models. Models can be used to prove the independence of a statement from given axioms; i.e., models can be used to demonstrate the impossibility of proving or disproving a statement from the axioms. Moreover, if an axiom system has many models that are essentially different from each other, as the models in Examples 1, 3, and 4 are essentially different from each other, then that system has a wide range of applicability. Propositions proved from the axioms of such a system are automatically correct statements within *any* of the models. Mathematicians often discover

---

[3] An incidence geometry with only finitely many points is called a *finite geometry*. There is an entertaining discussion of finite geometries (with applications to growing tomato plants) in Chapter 4 of Beck, Bleicher, and Crowe (1969). For an advanced treatment, see Dembowski (1968) or Stevenson (1972). See the exercises at the end of this chapter for more examples.

that an axiom system constructed with one particular model in mind has applications to completely different models never dreamed of.

At the other extreme, when all models of an axiom system are isomorphic to one another, the axioms are called *categorical*. (The axioms for Euclidean and hyperbolic geometries given later in the book are categorical.) The advantage of categorical axioms is that they completely describe all properties of the model that are expressible in the language of the system.[4] (For a simple example of a categorical system, suppose we add to the three incidence axioms a fourth axiom asserting that there do not exist four distinct points. Obviously, the three-point model in Example 1 is the only model, up to isomorphism, for this expanded axiom system.)

Finally, models provide evidence for the consistency of the axiom system. For example, if incidence geometry were inconsistent, the supposed proof of a contradiction could be translated into proof of a contradiction in the utterly trivial set theory for the set of three letters A, B, and C (Example 1).

ISOMORPHISM OF MODELS

We want to make precise the notion of two models being "essentially the same" or *isomorphic:* for incidence geometries, this will mean that there exists a one-to-one correspondence $P \leftrightarrow P'$ between the points of the models and a one-to-one correspondence $l \leftrightarrow l'$ between the lines of the models such that P lies on l if and only if P' lies on l'; such a correspondence is called an *isomorphism* from one model onto the other.

Example 5.   Consider a set $\{a, b, c\}$ of three letters, which we will call "lines" now. "Points" will be those subsets that contain exactly two letters — $\{a, b\}$, $\{a, c\}$, and $\{b, c\}$. Let incidence be set membership; for example, "point" $\{a, b\}$ is incident with "line" a and "line" b, not

---

[4] This is a nontrivial (and nonconstructive) theorem of mathematical logic called Gödel's completeness theorem, which says (modulo cardinality considerations) that if the system is categorical, then for every sentence S, there exists either a proof of S or a proof of $\sim S$.

with "line" c. This model certainly seems to be structurally the same as the three-point model in Example 1 — all we've changed is the notation. An explicit isomorphism is given by the following correspondences:

$$A \leftrightarrow \{a, b\} \qquad \{A, B\} \leftrightarrow b$$
$$B \leftrightarrow \{b, c\} \qquad \{B, C\} \leftrightarrow c$$
$$C \leftrightarrow \{a, c\} \qquad \{A, C\} \leftrightarrow a$$

Note that A lies on $\{A, B\}$ and $\{A, C\}$ only; its corresponding "point" $\{a, b\}$ lies on the corresponding "lines" b and a only. Similar checking with B and C shows that incidence is preserved by our correspondence. On the other hand, if we used a correspondence such as

$$\{A, B\} \leftrightarrow a$$
$$\{B, C\} \leftrightarrow b$$
$$\{A, C\} \leftrightarrow c$$

for the "lines," keeping the same correspondence for the "points," we would not have an isomorphism because, for example, A lies on $\{A, C\}$ but the corresponding "point" $\{a, b\}$ does not lie on the corresponding "line" c.

To further illustrate the idea that isomorphic models are "essentially the same," consider two models with different parallelism properties, such as one with the elliptic property and one with the Euclidean. We claim that these models are not isomorphic: suppose, on the contrary, that an isomorphism could be set up. Given line l and point P not on it; then every line through P meets l, by the elliptic property. Hence every line through the corresponding point P′ meets the corresponding line l', but that contradicts the Euclidean property of the second model.

Later on, we will need to use the concept of "isomorphism" for models of a geometry more complicated than incidence geometry — neutral geometry. In neutral geometry we will have betweenness and congruence relations, in addition to the incidence relation, and we will require an "isomorphism" to preserve those relations as well.

The general idea is that *an isomorphism of two models of an axiom system is a one-to-one correspondence between the basic objects of the system that preserves all the basic relations of the system.*

Another example to be discussed in Chapter 9 is the axiom system for a "group." Roughly speaking, a group is a set with a multiplication for its elements satisfying a few familiar axioms of algebra. An "isomorphism" of groups will then be a one-to-one mapping $x \rightarrow x'$ of one set onto the other which preserves the multiplication, i.e., for which $(xy)' = x'y'$.

PROJECTIVE AND AFFINE PLANES

We now very briefly discuss two types of models of incidence geometry that are particularly significant. During the Renaissance, around the fifteenth century, artists developed a theory of perspective in order to realistically paint two-dimensional representations of three-dimensional scenes. The theory described the projection of points in the scene onto the artist's canvas by lines from those points to a fixed viewing point in the artist's eye; the intersection of those lines with the plane of the canvas was used to construct the painting. The mathematical formulation of this theory was called *projective geometry*.

In this technique of projection, parallel lines that lie in a plane cutting the plane of the canvas are painted as meeting (visually, they appear to meet at a point on the horizon). This suggested an extension of Euclidean geometry in which parallel lines "meet at infinity," so that the Euclidean parallel property is replaced by the elliptic parallel property in the extended plane. We will carry out this extension rigorously. First, some definitions.

DEFINITION. A *projective plane* is a model of the incidence axioms having the elliptic parallel property (any two lines meet) and such that every line has at least three distinct points lying on it (strengthened Incidence Axiom 2).

Our proposed extension of the Euclidean plane uses only its incidence properties (not its betweenness and congruence properties); the purely incidence part of Euclidean geometry is called *affine* geometry, which leads to the next definition.

DEFINITION. An *affine plane* is a model of incidence geometry having the Euclidean parallel property.

Example 3 in this chapter illustrated the smallest affine plane (four points, six lines).

Let $\mathcal{A}$ be any affine plane. We introduce a relation $l \sim m$ on the lines of $\mathcal{A}$ to mean "$l = m$ or $l \parallel m$." This relation is obviously *reflexive* ($l \sim l$) and *symmetric* ($l \sim m \Rightarrow m \sim l$). Let us prove that it is *transitive* ($l \sim m$ and $m \sim n \Rightarrow l \sim n$): if any pair of these lines are equal, the conclusion is immediate, so assume that we have three distinct lines such that $l \parallel m$ and $m \parallel n$. Suppose, on the contrary, that l meets n at point P. P does not lie on m, because $l \parallel m$. Hence we have two distinct parallels n and l to m through P, which contradicts the Euclidean parallel property of $\mathcal{A}$.

A relation which is reflexive, symmetric, and transitive is called an *equivalence relation*. Such relations occur frequently in mathematics and are very important. Whenever they occur, we consider the equivalence classes determined by the relation: for example, the *equivalence class* $[l]$ of l is defined to be the set consisting of all lines equivalent to l—i.e., of l and all the lines in $\mathcal{A}$ parallel to l. In the familiar Cartesian model of the Euclidean plane, the set of all horizontal lines is one equivalence class, the set of verticals is another, the set of lines with slope 1 is a third, and so on. Equivalence classes take us from equivalence to equality: $l \sim m \Longleftrightarrow [l] = [m]$.

For historical and visual reasons, we call these equivalence classes *points at infinity;* we have made this vague idea precise within modern set theory. We now enlarge the model $\mathcal{A}$ to a new model $\mathcal{A}^*$ by adding these points, calling the points of $\mathcal{A}$ "ordinary" points for emphasis. We further enlarge the incidence relation by specifying that each of these equivalence classes lies on every one of the lines in that class: $[l]$ lies on l and on every line m such that $l \parallel m$. Thus, in the enlarged plane $\mathcal{A}^*$, l and m are no longer parallel, but they meet at $[l]$.

We want $\mathcal{A}^*$ to be a model of incidence geometry also, which requires one more step. To satisfy Euclid's Postulate I, we need to add one new line on which all (and only) the points at infinity lie: define *the line at infinity* l_∞ to be the set of all points at infinity. Let us now check that $\mathcal{A}^*$ is a projective plane, called the *projective completion* of $\mathcal{A}$:

Verification of I-1.
If P and Q are ordinary points, they lie on a unique line of $\mathcal{A}$ (since I-1 holds in $\mathcal{A}$) and they do not lie on l_∞. If P is

ordinary and Q is a point at infinity $[m]$, then either P lies on m and $\overleftrightarrow{PQ} = m$, or, by the Euclidean parallel property, P lies on a unique parallel n to m and Q also lies on n (by definition of incidence for points at infinity), so $\overleftrightarrow{PQ} = n$. If both P and Q are points at infinity, then $\overleftrightarrow{PQ} = l_\infty$.

Verification of Strengthened I-2. Each line m of $\mathscr{A}$ has at least two points on it (by I-2 in $\mathscr{A}$), and now we've added a third point $[m]$ at infinity. That l_∞ has at least three points on it follows from the existence in $\mathscr{A}$ of three lines that intersect in pairs (such as the lines joining the three noncollinear points furnished by Axiom I-3); the equivalence classes of those three lines do the job.

Verification of I-3. It holds already in $\mathscr{A}$.

Verification of the Elliptic Parallel Property. If two ordinary lines do not meet in $\mathscr{A}$, then they belong to the same equivalence class and meet at that point at infinity. An ordinary line m meets l_∞ at $[m]$. ■

Example 6. Figure 2.7 illustrates the smallest projective plane, projective completion of the smallest affine plane; it has seven points and seven lines. The dashed line could represent the line at infinity, for removing it and the three points C, B, and E that lie on it leaves us with a four-point, 6-line affine plane isomorphic to the one in Example 3, Figure 2.5.

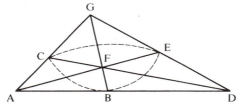

FIGURE 2.7 The smallest projective plane (7 points).

The usual Euclidean plane, regarded just as a model of incidence geometry (ignoring its betweenness and congruence structures), is called *the real affine plane*, and its projective completion is called *the real projective plane*. Coordinate descriptions of these planes are given in Major Exercises 9 and 10; other models isomorphic to the real projective plane are described in Exercise 10(c), and a "curved" model isomorphic to the real affine plane is described in Major Exercise 5.

Example 7. To visualize the projective completion $\mathscr{A}^*$ of the real affine plane $\mathscr{A}$, picture $\mathscr{A}$ as the plane T tangent to a sphere S in Euclidean three-space at its north pole N (Figure 2.8). If O is the center of sphere S, we can join each point P of T to O by a Euclidean line that will intersect the northern hemisphere of S in a unique point P'; this gives a one-to-one correspondence between the points P of T and the points P' of the northern hemisphere of S (N corresponds to itself). Similarly, given any line m of T, we join m to O by a plane Π through O that cuts out a great circle on the sphere and a great

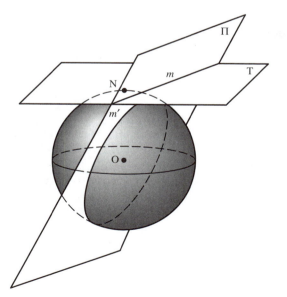

FIGURE 2.8

semicircle m' on the northern hemisphere; this gives a one-to-one correspondence between the lines m of T and the great semicircles m' of the northern hemisphere, a correspondence that clearly preserves incidence.

Now if $l \| m$ in T, the planes through O determined by these parallel lines will meet in a line lying in the plane of the equator, a line which (since it goes through O) cuts out a pair of antipodal points on the equator. Thus the line at infinity of $\mathscr{A}^*$ can be visualized under our isomorphism as the equator of S with antipodal points identified (they must be identified, or else Axiom I-1 will fail). In other words, $\mathscr{A}^*$ can be described as the northern hemisphere with antipodal points on the equator pasted to each other; however, we can't visualize this pasting very well, because it can be proved that the pasting cannot be done in Euclidean three-space without tearing the hemisphere.

Projective planes are the most important models of pure incidence geometry. We will see later on that Euclidean, hyperbolic, and, of course, elliptic geometry can all be considered "subgeometries" of projective geometry. This discovery by Cayley led him to exclaim that "projective geometry is all of geometry," which turned out to be an oversimplification.

REVIEW EXERCISE

Which of the following statements are correct?

(1) The "hypothesis" of a theorem is an assumption that implies the conclusion.

(2) A theorem may be proved by drawing an accurate diagram.

(3) To say that a step is "obvious" is an allowable justification in a rigorous proof.

(4) There is no way to program a computer to prove or disprove every statement in mathematics.

(5) To "disprove" a statement means to prove the negation of that statement.

(6) A "model" of an axiom system is the same as an "interpretation" of the system.

(7) The Pythagoreans discovered the existence of irrational lengths by an RAA proof.

(8) The negation of the statement "If 3 is an odd number, then 9 is even" is the statement "If 3 is an odd number, then 9 is odd."

(9) The negation of a conjunction is a disjunction.

(10) The statement "$1 = 2$ and $1 \neq 2$" is an example of a contradiction.

(11) The statement "Base angles of an isosceles triangle are congruent" has no hidden quantifiers.

(12) The statements "Some triangles are equilateral" and "There exists an equilateral triangle" have the same meaning.

(13) The converse of the statement "If you push me, then I will fall" is the statement "If you push me, then I won't fall."

(14) The following two statements are logically equivalent: If $l \parallel m$, then l and m have no point in common. If l and m have a point in common, then l and m are not parallel."

(15) Whenever a conditional statement is valid, its converse is also valid.

(16) If one statement implies a second statement and the second statement implies a third statement, then the first statement implies the third statement.

(17) The negation of "All triangles are isosceles" is "No triangles are isosceles."

(18) The hyperbolic parallel property is defined as "For every line l and every point P not on l there exist at least two lines through P parallel to l."

(19) The statement "Every point has at least two lines passing through it" is independent of the axioms for incidence geometry.

(20) "If $l \parallel m$ and $m \parallel n$, then $l \parallel n$" is independent of the axioms of incidence geometry.

EXERCISES

1. Let S be the following self-referential statement: "Statement S is false." Show that if S is either true or false then there is a contradiction in our language. (This is the *liar paradox*. Kurt Gödel used a variant of it as the starting point for his famous incompleteness theorem in logic; see De-Long, 1970)

2. (a) What is the negation of $[P \text{ or } Q]$?

   (b) What is the negation of $[P \, \& \sim Q]$?

   (c) Using the rules of logic given in the text, show that $P \Rightarrow Q$ means the same as $[\sim P \text{ or } Q]$. (Hint: Show they are both negations of the same thing.)

    (d) A symbolic way of writing Rule 2 for RAA proofs is $[[H \& \sim C] \Rightarrow [S \& \sim S]] \Rightarrow [H \Rightarrow C]$. Explain this.

3. Negate Euclid's fourth postulate.

4. Negate the Euclidean parallel postulate.

5. Write out the converse to the following statements:
   (a) "If lines l and m are parallel, then a transversal t to lines l and m cuts out congruent alternate interior angles."
   (b) "If the sum of the degree measures of the interior angles on one side of transversal t is less than $180°$, then lines l and m meet on that side of transversal t."

6. Prove all five propositions in incidence geometry as stated in this chapter. Don't use Incidence Axiom 2 in your proofs.

7. For each pair of axioms of incidence geometry, construct an interpretation in which those two axioms are satisfied but the third axiom is not. (This will show that the three axioms are *independent*, in the sense that it is impossible to prove any one of them from the other two.)

8. Show that the interpretations in Examples 3 and 4 in this chapter are models of incidence geometry and that the Euclidean and hyperbolic parallel properties, respectively, hold.

9. In each of the following interpretations of the undefined terms, which of the axioms of incidence geometry are satisfied and which are not? Tell whether each interpretation has the elliptic, Euclidean, or hyperbolic parallel property.
   (a) "Points" are dots on a sheet of paper, "lines" are circles drawn on the paper, "incidence" means that the dot lies on the circle.
   (b) "Points" are lines in Euclidean three-dimensional space, "lines" are planes in Euclidean three-space, "incidence" is the usual relation of a line lying in a plane.
   (c) Same as in (b), except that we restrict ourselves to lines and planes that pass through a fixed ordinary point O.
   (d) Fix a circle in the Euclidean plane. Interpret "point" to mean an ordinary Euclidean point *inside* the circle, interpret "line" to mean a chord of the circle, and let "incidence" mean that the point lies on the chord in the usual sense. (A *chord* of a circle is a segment whose endpoints lie on the circle.)
   (e) Fix a sphere in Euclidean three-space. Two points on the sphere are called *antipodal* if they lie on a diameter of the sphere; e.g., the north and south poles are antipodal. Interpret a "point" to be a set {P, P'} consisting of two antipodal points on the sphere. Interpret a "line" to be a great circle C on the sphere. Interpret a "point" {P, P'} to "lie on" a "line" C if one of the points P, P' lies on the great circle C (then the other point also lies on C).

10. (a) Prove that when each of two models of incidence geometry has exactly three "points" in it, the models are isomorphic.
    (b) Must two models having exactly four "points" be isomorphic? If you think so, prove this; if you think not, give a counterexample.
    (c) Show that the models in Exercises 9(c) and 9(e) are isomorphic. (Hint: Take the point O of Exercise 9(c) to be the center of the sphere in Exercise 9(e), and cut the sphere with lines and planes through point O to get the isomorphism.)
11. Construct a model of incidence geometry that has neither the elliptic, hyperbolic, nor Euclidean parallel properties. (These properties refer to any line *l* and any point P not on *l*. Construct a model that has different parallelism properties for different choices of *l* and P. Five points suffice.)
12. Suppose that in a given model for incidence geometry every "line" has at least three distinct "points" lying on it. What are the least number of "points" and the least number of "lines" such a model can have? Suppose further that the model has the Euclidean parallel property. Show that 9 is now the least number of "points" and 12 the least number of "lines" such a model can have.
13. The following syllogisms are by Lewis Carroll. Which of them are correct arguments?
    (a) No frogs are poetical; some ducks are unpoetical. Hence, some ducks are not frogs.
    (b) Gold is heavy; nothing but gold will silence him. Hence, nothing light will silence him.
    (c) All lions are fierce; some lions do not drink coffee. Hence, some creatures that drink coffee are not fierce.
    (d) Some pillows are soft; no pokers are soft. Hence, some pokers are not pillows.
14. Comment on the following example of isomorphic structures given by a music student: *Romeo and Juliet* and *West Side Story*.
15. Comment on the following statement by the artist David Hunter: "The only use for Logic is writing books on Logic and teaching courses in Logic; it has no application to human behavior."

MAJOR EXERCISES

1. Let *M* be a projective plane. Define a new interpretation *M'* by taking as "points" of *M'* the lines of *M* and as "lines" of *M'* the points of *M*, with the same incidence relation. Prove that *M'* is also a projective plane

(called the *dual plane* of $\mathscr{M}$). Suppose further that $\mathscr{M}$ has only finitely many points. Prove that all the lines in $\mathscr{M}$ have the same number of points lying on them. (Hint: See Figure 7.43 in Chapter 7.)

2. Let us add to the axioms of incidence geometry the following axioms:
   (i)   The Euclidean parallel property.
   (ii) The existence of only a finite number of points.
   (iii) The existence of lines l and m such that the number of points lying on l is different from the number of points lying on m.

   Show that this expanded axiom system is inconsistent. (Hint: Prove that (i) and (ii) imply the negation of (iii).)

3. Prove that every projective plane $\mathscr{B}$ is isomorphic to the projective completion of some affine plane $\mathscr{A}$. (Hint: As was done in Example 6, pick any line m in $\mathscr{B}$, pretend that m is "the line at infinity," remove m and the points lying on it, and prove that what's left is an affine plane $\mathscr{A}$ and that $\mathscr{B}$ is isomorphic to the completion $\mathscr{A}^*$.) A surprising discovery is that $\mathscr{A}$ need not be unique up to isomorphism (see Hartshorne, 1967).

4. Provide another solution to Major Exercise 2 by embedding the affine plane of that exercise in its completion and invoking Major Exercise 1.

5. Consider the following interpretation of incidence geometry. Begin with a punctured sphere in Euclidean three-space, i.e., a sphere with one point N removed. Interpret "points" as points on the punctured sphere. For each circle on the original sphere passing through N, interpret the punctured circle obtained by removing N as a "line." Interpret "incidence" in the Euclidean sense of a point lying on a punctured circle. Is this interpretation a model? If so, what parallel property does it have? Is it isomorphic to any other model you know? (Hint: If N is the north pole, project the punctured sphere from N onto the plane Π tangent to the sphere at the south pole, as in Figure 2.9. Use the fact that planes through N cut out circles on the sphere and lines in Π. For a hilarious discussion of this interpretation, refer to Chapter 3 of Sved, 1991.)

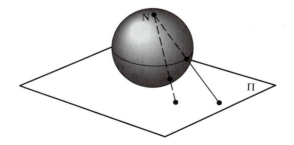

FIGURE 2.9

6. Consider the following statement in incidence geometry: "For any two lines l and m there exists a one-to-one correspondence between the set of points lying on l and the set of points lying on m." Prove that this statement is independent of the axioms of incidence geometry.

7. Let $\mathcal{M}$ be a finite projective plane so that, according to Major Exercise 1, all lines in $\mathcal{M}$ have the same number of points lying on them; call this number $n + 1$. Prove the following:
   (a) Each point in $\mathcal{M}$ has $n + 1$ lines passing through it.
   (b) The total number of points in $\mathcal{M}$ is $n^2 + n + 1$.
   (c) The total number of lines in $\mathcal{M}$ is $n^2 + n + 1$.

8. Let $\mathcal{A}$ be a finite affine plane so that, according to Major Exercise 2, all lines in $\mathcal{A}$ have the same number of points lying on them; call this number n. Prove the following:
   (a) Each point in $\mathcal{A}$ has $n + 1$ lines passing through it.
   (b) The total number of points in $\mathcal{A}$ is n^2.
   (c) The total number of lines in $\mathcal{A}$ is $n(n + 1)$.
   (Hint: Use Major Exercise 7.)

9. *The real affine plane* has as its "points" all ordered pairs (x, y) of real numbers. A "line" is determined by an ordered triple (u, v, w) of real numbers such that either $u \neq 0$ or $v \neq 0$, and it is defined as the set of all "points" (x, y) satisfying the linear equation $ux + vy + w = 0$. "Incidence" is defined as set membership. Verify that all axioms for an affine plane are satisfied by this interpretation.

10. A "point" $[x, y, z]$ in *the real projective plane* is determined by an ordered triple (x, y, z) of real numbers that are not all zero, and it consists of all the ordered triples of the form (kx, ky, kz) for all real numbers $k \neq 0$; thus, $[kx, ky, kz] = [x, y, z]$. A "line" in the real projective plane is determined by an ordered triple (u, v, w) of real numbers that are not all zero, and it is defined as the set of all "points" $[x, y, z]$ whose coordinates satisfy the linear equation $ux + vy + wz = 0$. "Incidence" is defined as set membership. Verify that all the axioms for a projective plane are satisfied by this interpretation. Prove that by taking $z = 0$ as the equation of the "line at infinity," by assigning the affine "point" (x, y) the "homogeneous coordinates" $[x, y, 1]$, and by assigning affine "lines" to projective "lines" in the obvious way, the real projective plane becomes isomorphic to the projective completion of the real affine plane. Prove that the models in Exercise 10(c) are also isomorphic to the real projective plane.

11. (a) Given an interpretation of some axioms, in order to show that the interpretation is a model, you must verify that the interpretations of the axioms hold. If you execute that verification precisely rather

than casually, you are actually giving proofs. In what axiomatic theory are those proofs given? Consider this question more specifically for the models presented in the text and exercises of this chapter.

(b) Some of the interpretations refer to a "sphere" in "Euclidean space," presuming that you already know the theory of such things, yet we are carefully laying the axiomatic foundations of the simpler theory of the Euclidean plane. Does this bother you? Comment.

(c) Can an inconsistent system (such as the one in Major Exercise 2) have a model? Explain.

12. Just because every step in a proof has been justified, that doesn't guarantee the correctness of the proof: the justifications may be in error. For example, the justification may not be one of the six types allowed by Logic Rule 1, or it may refer to a previous theorem that is not applicable, or it may draw erroneous inferences from a definition (such as "parallel lines are equidistant"). Thus a second "proof" should be given to verify the correctness of the justifications in the first proof. But then how can we be certain the second "proof" is correct? Do we have to give a third "proof" and so on ad infinitum? Discuss.

PROJECTS

1. The following statement is by the French mathematician G. Desargues: "If the vertices of two triangles correspond in such a way that the lines joining corresponding vertices are concurrent, then the intersections of corresponding sides are collinear." (See Figure 2.10.) This statement is independent of the axioms for projective planes: it holds in the real projective plane, but there exist other projective planes in which it fails. Report on this independence result (see Artzy, 1965, or Stevenson, 1972).

2. An isomorphism of a projective plane $\mathcal{M}$ onto its dual plane $\mathcal{M}'$ (see Major Exercise 1) is called a *polarity* of $\mathcal{M}$. By definition of "isomorphism," it assigns to each point A of $\mathcal{M}$ a line $p(\text{A})$ of $\mathcal{M}$ called the *polar* of A, and to each line m of $\mathcal{M}$ a point $\text{P}(m)$ of $\mathcal{M}$ called its *pole*, in such a way that A lies on m if and only if $\text{P}(m)$ lies on $p(\text{A})$. The *conic* γ determined by this polarity is defined to be the set of all points A such that A lies on its polar $p(\text{A})$; $p(\text{A})$ is defined to be the *tangent line* to the conic at A. Point B is defined to be *interior* to γ if every line through B intersects γ in two

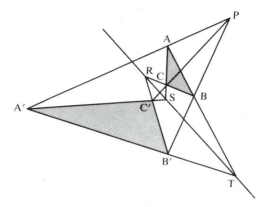

FIGURE 2.10 Desargues's theorem.

points. This very abstract definition[5] of "conic" can be reconciled with more familiar descriptions, such as (using coordinates) the solution set to a homogeneous quadratic equation in three variables. The theory of conics is one of the most important topics in plane projective geometry. Report on this, using some good projective geometry text such as Coxeter (1960). A polarity will play a crucial role in Chapter 7 (see also Major Exercise 13, Chapter 6).

3. Aristotle is considered the founder of classical logic. Up through the 1930s, some important logicians were Leibniz, Boole, Frege, Russell, Whitehead, Hilbert, Ackermann, Skolem, Gödel, Church, Tarski, and Kleene. Report on some of the history of logic, using DeLong (1970) and his bibliography as references.

---

[5] The poet Goethe said: "Mathematicians are like Frenchmen: whatever you say to them, they translate it into their own language and forthwith it is something entirely different."

CHAPTER 3 HILBERT'S AXIOMS

> The value of Euclid's work as a masterpiece of logic has been very grossly exaggerated.
>
> BERTRAND RUSSELL

FLAWS IN EUCLID

Having clarified our rules of reasoning (Chapter 2), let us return to the postulates of Euclid. In Exercises 9 and 10 of Chapter 1 we saw that Euclid neglected to state his assumptions that points and lines exist, that not all points are collinear, and that every line has at least two points lying on it. We made these assumptions explicit in Chapter 2 by adding two more axioms of incidence to Euclid's first postulate.

In Exercises 6 and 7, Chapter 1, we saw that some assumptions about "betweenness" are needed. In fact, Euclid never mentioned this notion explicitly, but tacitly assumed certain facts about it that are obvious in diagrams. In Chapter 1 we saw the danger of reasoning from diagrams, so these tacit assumptions will have to be made explicit.

Quite a few of Euclid's proofs are based on reasoning from diagrams. To make these proofs rigorous, a much larger system of explicit axioms is needed. Many such axiom systems have been proposed. We will present a modified version of David Hilbert's system of axioms.

David Hilbert

Hilbert's system was not the first, but his axioms are perhaps the most intuitive and are certainly the closest in spirit to Euclid's.[1]

During the first quarter of the twentieth century Hilbert was considered the leading mathematician of the world.[2] He made outstanding, original contributions to a wide range of mathematical fields as well as to physics. He is perhaps best known for his research in the foundations of geometry as well as the foundations of algebraic number theory, infinite-dimensional spaces, and mathematical logic. A

[1] Let us not forget that no serious work toward constructing new axioms for Euclidean geometry had been done until the discovery of non-Euclidean geometry shocked mathematicians into reexamining the foundations of the former. We have the paradox of non-Euclidean geometry helping us to better understand Euclidean geometry!

[2] I heartily recommend the warm and colorful biography of Hilbert by Constance Reid (1970). It is nontechnical and conveys the excitement of the time when Göttingen was the capital of the mathematical world.

great champion of the axiomatic method, he "axiomatized" all of the above subjects except for physics (although he did succeed in providing physicists with very valuable mathematical techniques). He was also a mathematical prophet; in 1900 he predicted 23 of the most important mathematical problems of this century.

He has been quoted as saying: "One must be able to say at all times — instead of points, lines and planes — tables, chairs and beer mugs." In other words, since no properties of points, lines, and planes may be used in a proof other than the properties given by the axioms, you may as well call these undefined entities by other names.

Hilbert's axioms are divided into five groups: incidence, betweenness, congruence, continuity, and parallelism. We have already seen the three axioms of incidence in Chapter 2. In the next sections we will deal successively with the other groups of axioms.

AXIOMS OF BETWEENNESS

To further illustrate the need for axioms of betweenness, consider the following attempted proof of the theorem that base angles of an isosceles triangle are congruent. This is not Euclid's proof, which is flawed in other ways (see Golos, 1968, p. 57), but is an argument found in some high school geometry texts.

Proof:
Given $\triangle ABC$ with $AC \cong BC$. To prove $\angle A \cong \angle B$ (see Figure 3.1):

(1) Let the bisector of $\angle C$ meet AB at D (every angle has a bisector).

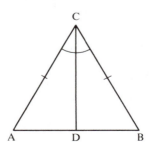

FIGURE 3.1

(2) In triangles $\triangle ACD$ and $\triangle BCD$, $AC \cong BC$ (hypothesis).
(3) $\sphericalangle ACD \cong \sphericalangle BCD$ (definition of bisector of an angle).
(4) $CD \cong CD$ (things that are equal are congruent).
(5) $\triangle ACD \cong \triangle BCD$ (SAS).
(6) Therefore, $\sphericalangle A \cong \sphericalangle B$ (corresponding angles of congruent triangles). ∎

Consider the first step, whose justification is that every angle has a bisector. This is a correct statement and can be proved separately. But how do we know that the bisector of $\sphericalangle C$ meets $\overleftrightarrow{AB}$, or if it does, how do we know that the point of intersection D lies *between* A and B? This may seem obvious, but if we are to be rigorous, it requires proof. For all we know, the picture might look like Figure 3.2. If this were the case, steps 2 – 5 would still be correct, but we could conclude only that $\sphericalangle B$ is congruent to $\sphericalangle CAD$, not to $\sphericalangle CAB$, since $\sphericalangle CAD$ is the angle in $\triangle ACD$ that corresponds to $\sphericalangle B$.

Once we state our four axioms of betweenness, it will be possible to prove (after a considerable amount of work) that the bisector of $\sphericalangle C$ does meet $\overleftrightarrow{AB}$ in a point D between A and B, so the above argument will be repaired (see the crossbar theorem, later in this section). There is, however, an easier proof of the theorem (given in the next section). We will use the shorthand notation

$$A * B * C$$

to abbreviate the statement "point B is between point A and point C."

BETWEENNESS AXIOM 1. If $A * B * C$, then A, B, and C are three distinct points all lying on the same line, and $C * B * A$.

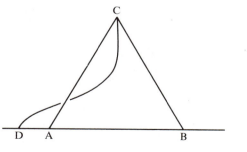

FIGURE 3.2

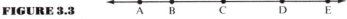

FIGURE 3.3

The first part of this axiom fills the gap mentioned in Exercise 6, Chapter 1. The second part (C * B * A) makes the obvious remark that "between A and C" means the same as "between C and A" — it doesn't matter whether A or C is mentioned first.

BETWEENNESS AXIOM 2. Given any two distinct points B and D, there exist points A, C, and E lying on $\overleftrightarrow{BD}$ such that A * B * D, B * C * D, and B * D * E (Figure 3.3).

This axiom ensures that there are points between B and D and that the line $\overleftrightarrow{BD}$ does not end at either B or D.

BETWEENNESS AXIOM 3. If A, B, and C are three distinct points lying on the same line, then one and only one of the points is between the other two.

This axiom ensures that a line is not circular; if the points were on a circle, you would then have to say that each is between the other two (or none is between the other two — it would depend on which of the two arcs you look at — see Figure 3.4).

Before stating the last betweenness axiom, let us examine some consequences of the first three. Recall that the *segment* AB is defined as the set of all points between A and B together with the endpoints A and B. The *ray* $\overrightarrow{AB}$ is defined as the set of all points on the segment AB together with all points C such that A * B * C. The second axiom ensures that such points as C exist, so the ray $\overrightarrow{AB}$ is larger than the segment AB. We can now prove the formulas you encountered in Exercise 7, Chapter 1.

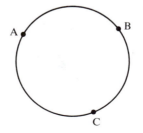

FIGURE 3.4

PROPOSITION 3.1. For any two points A and B: (i) $\overrightarrow{AB} \cap \overrightarrow{BA} = AB$, and (ii) $\overrightarrow{AB} \cup \overrightarrow{BA} = \{\overleftrightarrow{AB}\}$.

Proof of (i):

(1) By definition of segment and ray, $AB \subset \overrightarrow{AB}$ and $AB \subset \overrightarrow{BA}$, so by definition of intersection, $AB \subset \overrightarrow{AB} \cap \overrightarrow{BA}$.

(2) Conversely, let the point C belong to the intersection of $\overrightarrow{AB}$ and $\overrightarrow{BA}$; we wish to show that C belongs to AB.

(3) If C = A or C = B, C is an endpoint of AB. Otherwise, A, B, and C are three collinear points (by definition of ray and Axiom 1), so exactly one of the relations A * C * B, A * B * C, or C * A * B holds (Axiom 3).

(4) If A * B * C holds, then C is not on $\overrightarrow{BA}$; if C * A * B holds, then C is not on $\overrightarrow{AB}$. In either case, C does not belong to both rays.

(5) Hence, the relation A * C * B must hold, so C belongs to AB. ■

The proof of (ii) is similar and is left as an exercise. (Recall that $\{\overleftrightarrow{AB}\}$ is the set of points lying on the line $\overleftrightarrow{AB}$.)

Recall next that if C * A * B, then $\overrightarrow{AC}$ is said to be *opposite* to $\overrightarrow{AB}$ (see Figure 3.5). By Axiom 1, points A, B, and C are collinear, and by Axiom 3, C does not belong to $\overrightarrow{AB}$, so rays $\overrightarrow{AB}$ and $\overrightarrow{AC}$ are distinct. This definition is therefore in agreement with the definition given in Chapter 1 (see Proposition 3.6). Axiom 2 guarantees that every ray $\overrightarrow{AB}$ has an opposite ray $\overrightarrow{AC}$.

It seems clear from Figure 3.5 that every point P lying on the line *l* through A, B, C must belong either to ray $\overrightarrow{AB}$ or to an opposite ray $\overrightarrow{AC}$. This statement seems similar to the second assertion of Proposition 3.1, but it is actually more complicated; we are now discussing *four* points A, B, C, and P, whereas previously we had to deal with only three points at a time. In fact, we encounter here another "pictorially obvious" assertion that cannot be proved without introducing another axiom (see Exercise 17).

FIGURE 3.5

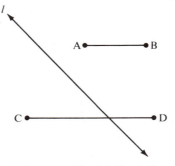

FIGURE 3.6   A and B are on the same side of *l*; C and D are on opposite sides of *l*.

Suppose we call the assertion "C * A * B and P collinear with A, B, C $\Rightarrow$ P $\in$ $\overrightarrow{AC}$ $\cup$ $\overrightarrow{AB}$" *the line separation property*. Some mathematicians take this property as another axiom. However, it is considered inelegant in mathematics to assume more axioms than are necessary (although we pay for elegance by having to work harder to prove results that appear obvious). So we will not assume the line separation property as an axiom; instead, we will prove it as a consequence of our previous axioms and our last betweenness axiom, called *the plane separation axiom*.

DEFINITION.   Let *l* be any line, A and B any points that do not lie on *l*. If A = B or if segment AB contains no point lying on *l*, we say A and B are *on the same side of l*, whereas if A $\neq$ B and segment AB does intersect *l*, we say that A and B are *on opposite sides* of *l* (see Figure 3.6). The law of the excluded middle (Rule 10) tells us that A and B are either on the same side or on opposite sides of *l*.

BETWEENNESS AXIOM 4 (Plane Separation).   For every line *l* and for any three points A, B, and C not lying on *l*:

(i)   If A and B are on the same side of *l* and B and C are on the same side of *l*, then A and C are on the same side of *l* (see Figure 3.7).

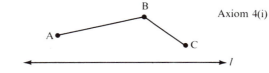

FIGURE 3.7

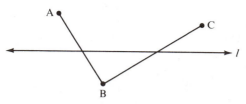

FIGURE 3.8

(ii) If A and B are on opposite sides of *l* and B and C are on opposite sides of *l*, then A and C are on the same side of *l* (see Figure 3.8).

COROLLARY. (iii) If A and B are on opposite sides of *l* and B and C are on the same side of *l*, then A and C are on opposite sides of *l*.

Axiom 4(i) indirectly guarantees that our geometry is two-dimensional, since it does not hold in three-space. (Line *l* could be outside the plane of this page and cut through segment AC; this interpretation shows that if we assumed the line separation property as an axiom, we could not prove the plane separation property.) Betweenness Axiom 4 is also needed to make sense out of Euclid's fifth postulate, which talks about two lines meeting on one "side" of a transversal. We can now define a *side* of a line *l* as the set of all points that are on the same side of *l* as some particular point A not lying on *l*. If we denote this side by H_A, notice that if C is on the same side of *l* as A, then by Axiom 4(i), $H_C = H_A$. (The definition of a *side* may seem circular because we use the word "side" twice, but it is not; we have already defined the compound expression "on the same side.") Another expression commonly used for a "side of *l*" is a *half-plane bounded by l*.

PROPOSITION 3.2. Every line bounds exactly two half-planes and these half-planes have no point in common.

Proof:
(1) There is a point A not lying on *l* (Proposition 2.3).
(2) There is a point O lying on *l* (Incidence Axiom 2).
(3) There is a point B such that B * O * A (Betweenness Axiom 2).
(4) Then A and B are on opposite sides of *l* (by definition), so *l* has at least two sides.

(5) Let C be any point distinct from A and B and not lying on l. If C and B are not on the same side of l, then C and A are on the same side of l (by the law of excluded middle and Betweenness Axiom 4(ii)). So the set of points not on l is the union of the side H_A of A and the side H_B of B.

(6) If C were on both sides (RAA hypothesis), then A and B would be on the same side (Axiom 4(i)), contradicting step 4; hence the two sides are disjoint (RAA conclusion). ∎

We next apply the plane separation property to study betweenness relations among four points.

PROPOSITION 3.3. Given A * B * C and A * C * D. Then B * C * D and A * B * D. (See Figure 3.9.)

Proof:

(1) A, B, C, and D are four distinct collinear points (see Exercise 1).

(2) There exists a point E not on the line through A, B, C, D (Proposition 2.3).

(3) Consider line $\overleftrightarrow{EC}$. Since (by hypothesis) AD meets this line in point C, A and D are on opposite sides of $\overleftrightarrow{EC}$.

(4) We claim A and B are on the same side of $\overleftrightarrow{EC}$. Assume on the contrary that A and B are on opposite sides of $\overleftrightarrow{EC}$ (RAA hypothesis).

(5) Then $\overleftrightarrow{EC}$ meets $\overleftrightarrow{AB}$ in a point between A and B (definition of "opposite sides").

(6) That point must be C (Proposition 2.1).

(7) Thus, A * B * C and A * C * B, which contradicts Betweenness Axiom 3.

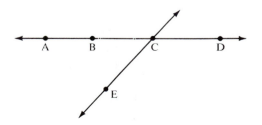

FIGURE 3.9

(8) Hence, A and B are on the same side of $\overleftrightarrow{EC}$ (RAA conclusion).

(9) B and D are on opposite sides of $\overleftrightarrow{EC}$ (steps 3 and 8 and the corollary to Betweenness Axiom 4).

(10) Hence, the point C of intersection of lines $\overleftrightarrow{EC}$ and $\overleftrightarrow{BD}$ lies between B and D (definition of "opposite sides"; Proposition 2.1, i.e., that the point of intersection is unique).

A similar argument involving $\overleftrightarrow{EB}$ proves that A * B * D (Exercise 2(b)). ∎

COROLLARY. Given A * B * C and B * C * D. Then A * B * D and A * C * D.

Finally we prove the *line separation property*.

PROPOSITION 3.4. If C * A * B and *l* is the line through A, B, and C (Betweenness Axiom 1), then for every point P lying on *l*, P lies either on ray $\overrightarrow{AB}$ or on the opposite ray $\overrightarrow{AC}$.

Proof:

(1) Either P lies on $\overrightarrow{AB}$ or it does not (law of excluded middle).

(2) If P does lie on $\overrightarrow{AB}$, we are done, so assume it doesn't; then P * A * B (Betweenness Axiom 3).

(3) If P = C then P lies on $\overrightarrow{AC}$ (by definition), so assume P ≠ C; then exactly one of the relations C * A * P, C * P * A, and P * C * A holds (Betweenness Axiom 3 again).

(4) Suppose the relation C * A * P holds (RAA hypothesis).

(5) We know (by Betweenness Axiom 3) that exactly one of the relations P * C * B, C * P * B, and C * B * P holds.

(6) If P * B * C, then combining this with P * A * B (step 2) gives A * B * C (Proposition 3.3), contradicting the hypothesis.

(7) If C * P * B, then combining this with C * A * P (step 4) gives A * P * B (Proposition 3.3), contradicting step 2.

(8) If B * C * P, then combining this with B * A * C (hypothesis and Betweenness Axiom 1) gives A * C * P (Proposition 3.3), contradicting step 4.

(9) Since we obtain a contradiction in all three cases, C * A * P does not hold (RAA conclusion).

(10) Therefore, C * P * A or P * C * A (step 3), which means that P lies on the opposite ray $\overrightarrow{AC}$. ∎

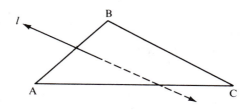

FIGURE 3.10

The next theorem states a visually obvious property that Pasch discovered Euclid to be using without proof.

PASCH'S THEOREM. If A, B, C are distinct noncollinear points and *l* is any line intersecting AB in a point between A and B, then *l* also intersects either AC or BC (see Figure 3.10). If C does not lie on *l*, then *l* does not intersect both AC and BC.

Intuitively, this theorem says that if a line "goes into" a triangle through one side, it must "come out" through another side.

Proof:
(1) Either C lies on *l* or it does not; if it does, the theorem holds (law of excluded middle).
(2) A and B do not lie on *l*, and the segment AB does intersect *l* (hypothesis and Axiom 1).
(3) Hence, A and B lie on opposite sides of *l* (by definition).
(4) From step 1 we may assume that C does not lie on *l*, in which case C is either on the same side of *l* as A or on the same side of *l* as B (separation axiom).
(5) If C is on the same side of *l* as A, then C is on the opposite side from B, which means that *l* intersects BC and does not intersect AC; similarly if C is on the same side of *l* as B, then *l* intersects AC and does not intersect BC (separation axiom).
(6) The conclusions of Pasch's theorem hold (Logic Rule 11 — proof by cases). ∎

Here are some more results on betweenness and separation that you will be asked to prove in the exercises.

PROPOSITION 3.5. Given A * B * C. Then AC = AB ∪ BC and B is the only point common to segments AB and BC.

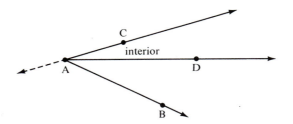

FIGURE 3.11

PROPOSITION 3.6. Given A * B * C. Then B is the only point common to rays $\overrightarrow{BA}$ and $\overrightarrow{BC}$, and $\overrightarrow{AB} = \overrightarrow{AC}$.

DEFINITION. Given an angle ∢CAB, define a point D to be in the *interior* of ∢CAB if D is on the same side of $\overleftrightarrow{AC}$ as B and if D is also on the same side of $\overleftrightarrow{AB}$ as C. (Thus, the interior of an angle is the intersection of two half-planes.) See Figure 3.11.

PROPOSITION 3.7. Given an angle ∢CAB and point D lying on line $\overleftrightarrow{BC}$. Then D is in the interior of ∢CAB if and only if B * D * C (see Figure 3.12).

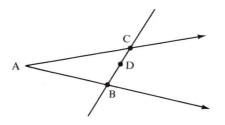

FIGURE 3.12

Warning. Do not assume that every point in the interior of an angle lies on a segment joining a point on one side of the angle to a point on the other side. In fact, this assumption is false in hyperbolic geometry (see Exercise 36).

PROPOSITION 3.8. If D is in the interior of ∢CAB; then: (a) so is every other point on ray $\overrightarrow{AD}$ except A; (b) no point on the opposite ray to $\overrightarrow{AD}$ is in the interior of ∢CAB; and (c) if C * A * E, then B is in the interior of ∢DAE (see Figure 3.13).

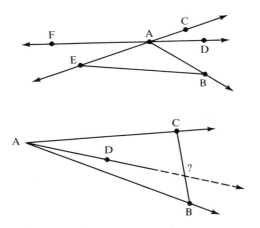

FIGURE 3.13

FIGURE 3.14

DEFINITION. Ray $\overrightarrow{AD}$ is *between* rays $\overrightarrow{AC}$ and $\overrightarrow{AB}$ if $\overrightarrow{AB}$ and $\overrightarrow{AC}$ are not opposite rays and D is interior to ∢CAB. (By Proposition 3.8(a), this definition does not depend on the choice of point D on $\overrightarrow{AD}$.)

CROSSBAR THEOREM. If $\overrightarrow{AD}$ is between $\overrightarrow{AC}$ and $\overrightarrow{AB}$, then $\overrightarrow{AD}$ intersects segment BC (see Figure 3.14).

DEFINITION. The *interior* of a triangle is the intersection of the interiors of its three angles. Define a point to be *exterior* to the triangle if it is not in the interior and does not lie on any side of the triangle.

PROPOSITION 3.9. (a) If a ray r emanating from an exterior point of △ABC intersects side AB in a point between A and B, then r also intersects side AC or side BC. (b) If a ray emanates from an interior point of △ABC, then it intersects one of the sides, and if it does not pass through a vertex, it intersects only one side.

AXIOMS OF CONGRUENCE

If we were more pedantic, "congruent," the last of our undefined terms, would be replaced by two terms, since it refers to either a relation between segments or a relation between angles. We are accustomed to congruence as a relation between triangles, but we can now define this as follows: two triangles are *congruent* if a one-to-one correspondence can be set up between their vertices so that corresponding sides are congruent and corresponding angles are congruent. When we write △ABC ≅ △DEF we understand that A corresponds to D, B to E, and C to F. Similar definitions can be given for congruence of quadrilaterals, pentagons, and so forth.

CONGRUENCE AXIOM 1. If A and B are distinct points and if A′ is any point, then for each ray *r* emanating from A′ there is a *unique* point B′ on *r* such that B′ ≠ A′ and AB ≅ A′B′. (See Figure 3.15.)

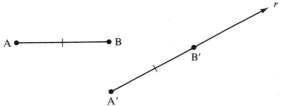

FIGURE 3.15

Intuitively speaking, this axiom says you can "move" the segment AB so that it lies on the ray *r* with A superimposed on A′, and B superimposed on B′. (In Major Exercise 2, Chapter 1, you showed how to do this with a straightedge and collapsible compass.)

CONGRUENCE AXIOM 2. If AB ≅ CD and AB ≅ EF, then CD ≅ EF. Moreover, every segment is congruent to itself.

This axiom replaces Euclid's first common notion, since it says that segments congruent to the same segment are congruent to each other. It also replaces the fourth common notion, since it says that segments that coincide are congruent.

CONGRUENCE AXIOM 3. If A * B * C, A′ * B′ * C′, AB ≅ A′B′, and BC ≅ B′C′, then AC ≅ A′C′. (See Figure 3.16.)

This axiom replaces the second common notion, since it says that if congruent segments are "added" to congruent segments, the sums are congruent. Here, "adding" means juxtaposing segments along the same line. For example, using Congruence Axioms 1 and 3, you can

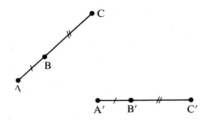

FIGURE 3.16

lay off a copy of a given segment AB two, three, . . . , n times, to get a new segment $n \cdot$ AB. (See Figure 3.17.)

CONGRUENCE AXIOM 4. Given any ∢BAC (where, by definition of "angle," $\overrightarrow{AB}$ is not opposite to $\overrightarrow{AC}$), and given any ray $\overrightarrow{A'B'}$ emanating from a point A', then there is a *unique* ray $\overrightarrow{A'C'}$ on a given side of line $\overleftrightarrow{A'B'}$ such that ∢B'A'C' ≅ ∢BAC. (See Figure 3.18.)

This axiom can be paraphrased to state that a given angle can be "laid off" on a given side of a given ray in a unique way (see Major Exercise 1(g), Chapter 1).

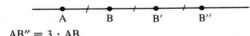

FIGURE 3.17   AB″ = 3 · AB.

CONGRUENCE AXIOM 5. If ∢A ≅ ∢B and ∢A ≅ ∢C, then ∢B ≅ ∢C. Moreover, every angle is congruent to itself.

This is the analogue for angles of Congruence Axiom 2 for segments; the first part asserts the transitivity and the second part the reflexivity of the congruence relation. Combining them, we can prove the symmetry of this relation: ∢A ≅ ∢B ⟹ ∢B ≅ ∢A.

Proof:
∢A ≅ ∢B (hypothesis) and ∢A ≅ ∢A (reflexivity) imply (substituting A for C in Congruence Axiom 5) ∢B ≅ ∢A (transitivity). ∎

(By the same argument, congruence of segments is a symmetric relation.)

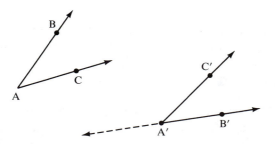

FIGURE 3.18

It would seem natural to assume next an "addition axiom" for congruence of *angles* analogous to Congruence Axiom 3 (the addition axiom for congruence of segments). We won't do this, however, because such a result can be proved using the next congruence axiom (see Proposition 3.19).

CONGRUENCE AXIOM 6 (SAS). If two sides and the included angle of one triangle are congruent respectively to two sides and the included angle of another triangle, then the two triangles are congruent (see Figure 3.19).

This side-angle-side criterion for congruence of triangles is a profound axiom. It provides the "glue" which binds the relation of congruence of segments to the relation of congruence of angles. It enables us to deduce all the basic results about triangle congruence with which you are presumably familiar. For example, here is one immediate consequence which states that we can "lay off" a given triangle on a given base and a given half-plane.

COROLLARY TO SAS. Given $\triangle ABC$ and segment $DE \cong AB$, there is a unique point F on a given side of line $\overleftrightarrow{DE}$ such that $\triangle ABC \cong \triangle DEF$.

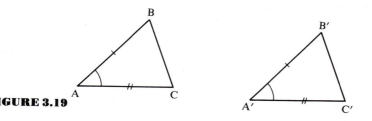

FIGURE 3.19

Proof:

There is a unique ray $\overrightarrow{DF}$ on the given side such that $\sphericalangle CAB \cong \sphericalangle FDE$, and F on that ray can be chosen to be the unique point such that $AC \cong DF$ (by Congruence Axioms 4 and 1). Then $\triangle ABC \cong \triangle DEF$ (SAS). ■

As we said, Euclid did not take SAS as an axiom but tried to prove it as a theorem. His argument was essentially as follows. Move $\triangle A'B'C'$ so as to place point A' on point A and $\overrightarrow{A'B'}$ on $\overrightarrow{AB}$. Since $AB \cong A'B'$, by hypothesis, point B' must fall on point B. Since $\sphericalangle A \cong \sphericalangle A'$, $\overrightarrow{A'C'}$ must fall on $\overrightarrow{AC}$, and since $AC \cong A'C'$, point C' must coincide with point C. Hence, B'C' will coincide with BC and the remaining angles will coincide with the remaining angles, so the triangles will be congruent.

This argument is called *superposition*. It derives from the experience of drawing two triangles on paper, cutting out one, and placing it on top of the other. Although this is a good way to convince a novice in geometry to accept SAS, it is not a proof, and Euclid reluctantly used it in only one other theorem. It is not a proof because Euclid never stated an axiom that allows figures to be moved around without changing their size and shape.

Some modern writers introduce "motion" as an undefined term and lay down axioms for this term. (In fact, in Pieri's foundations of geometry, "point" and "motion" are the only undefined terms.) Or else, the geometry is first built up on a different basis, "distances" introduced, and a "motion" defined as a one-to-one transformation of the plane onto itself that preserves distance. Euclid can be vindicated by either approach. In fact, Felix Klein, in his 1872 *Erlanger Programme*, defined a geometry as the study of those properties of figures that remain invariant under a particular group of transformations. This idea will be developed in Chapter 9.

You will show in Exercise 35 that it is impossible to prove SAS or any of the other criteria for congruence of triangles (SSS, ASA, SAA) from the preceding axioms. As usual, the method for proving the impossibility of proving some statement S is to invent a model for the preceding axioms in which S is false.

As an application of SAS, the simple proof of Pappus (A.D.300) for the theorem on base angles of an isosceles triangle follows.

PROPOSITION 3.10.   If in $\triangle ABC$ we have $AB \cong AC$, then $\sphericalangle B \cong \sphericalangle C$ (see Figure 3.20).

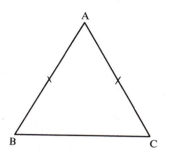

FIGURE 3.20

Proof:

(1) Consider the correspondence of vertices $A \leftrightarrow A$, $B \leftrightarrow C$, $C \leftrightarrow B$. Under this correspondence, two sides and the included angle of $\triangle ABC$ are congruent respectively to the corresponding sides and included angle of $\triangle ACB$ (by hypothesis and Congruence Axiom 5 that an angle is congruent to itself).

(2) Hence, $\triangle ABC \cong \triangle ACB$ (SAS), so the corresponding angles, $\sphericalangle B$ and $\sphericalangle C$, are congruent (by definition of congruence of triangles). ■

Here are some more familiar results on congruence. We will prove some of them; if the proof is omitted, see the exercises.

PROPOSITION 3.11 (Segment Subtraction). If $A * B * C$, $D * E * F$, $AB \cong DE$, and $AC \cong DF$, then $BC \cong EF$ (see Figure 3.21).

PROPOSITION 3.12. Given $AC \cong DF$, then for any point B between A and C, there is a unique point E between D and F such that $AB \cong DE$.

Proof:

(1) There is a unique point E on $\overrightarrow{DF}$ such that $AB \cong DE$ (Congruence Axiom 1).

(2) Suppose E were not between D and F (RAA hypothesis; see Figure 3.22).

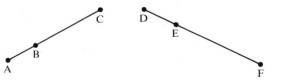

FIGURE 3.21

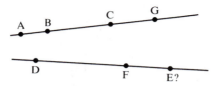

FIGURE 3.22

(3) Then either $E = F$ or $D * F * E$ (definition of $\overrightarrow{DF}$).

(4) If $E = F$, then B and C are two distinct points on $\overrightarrow{AC}$ such that $AC \cong DF \cong AB$ (hypothesis, step 1), contradicting the uniqueness part of Congruence Axiom 1.

(5) If $D * F * E$, then there is a point G on the ray opposite to $\overrightarrow{CA}$ such that $FE \cong CG$ (Congruence Axiom 1).

(6) Then $AG \cong DE$ (Congruence Axiom 3).

(7) Thus, there are two distinct points B and G on $\overrightarrow{AC}$ such that $AG \cong DE \cong AB$ (steps 1, 5, and 6), contradicting the uniqueness part of Congruence Axiom 1.

(8) $D * E * F$ (RAA conclusion). ∎

DEFINITION. $AB < CD$ (or $CD > AB$) means that there exists a point E between C and D such that $AB \cong CE$.

PROPOSITION 3.13 (Segment Ordering). (a) Exactly one of the following conditions holds (*trichotomy*): $AB < CD$, $AB \cong CD$, or $AB > CD$. (b) If $AB < CD$ and $CD \cong EF$, then $AB < EF$. (c) If $AB > CD$ and $CD \cong EF$, then $AB > EF$. (d) if $AB < CD$ and $CD < EF$, then $AB < EF$ (transitivity).

PROPOSITION 3.14. Supplements of congruent angles are congruent.

PROPOSITION 3.15. (a) Vertical angles are congruent to each other. (b) An angle congruent to a right angle is a right angle.

PROPOSITION 3.16. For every line l and every point P there exists a line through P perpendicular to l.

Proof:

(1) Assume first that P does not lie on l and let A and B be any two points on l (Incidence Axiom 2). (See Figure 3.23.)

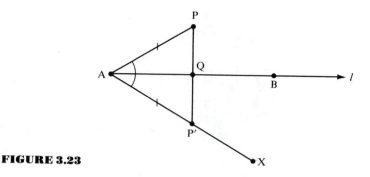

FIGURE 3.23

(2) On the opposite side of l from P there exists a ray $\overrightarrow{AX}$ such that ∢XAB ≅ ∢PAB (Congruence Axiom 4).

(3) There is a point P′ on $\overrightarrow{AX}$ such that AP′ ≅ AP (Congruence Axiom 1).

(4) PP′ intersects l in a point Q (definition of opposite sides of l).

(5) If Q = A, then $\overleftrightarrow{PP'}$ ⊥ l (definition of ⊥).

(6) If Q ≠ A, then △PAQ ≅ △P′AQ (SAS).

(7) Hence, ∢PQA ≅ ∢P′QA (corresponding angles), so $\overleftrightarrow{PP'}$ ⊥ l (definition of ⊥).

(8) Assume now that P lies on l. Since there are points not lying on l (Proposition 2.3), we can drop a perpendicular from one of them to l (steps 5 and 7), thereby obtaining a right angle.

(9) We can lay off an angle congruent to this right angle with vertex at P and one side on l (Congruence Axiom 4); the other side of this angle is part of a line through P perpendicular to l (Proposition 3.15(b)). ■

It is natural to ask whether the perpendicular to l through P constructed in Proposition 3.16 is unique. If P lies on l, Proposition 3.23 (later in this chapter) and the uniqueness part of Congruence Axiom 4 guarantee that the perpendicular is unique. If P does not lie on l, we will not be able to prove uniqueness for the perpendicular until the next chapter.

Note on Elliptic Geometry. Informally, elliptic geometry may be thought of as the geometry on a Euclidean sphere with antipodal points identified (the model of incidence geometry first described

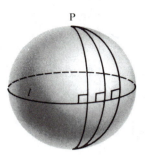

FIGURE 3.24

in Exercise 9(e), Chapter 2). Its "lines" are the great circles on the sphere. Given such a "line" l, there is a point P called the "pole" of l such that every line through P is perpendicular to l! To visualize this, think of l as the equator on a sphere and P as the north pole; every great circle through the north pole is perpendicular to the equator (Figure 3.24).

PROPOSITION 3.17 (ASA Criterion for Congruence). Given $\triangle ABC$ and $\triangle DEF$ with $\angle A \cong \angle D$, $\angle C \cong \angle F$, and $AC \cong DF$. Then $\triangle ABC \cong \triangle DEF$.

PROPOSITION 3.18 (Converse of Proposition 3.10). If in $\triangle ABC$ we have $\angle B \cong \angle C$, then $AB \cong AC$ and $\triangle ABC$ is isosceles.

PROPOSITION 3.19 (Angle Addition). Given $\overrightarrow{BG}$ between $\overrightarrow{BA}$ and $\overrightarrow{BC}$, $\overrightarrow{EH}$ between $\overrightarrow{ED}$ and $\overrightarrow{EF}$, $\angle CBG \cong \angle FEH$, and $\angle GBA \cong \angle HED$. Then $\angle ABC \cong \angle DEF$. (See Figure 3.25.)

> *Proof:*
> (1) By the crossbar theorem,[3] we may assume G is chosen so that $A * G * C$.
> (2) By Congruence Axiom 1, we assume D, F, and H chosen so that $AB \cong ED$, $GB \cong EH$, and $CB \cong EF$.
> (3) Then $\triangle ABG \cong \triangle DEH$ and $\triangle GBC \cong \triangle HEF$ (SAS).

---

[3] This renaming technique will be used frequently. G is just a label for any point $\neq$ B on the ray which intersects AC, so we may as well choose G to be the point of intersection rather than clutter the argument with a new label.

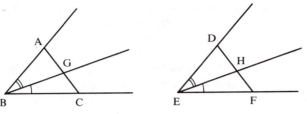

FIGURE 3.25

(4) ∡DHE ≅ ∡AGB, ∡FHE ≅ ∡CGB (step 3), and ∡AGB is supplementary to ∡CGB (step 1).

(5) D, H, F are collinear and ∡DHE is supplementary to ∡FHE (step 4, Proposition 3.14, and Congruence Axiom 4).

(6) D * H * F (Proposition 3.7, using the hypothesis on $\overrightarrow{EH}$).

(7) AC ≅ DF (steps 3 and 6, Congruence Axiom 3).

(8) ∡BAC ≅ ∡EDF (steps 3 and 6).

(9) △ABC ≅ △DEF (SAS; steps 2, 7, and 8).

(10) ∡ABC ≅ ∡DEF (corresponding angles). ∎

PROPOSITION 3.20 (Angle Subtraction). Given $\overrightarrow{BG}$ between $\overrightarrow{BA}$ and $\overrightarrow{BC}$, $\overrightarrow{EH}$ between $\overrightarrow{ED}$ and $\overrightarrow{EF}$, ∡CBG ≅ ∡FEH, and ∡ABC ≅ ∡DEF. Then ∡GBA ≅ ∡HED.

DEFINITION. ∡ABC < ∡DEF means there is a ray $\overrightarrow{EG}$ between $\overrightarrow{ED}$ and $\overrightarrow{EF}$ such that ∡ABC ≅ ∡GEF (see Figure 3.26).

PROPOSITION 3.21 (Ordering of Angles). (a) Exactly one of the following conditions holds (*trichotomy*): ∡P < ∡Q, ∡P ≅ ∡Q, or ∡Q < ∡P. (b) If ∡P < ∡Q and ∡Q ≅ ∡R, then ∡P < ∡R. (c) If ∡P < ∡Q and ∡Q ≅ ∡R, then ∡P > ∡R. (d) If ∡P < ∡Q and ∡Q < ∡R, then ∡P < ∡R.

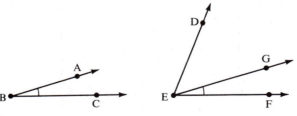

FIGURE 3.26

PROPOSITION 3.22 (SSS Criterion for Congruence). Given △ABC and △DEF. If AB ≅ DE, BC ≅ EF, and AC ≅ DF, then △ABC ≅ △DEF.

The AAS criterion for congruence will be given in the next chapter because its proof is more difficult. The next proposition was assumed as an axiom by Euclid, but can be proved from Hilbert's axioms.

PROPOSITION 3.23 (Euclid's Fourth Postulate). All right angles are congruent to each other. (See Figure 3.27.)

Proof:
(1) Given ∡BAD ≅ ∡CAD and ∡FEH ≅ ∡GEH (two pairs of right angles, by definition). Assume the contrary, that ∡BAD is not congruent to ∡FEH (RAA hypothesis).
(2) Then one of these angles is smaller than the other, e.g., ∡FEH < ∡BAD (Proposition 3.21(a)), so that by definition there is a ray $\overrightarrow{AJ}$ between $\overrightarrow{AB}$ and $\overrightarrow{AD}$ such that ∡BAJ ≅ ∡FEH.
(3) ∡CAJ ≅ ∡GEH (Proposition 3.14).
(4) ∡CAJ ≅ ∡FEH (steps 1 and 3, Congruence Axiom 5).
(5) There is a ray $\overrightarrow{AK}$ between $\overrightarrow{AD}$ and $\overrightarrow{AC}$ such that ∡BAJ ≅ ∡CAK (step 1 and Proposition 3.21(b)).
(6) ∡BAJ ≅ ∡CAJ (steps 2 and 4, and Congruence Axiom 5).
(7) ∡CAJ ≅ ∡CAK (steps 5 and 6, and Congruence Axiom 5).
(8) Thus, we have ∡CAD greater than ∡CAK (by definition) and less than its congruent angle ∡CAJ (step 7 and Proposition 3.8(c)), which contradicts Proposition 3.21.
(9) ∡BAD ≅ ∡FEH (RAA conclusion). ∎

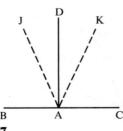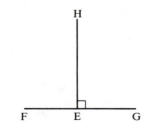

FIGURE 3.27

AXIOMS OF CONTINUITY

The axioms of continuity are needed to fill a number of gaps in Euclid's *Elements*. Consider the argument Euclid gives to justify his very first proposition.

EUCLID'S PROPOSITION 1. Given any segment, there is an equilateral triangle having the given segment as one of its sides.

Euclid's Proof:
(1) Let AB be the given segment. With center A and radius AB, let the circle BCD be described (Postulate III). (See Figure 3.28.)
(2) Again with center B and radius BA, let the circle ACE be described (Postulate III).
(3) From a point C in which the circles cut one another, draw the segments CA and CB (Postulate I).
(4) Since A is the center of the circle CDB, AC is congruent to AB (definition of circle).
(5) Again, since B is the center of circle CAE, BC is congruent to BA (definition of circle).
(6) Since CA and CB are each congruent to AB (steps 4 and 5), they are congruent to each other (first common notion).
(7) Hence, △ABC is an equilateral triangle (by definition) having AB as one of its sides. ■

Since very step has apparently been justified, you may not see the gap in the proof. It occurs in the first three steps, especially in the third step, which explicitly states that C is a point in which the circles cut

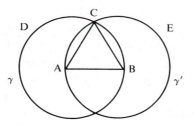

FIGURE 3.28

each other. (The second step states this implicitly by using the same letter "C" to denote part of the circle, as in the first step.) The point is: How do we know that such a point C exists?

If you believe it is obvious from the diagram that such a point C exists you are right — but you are not allowed to use the diagram to justify this! We aren't saying that the circles constructed do not cut each other; we're saying only that another axiom is needed to *prove* that they do.

The gap can be filled by assuming the following *circular continuity principle:*

CIRCULAR CONTINUITY PRINCIPLE. If a circle γ has one point inside and one point outside another circle γ', then the two circles intersect in two points.

Here a point P is defined as *inside* a circle with center O and radius OR if OP < OR (*outside* if OP > OR). In Figure 3.28, point B is inside circle γ', and the point B′ (not shown) such that A is the midpoint of BB′ is outside γ'. This principle is also needed to prove Euclid's 22nd proposition, the converse to the triangle inequality (see Major Exercise 4, Chapter 4). Another gap occurs in Euclid's method of dropping a perpendicular to a line (his 12th proposition, our Proposition 3.16). His construction tacitly assumes that if a line passes through a point inside a circle, then the line intersects the circle in two points — an assumption you can justify using the circular continuity principle (Major Exercise 1, Chapter 4; but our justification uses Proposition 3.16, so Euclid's argument must be discarded to avoid circular reasoning). Here is another useful consequence (see Major Exercise 2, Chapter 4).

ELEMENTARY CONTINUITY PRINCIPLE. If one endpoint of a segment is inside a circle and the other outside, then the segment intersects the circle.

Can you see why these are "continuity principles"? For example, in Figure 3.29, if you were drawing the segment with a pencil moving continuously from A to B, it would have to cross the circle (if it didn't, there would be "a hole" in the segment or the circle).

The next statement is not about continuity but rather about measurement. Archimedes was astute enough to recognize that a new

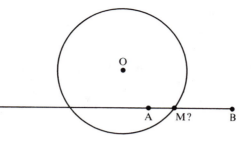

FIGURE 3.29

axiom was needed. It is listed here because we will show that it is a consequence of Dedekind's continuity axiom, given later in this section. It is needed so that we can assign a positive real number as the *length* $\overline{AB}$ of an arbitrary segment AB, as will be explained in Chapter 4.

ARCHIMEDES' AXIOM.   If CD is any segment, A any point, and r any ray with vertex A, then for every point B $\neq$ A on r there is a number n such that when CD is laid off n times on r starting at A, a point E is reached such that $n \cdot CD \cong AE$ and either B = E or B is between A and E.

Here we use Congruence Axiom 1 to begin laying off CD on r starting at A, obtaining a unique point A_1 on r such that $AA_1 \cong CD$, and we define $1 \cdot CD$ to be AA_1. Let r_1 be the ray emanating from A_1 that is contained in r. By the same method, we obtain a unique point A_2 on r_1 such that $A_1A_2 \cong CD$, and we define $2 \cdot CD$ to be AA_2. Iterating this process, you can define, by induction on n, the segment $n \cdot CD$ to be AA_n.

For example, if AB were π units long and CD of one unit length, you would have to lay off CD at least four times to get to a point E beyond the point B (see Figure 3.30).

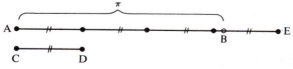

FIGURE 3.30

The intuitive content of Archimedes' axiom is that if you arbitrarily choose one segment CD as a unit of length, then every other segment has finite length with respect to this unit (in the notation of the axiom the length of AB with respect to CD as unit is at most n units). Another way to look at it is to choose AB as unit of length. The axiom says that no other segment can be infinitesimally small with respect to this unit (the length of CD with respect to AB as unit is at least $1/n$ units).

The next statement is a consequence of Archimedes' axiom and the previous axioms (as you will show in Exercise 6, Chapter 5), but if one wants to do geometry with segments of infinitesimal length allowed, this statement can replace Archimedes' axiom (see my note "Aristotle's Axiom in the Foundations of Hyperbolic Geometry," *Journal of Geometry*, vol. 33, 1988). Besides, Archimedes' axiom is not a purely geometric axiom, since it asserts the existence of a *number*.

ARISTOTLE'S AXIOM. Given any side of an acute angle and any segment AB, there exists a point Y on the given side of the angle such that if X is the foot of the perpendicular from Y to the other side of the angle, XY > AB.

Informally, if we start with any point Y on the given side, then as Y "recedes endlessly" from the vertex V of the angle, perpendicular segment XY "increases indefinitely" (because it is eventually bigger than any previously given segment AB). This principle will be valuable in Chapter 5 when we examine Proclus' attempt to prove Euclid's parallel postulate (see Figure 5.2). The idea of the proof from Archimedes' axiom is that if the starting XY is not already greater than the given segment AB, one simply lays off enough copies of VY on ray $\overrightarrow{VY}$ until point Y′ is reached such that the perpendicular segment dropped from Y′ is greater than AB (see Exercise 6, Chapter 5).

IMPORTANT COROLLARY. Let $\overrightarrow{AB}$ be any ray, P any point not collinear with A and B, and ⦨XVY any acute angle. Then there exists a point R on ray $\overrightarrow{AB}$ such that ⦨PRA < ⦨XVY.

Informally, if we start with any point R on $\overrightarrow{AB}$, then as R "recedes endlessly" from the vertex A of the ray, ⦨PRA decreases to zero (because it is eventually smaller than any previously given angle ⦨XVY). This result will be used in Chapter 6. Its proof uses Theorem 4.2 of Chapter 4 (the exterior angle theorem) and so it should be given

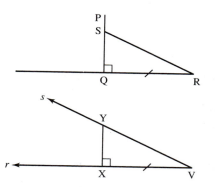

FIGURE 3.31

after that theorem is proved, but we sketch the proof now for convenience of reference. You may skip it now and return when needed.

Proof:

Let Q be the foot of the perpendicular from P to $\overleftrightarrow{AB}$. Since point B is just a label, we choose it so that $Q \neq B$ and Q lies on ray $\overrightarrow{BA}$. X and Y are arbitrary points on the rays r and s that are the sides of $\sphericalangle XVY$. Let X' be the foot of the perpendicular from Y to the line containing r. By the hypothesis that the angle is acute and the exterior angle theorem, we can show (by an RAA argument) that X' actually lies on r, and so we can choose X to be X'. Aristotle's axiom guarantees that Y can be chosen such that $XY > PQ$. By Congruence Axiom 1, there is one point R on $\overrightarrow{QB}$ such that $QR \cong XV$. We claim that $\sphericalangle PRQ < \sphericalangle XVY$. Assume the contrary. By trichotomy, there is a ray $\overrightarrow{RS}$ such that $\sphericalangle QRS \cong \sphericalangle XVY$ and $\overrightarrow{RS}$ either equals $\overrightarrow{RP}$ or is between $\overrightarrow{RP}$ and $\overrightarrow{RQ}$. By the crossbar theorem, point S (which thus far is also merely a label) can be chosen to lie on segment PQ; then SQ is not greater than PQ. By the ASA congruence criterion, $SQ \cong XY$. Hence XY is not greater than PQ, contradicting our choice of Y. Thus $\sphericalangle PRQ < \sphericalangle XVY$, as claimed. If R lies on ray $\overrightarrow{AB}$, then $\sphericalangle PRQ = \sphericalangle PRA$ and we are done. If not, R and Q lie on the opposite ray. By the exterior angle theorem, if R' is any point such that $Q * R * R'$, then $\sphericalangle PR'Q < \sphericalangle PRQ < XVY$. We get $\sphericalangle PBA = \sphericalangle PBQ < \sphericalangle XVY$ by taking R' = B. ∎

All four principles thus far stated are in the spirit of ancient Greek geometry. They are all consequences of the next axiom, which is utterly modern.

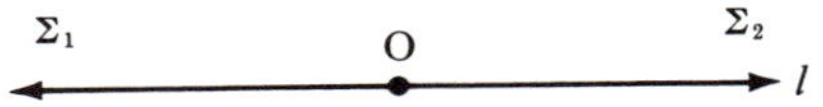

FIGURE 3.32

$$\Sigma_1 \qquad\qquad O \qquad\qquad \Sigma_2$$
$$\longleftarrow\!\!\bullet\!\!\longrightarrow l$$

DEDEKIND'S AXIOM.[4]   Suppose that the set $\{l\}$ of all points on a line l is the disjoint union $\Sigma_1 \cup \Sigma_2$ of two nonempty subsets such that no point of either subset is between two points of the other. Then there exists a unique point O on l such that one of the subsets is equal to a ray of l with vertex O and the other subset is equal to the complement.

Dedekind's axiom is a sort of converse to the line separation property stated in Proposition 3.4. That property says that any point O on l separates all the other points on l into those to the left of O and those to the right (see Figure 3.32; more precisely, $\{l\}$ is the union of the two rays of l emanating from O). Dedekind's axiom says that, conversely, any separation of points on l into left and right is produced by a unique point O. A pair of subsets Σ_1 and Σ_2 with the properties in Dedekind's axiom is called a *Dedekind cut* of the line.

Loosely speaking, the purpose of Dedekind's axiom is to ensure that a line l has no "holes" in it, in the sense that for any point O on l and any positive real number x there exist unique points P_{-x} and P_x on l such that $P_{-x} * O * P_x$ and segments $P_{-x}O$ and OP_x both have length x (with respect to some unit segment of measurement); see Figure 3.33.

Without Dedekind's axiom there would be no guarantee, for example, of the existence of a segment of length π. With it, we can introduce a rectangular coordinate system into the plane and do geometry analytically, as Descartes and Fermat discovered in the seventeenth century. This coordinate system enables us to prove that our axioms for Euclidean geometry are *categorical* in the sense that the system has a unique model (up to isomorphism — see the section Isomorphism of Models in Chapter 2), namely, the usual Cartesian coordinate plane of all ordered pairs of real numbers.

If we omitted Dedekind's axiom, then another model would be the so-called *surd plane*, a plane that is used to prove the impossibility of

[4] This axiom was proposed by J. W. R. Dedekind in 1871; an analogue of it is used in analysis texts to express the completeness of the real number system. It implies that every Cauchy sequence converges, that continuous functions satisfy the intermediate value theorem, that the definite integral of a continuous function exists, and other important conclusions. Dedekind actually defined a "real number" as a Dedekind cut on the set of rational numbers, an idea Eudoxus had 2000 years earlier (see Moise, 1990, Chapter 20).

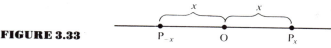

FIGURE 3.33

trisecting every angle with a straightedge and compass (see Moise, 1990, p. 282 ff.). The categorical natural of all the axioms is proved in Borsuk and Szmielew (1960, p. 276 ff.).

Warning. If you have never seen Dedekind's axiom before, arguments using it may be difficult to follow. Don't be discouraged. With the exception of Theorem 6.6 in hyperbolic geometry, it is not needed for studying the main theme of this book. I advise the beginning student to skip to the next section, Axiom of Parallelism.

Although Dedekind's axiom implies the other four principles and is the only continuity axiom we need assume, we still refer to the others as "axioms." Let us sketch a proof that *Archimedes' axiom is a consequence of Dedekind's* (and the axioms preceding this section).

Proof:
Given a segment CD and a point A on line l, with a ray r of l emanating from A. In the terminology of Archimedes' axiom, let Σ_1 consist of A and all points B on r reached by laying off copies of segment CD on r starting from A. Let Σ_2 be the complement of Σ_1 in r. We wish to prove that Σ_2 is empty, so assume the contrary.

In that case, let us show that we have defined a Dedekind cut of r (see Exercise 7(a)). Start with two points P, Q in Σ_2 and say A * P * Q. We must show that PQ $\subset \Sigma_2$. Let B be between P and Q. Suppose B could be reached, so that n and E are as in the statement of Archimedes' axiom; then, by Proposition 3.3, P is reached by the same n and E, contradicting P $\in \Sigma_2$. Thus PQ $\subset \Sigma_2$. Similarly, you can show that when P and Q are two points in Σ_1, PQ $\subset \Sigma_1$ (Exercise 7(b)). So we have a Dedekind cut. Let O be the point of r furnished by Dedekind's axiom.

Case 1. $O \in \Sigma_1$. Then for some number n, O can be reached by laying off n copies of segment CD on r starting from A. By laying off one more copy of CD, we can reach a point in Σ_2, but by definition of Σ_2, that is impossible.

Case 2. $O \in \Sigma_2$. Lay off a copy of CD on the ray opposite to Σ_2 starting at O, obtaining a point P; P lies on r (Exercise 7(b)), so $P \in \Sigma_1$. Then for some number n, P can be reached by laying off n copies of segment CD on r starting from A. By laying off one more copy of CD, we can reach O. That contradicts $O \in \Sigma_2$.

So in either case, we obtain a contradiction, and we can reject the RAA hypothesis that Σ_2 is nonempty. ■

To further get an idea of how Dedekind's axiom gives us continuity results, we sketch a proof now of the elementary continuity principle from Dedekind's axiom (logically, this proof should be given later, because it uses results from Chapter 4). Refer to Figure 3.29, p. 95.

Proof:
By the definitions of "inside" and "outside" of a circle γ with center O and radius OR, we have $OA < OR < OB$. Let Σ_2 be the set of all points P on the ray $\overrightarrow{AB}$ that either lie on γ or are outside γ, and let Σ_1 be its complement in $\overrightarrow{AB}$. By trichotomy (Proposition 3.13(a)), Σ_1 consists of all points of the segment AB that lie inside γ. Applying Exercise 27 of Chapter 4, you can convince yourself that (Σ_1, Σ_2) is a Dedekind cut. Let M be the point on $\overrightarrow{AB}$ furnished by Dedekind's axiom. Assume M does not lie on γ (RAA hypothesis).

Case 1. $OM < OR$. Then $M \in \Sigma_1$. Let m and r be the lengths (defined in Chapter 4) of OM and OR, respectively. Since Σ_2 with M is a ray, there is a point $N \in \Sigma_2$ such that the length of MN is $\frac{1}{2}(r - m)$ (e.g., by laying off a segment whose length is $\frac{1}{2}(r - m)$, using Theorem 4.3(11)). But by the *triangle inequality* (Corollary 2 to Theorem 4.3), the length of ON is less than $m + \frac{1}{2}(r - m) < m + (r - m) = r$, which contradicts $N \in \Sigma_2$.

Case 2. OM > OR. The same argument applies, interchanging the roles of Σ_2 and Σ_1.

So in either case, we obtain a contradiction, and M must lie on γ. ∎

You will find a lovely proof of the circular continuity principle from Dedekind's axiom on pp. 238–240 of Heath's translation and commentary on Euclid's *Elements* (1956). It assumes that Dedekind's axiom holds for semicircles, which you can easily prove in Major Exercise 4, and also uses the triangle inequality and the fact that the hypotenuse is greater than the leg (proved in Chapter 4).

Euclid's tacit use of continuity principles can often be avoided. We did not use them in our proof of the existence of perpendiculars (Proposition 3.16). We did use the circular continuity principle to prove the existence of equilateral triangles on a given base, and Euclid used that to prove the existence of midpoints, as in your straightedge-and-compass solution to Major Exercise 1 (a) of Chapter 1. But there is an ingenious way to prove the existence of midpoints using only the very mild continuity given by Pasch's theorem (see Exercise 12, Chapter 4).

Figure 3.34 shows the implications discussed (assuming all the incidence, betweenness, and congruence axioms — especially SAS).

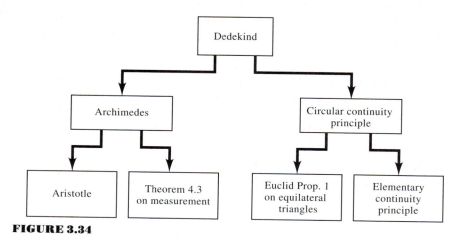

FIGURE 3.34

AXIOM OF PARALLELISM

If we were to stop with the axioms we now have, we could do quite a bit of geometry, but we still couldn't do all of Euclidean geometry. We would be able to do what J. Bolyai called "absolute geometry." This name is misleading because it does not include elliptic geometry and other geometries (see Appendix B). Preferable is the name suggested by W. Prenowitz and M. Jordan (1965), *neutral geometry,* so called because in doing this geometry we remain neutral about the one axiom from Hilbert's list left to be considered — historically the most controversial axiom of all.

HILBERT'S AXIOM OF PARALLELISM. For every line *l* and every point P not lying on *l* there is at most one line *m* through P such that *m* is parallel to *l* (Figure 3.35).

Note that this axiom is weaker than the Euclidean parallel postulate introduced in Chapter 1. This axiom asserts only that *at most* one line through P is parallel to *l*, whereas the Euclidean parallel postulate asserts in addition that *at least* one line through P is parallel to *l*. The reason "at least" is omitted from Hilbert's axiom is that it can be proved from the other axioms (see Corollary 2 to Theorem 4.1 in Chapter 4); it is therefore unnecessary to assume this as part of an axiom. This observation is important because it implies that the elliptic parallel property (no parallel lines exist) is inconsistent with the axioms of neutral geometry. Thus, a different set of axioms is needed for the foundation of elliptic geometry (see Appendix A).

The axiom of parallelism completes our list of 16 axioms for Euclidean geometry. A *Euclidean plane* is a model of these axioms. In referring to these axioms we will use the following shorthand: the incidence axioms will be denoted by I-1, I-2, and I-3; the betweenness axioms by B-1, B-1, B-3, and B-4; the congruence axioms by C-1, C-2, C-3, C-4, C-5, and C-6 (or SAS). The continuity axioms and the parallelism axiom will be referred to by name.

FIGURE 3.35

REVIEW EXERCISE

Which of the following statements are correct?

(1) Hilbert's axiom of parallelism is the same as the Euclidean parallel postulate given in Chapter 1.

(2) A * B * C is logically equivalent to C * B * A.

(3) In Axiom B-2 it is unnecessary to assume the existence of a point E such that B * D * E because this can be proved from the rest of the axiom and Axiom B-1, by interchanging the roles of B and D and taking E to be A.

(4) If A, B, and C are distinct collinear points, it is possible that *both* A * B * C *and* A * C * B.

(5) The "line separation property" asserts that a line has two sides.

(6) If points A and B are on opposite sides of a line *l*, then a point C not on *l* must be either on the same side of *l* as A or on the same side of *l* as B.

(7) If line *m* is parallel to line *l*, then all the points on *m* lie on the same side of *l*.

(8) If we were to take Pasch's theorem as an axiom instead of the separation axiom B-4, then B-4 could be proved as a theorem.

(9) The notion of "congruence" for two triangles is not defined in this chapter.

(10) It is an immediate consequence of Axiom C-2 that if AB ≅ CD, then CD ≅ AB.

(11) One of the congruence axioms asserts that if congruent segments are "subtracted" from congruent segments, the differences are congruent.

(12) In the statement of Axiom C-4 the variables A, B, C, A′, and B′ are quantified universally, and the variable C′ is quantified existentially.

(13) One of the congruence axioms is the side-side-side (SSS) criterion for congruence of triangles.

(14) Euclid attempted unsuccessfully to prove the side-angle-side criterion (SAS) for congruence by a method called "superposition."

(15) We can use Pappus' method to prove the converse of the theorem on base angles of an isosceles triangle if we first prove the angle-side-angle (ASA) criterion for congruence.

(16) Archimedes' axiom is independent of the other 15 axioms for Euclidean geometry given in this book.

(17) AB < CD means that there is a point E between C and D such that AB ≅ CE.

(18) Neutral geometry used to be called *absolute geometry;* it is the geometry you have when the axiom of parallelism is excluded from the system of axioms given here.

EXERCISES ON BETWEENNESS

1. Given $A * B * C$ and $A * C * D$.
   (a) Prove that A, B, C, and D are four distinct points (the proof requires an axiom).
   (b) Prove that A, B, C, and D are collinear.
   (c) Prove the corollary to Axiom B-4.
2. (a) Finish the proof of Proposition 3.1 by showing that $\overrightarrow{AB} \cup \overrightarrow{BA} = \overleftrightarrow{AB}$.
   (b) Finish the proof of Proposition 3.3 by showing that $A * B * D$.
   (c) Prove the converse of Proposition 3.3 by applying Axiom B-1.
   (d) Prove the corollary to Proposition 3.3.
3. Given $A * B * C$.
   (a) Use Proposition 3.3 to prove that $AB \subset AC$. Interchanging A and C, deduce $CB \subset CA$; which axiom justifies this interchange?
   (b) Use Axiom B-4 to prove that $AC \subset AB \cup BC$. (Hint: If P is a fourth point on AC, use another line through P to show $P \in AB$ or $P \in BC$.)
   (c) Finish the proof of Proposition 3.5. (Hint: If $P \neq B$ and $P \in AB \cap BC$, use another line through P to get a contradiction.)
4. Given $A * B * C$.
   (a) If P is a fourth point collinear with A, B, and C, use Proposition 3.3 and an axiom to prove that $\sim A * B * P \Rightarrow \sim A * C * P$.
   (b) Deduce that $\overrightarrow{BA} \subset \overrightarrow{CA}$ and, symmetrically, $\overrightarrow{BC} \subset \overrightarrow{AC}$.
   (c) Use this result, Proposition 3.1(a), Proposition 3.3, and Proposition 3.5 to prove that B is the only point that $\overrightarrow{BA}$ and $\overrightarrow{BC}$ have in common.
5. Given $A * B * C$. Prove that $\overrightarrow{AB} = \overrightarrow{AC}$, completing the proof of Proposition 3.6. Deduce that every ray has a *unique* opposite ray.
6. In Axiom B-2 we were given distinct points B and D and we asserted the existence of points A, C, and E such that $A * B * D$, $B * C * D$, and $B * D * E$. We can now show that it was not necessary to assume the existence of a point C between B and D because we can prove from our other axioms (including the rest of Axiom B-2) and from Pasch's theorem (which was proved without using Axiom B-2) that C exists.[5] Your job is to justify each step in the proof (some of the steps require a separate RAA argument).

---

[5] Regarding superfluous hypotheses, there is a story that Napoleon, after examining a copy of Laplace's *Celestial Mechanics,* asked Laplace why there was no mention of God in the work. The author replied, "I have no need of this hypothesis."

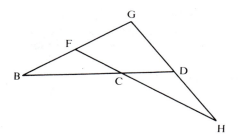

FIGURE 3.36

Proof (see Figure 3.36):

(1) There exists a line $\overrightarrow{BD}$ through B and D.
(2) There exists a point F not lying on $\overrightarrow{BD}$.
(3) There exists a line $\overrightarrow{BF}$ through B and F.
(4) There exists a point G such that B * F * G.
(5) Points B, F, and G are collinear.
(6) G and D are distinct points and D, B, and G are not collinear.
(7) There exists a point H such that G * D * H.
(8) There exists a line $\overrightarrow{GH}$.
(9) H and F are distinct points.
(10) There exists a line $\overleftrightarrow{FH}$.
(11) D does not lie on $\overleftrightarrow{FH}$.
(12) B does not lie on $\overleftrightarrow{FH}$.
(13) G does not lie on $\overleftrightarrow{FH}$.
(14) Points D, B, and G determine $\triangle DBG$ and $\overleftrightarrow{FH}$ intersects side BG in a point between B and G.
(15) H is the only point lying on both $\overleftrightarrow{FH}$ and $\overleftrightarrow{GH}$.
(16) No point between G and D lies on $\overleftrightarrow{FH}$.
(17) Hence, $\overleftrightarrow{FH}$ intersects side BD in a point C between D and B.
(18) Thus, there exists a point C between D and B. ■

7. (a) Define a Dedekind cut on a ray r the same way a Dedekind cut is defined for a line. Prove that the conclusion of Dedekind's axiom also holds for r. (Hint: One of the subsets, say, Σ_1, contains the vertex A of r; enlarge this set so as to include the ray opposite to r and show that a Dedekind cut of the line l containing r is obtained.) Similarly, state and prove a version of Dedekind's axiom for a cut on a segment.

   (b) Supply the indicated arguments left out of the proof of Archimedes' axiom from Dedekind's axiom.

8. From the three-point model (Example 1 in Chapter 2) we saw that if we used only the axioms of incidence we could not prove that a line has

more than two points lying on it. Using the betweenness axioms as well, prove that every line has at least five points lying on it. Give an informal argument to show that every segment (a fortiori, every line) has an infinite number of points lying on it (a formal proof requires the technique of mathematical induction).

9. Given a line l, a point A on l, and a point B not on l. Then every point of the ray $\overrightarrow{AB}$ (except A) is on the same side of l as B. (Hint: Use an RAA argument).

10. Prove Proposition 3.7.

11. Prove Proposition 3.8. (Hint: For Proposition 3.8(c) prove in two steps that E and B lie on the same side of $\overleftrightarrow{AD}$, first showing that EB does not meet $\overleftrightarrow{AD}$, then showing that EB does not meet the opposite ray $\overrightarrow{AF}$. Use Exercise 9.)

12. Prove the crossbar theorem. (Hint: Assume the contrary, and show that B and C lie on the same side of $\overleftrightarrow{AD}$. Use Proposition 3.8(c) to derive a contradiction.)

13. Prove Proposition 3.9. (Hint: For Proposition 3.9(a) use Pasch's theorem and Proposition 3.7; see Figure 3.37. For Proposition 3.9(b) let the ray emanate from point D in the interior of $\triangle ABC$. Use the crossbar theorem and Proposition 3.7 to show that $\overrightarrow{AD}$ meets BC in a point E such that $A * D * E$. Apply Pasch's theorem to $\triangle ABE$ and $\triangle AEC$; see Figure 3.38.)

14. Prove that a line cannot be contained in the interior of a triangle.

15. If a, b, and c are rays, let us say that they are *coterminal* if they emanate from the same point, and let us use the notation $a * b * c$ to mean that b is between a and c (as defined on p. 82). The analogue of Axiom B-1 states that if $a * b * c$, then a, b, c are distinct and coterminal and $c * b * a$; this analogue is obviously correct. State the analogues of Axioms B-2 and B-3 and Proposition 3.3 and tell which parts of these analogues are correct. (Beware of opposite rays!)

16. Find an interpretation in which the incidence axioms and the first two betweenness axioms hold but Axiom B-3 fails in the following way: there

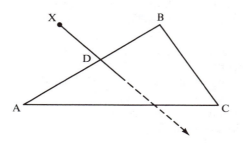

FIGURE 3.37

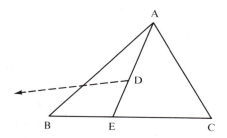

FIGURE 3.38

exist three collinear points, no one of which is between the other two. (Hint: In the usual Euclidean model, introduce a new betweenness relation A * B * C to mean that B is the midpoint of AC.)

17. Find an interpretation in which the incidence axioms and the first three betweenness axioms hold but the line separation property (Proposition 3.4) fails. (Hint: In the usual Euclidean model, pick a point P that is between A and B in the usual Euclidean sense and specify that A will now be considered to be between P and B. Leave all other betweenness relations among points alone. Show that P lies neither on ray $\overrightarrow{AB}$ nor on its opposite ray $\overrightarrow{AC}$.)

18. A rational number of the form $a/2^n$ (with a, n integers) is called *dyadic*. In the interpretations of Project 2 for this chapter, restrict to those points which have dyadic coordinates and to those lines which pass through several dyadic points. The incidence axioms, the first three betweenness axioms, and the line separation property all hold in this dyadic rational plane; show that Pasch's theorem fails. (Hint: The lines $3x + y = 1$ and $y = 0$ do not meet in this plane.)

19. A set of points S is called *convex* if whenever two points A and B are in S, the entire segment AB is contained in S. Prove that a half-plane, the interior of an angle, and the interior of a triangle are all convex sets, whereas the exterior of a triangle is not convex. Is a triangle a convex set?

EXERCISES ON CONGRUENCE

20. Justify each step in the following proof of Proposition 3.11:

Proof:
(1) Assume on the contrary that BC is not congruent to EF.
(2) Then there is a point G on $\overrightarrow{EF}$ such that BC $\cong$ EG.

(3) G ≠ F.

(4) Since AB ≅ DE, adding gives AC ≅ DG.

(5) However, AC ≅ DF.

(6) Hence, DF ≅ DG.

(7) Therefore, F = G.

(8) Our assumption has led to a contradiction; hence, BC ≅ EF. ■

21. Prove Proposition 3.13(a). (Hint: In case AB and CD are not congruent, there is a unique point F ≠ D on $\overrightarrow{CD}$ such that AB ≅ CF (reason ?). In case C * F * D, show that AB < CD. In case C * D * F, use Proposition 3.12 and some axioms to show that CD < AB.)

22. Use Proposition 3.12 to prove Proposition 3.13(b) and (c).

23. Use the previous exercise and Proposition 3.3 to prove Proposition 3.13(d).

24. Justify each step in the following proof of Proposition 3.14 (see Figure 3.39).

Proof:

Given ∢ABC ≅ ∢DEF. To prove ∢CBG ≅ ∢FEH:

(1) The points A, C, and G being given arbitrarily on the sides of ∢ABC and the supplement ∢CBG of ∢ABC, we can choose the points D, F, and H on the sides of the other angle and its supplement so that AB ≅ DE, CB ≅ FE, and BG ≅ EH.

(2) Then, △ABC ≅ △DEF.

(3) Hence, AC ≅ DF and ∢A ≅ ∢D.

(4) Also, AG ≅ DH.

(5) Hence, △ACG ≅ △DFH.

(6) Therefore, CG ≅ FH and ∢G ≅ ∢H.

(7) Hence, △CBG ≅ △FEH.

(8) It follows that ∢CBG ≅ ∢FEH, as desired. ■

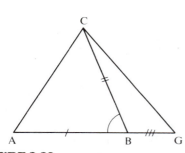
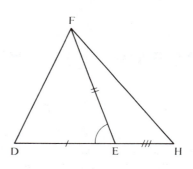

FIGURE 3.39

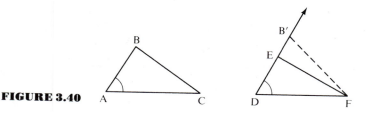

FIGURE 3.40

25. Deduce Proposition 3.15 from Proposition 3.14.
26. Justify each step in the following proof of Proposition 3.17 (see Figure 3.40):

Proof:
Given △ABC and △DEF with ∢A ≅ ∢D, ∢C ≅ ∢F, and AC ≅ DF. To prove △ABC ≅ △DEF:
(1) There is a unique point B′ on ray $\overrightarrow{DE}$ such that DB′ ≅ AB.
(2) △ABC ≅ △DB′F.
(3) Hence, ∢DFB′ ≅ ∢C.
(4) This implies $\overrightarrow{FE} = \overrightarrow{FB'}$.
(5) In that case, B′ = E.
(6) Hence, △ABC ≅ △DEF. ∎

27. Prove Proposition 3.18.
28. Prove that an equiangular triangle (all angles congruent to one another) is equilateral.
29. Prove Proposition 3.20. (Hint: Use Axiom C-4 and Proposition 3.19.)
30. Given ∢ABC ≅ ∢DEF and $\overrightarrow{BG}$ between $\overrightarrow{BA}$ and $\overrightarrow{BC}$. Prove that there is a unique ray $\overrightarrow{EH}$ between $\overrightarrow{ED}$ and $\overrightarrow{EF}$ such that ∢ABG ≅ ∢DEH. (Hint: Show that D and F can be chosen so that AB ≅ DE and BC ≅ EF, and that G can be chosen so that A * G * C. Use Propositions 3.7 and 3.12 and SAS to get H; see Figure 3.25.)
31. Prove Proposition 3.21 (imitate Exercises 21–23).
32. Prove Proposition 3.22. (Hint: Use the corollary to SAS to reduce to the case where A = D, C = F, and the points B and E are on opposite sides of $\overleftrightarrow{AC}$. Then consider the three cases in Figure 3.41 separately.)
33. If AB < CD, prove that 2AB < 2CD.
34. Let $\mathbb{Q}^2$ be the *rational plane* of all ordered pairs (x, y) of rational numbers with the usual interpretations of the undefined geometric terms used in analytic geometry. Show that Axiom C-1 and the elementary continuity principle fail in $\mathbb{Q}^2$. (Hint: The setgment from $(0, 0)$ to $(1, 1)$ cannot be laid off on the x axis from the origin.)

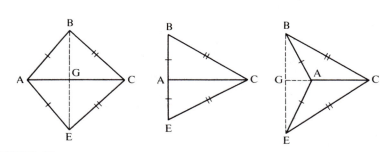

FIGURE 3.41

35. In the usual Euclidean plane we are all familiar with, there is a notion of length of a segment. Let us agree to measure all lengths in inches except for segments on one particular line called the *x* axis, where we will measure lengths in feet, and let us now interpret *congruence of segments* to mean that two segments have the same "length" in this perverse way of measuring. Incidence, betweenness, and congruence of angles will have their usual meaning. Show informally that the first five congruence axioms and angle addition (Proposition 3.19) still hold in this interpretation but that SAS fails (see Figure 3.42). Draw a picture of a "circle" with center on the *x* axis in this interpretation and use that picture to show that the circular continuity principle and the elementary continuity principle fail. Show that Dedekind's axiom still holds. Draw other pictures to show that SSS, ASA, and SAA all fail.

36. In Chapter 2 we displayed many models of the incidence axioms. As soon as we add the betweenness axioms, most of those interpretations are no longer models (for example, we lose all the finite models and the models in which "lines" are circles). Show, however, that the model in Exercise 9(d), which has the hyperbolic parallel property, is still a model under the natural interpretation of betweenness. It is called *the Klein model* and will be further studied in Chapter 7. Draw a picture to show that in this model, a point in the interior of an angle need not lie on a segment joining a point on one ray of the angle to a point on the other ray.

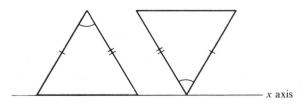

FIGURE 3.42

MAJOR EXERCISES

1. Let γ be a circle with center A and radius of length r. Let γ' be another circle with center A' and radius of length r', and let d be the distance from A to A' (see Figure 3.43). There is a hypothesis about the numbers r, r', and d that ensures that the circles γ and γ' intersect in two distinct points. Figure out what this hypothesis is. (Hint: It's statement that certain numbers obtained from r, r', and d are less than certain others.)

   What hypothesis on r, r', and d ensures that γ and γ' intersect in only one point, i.e., that the circles are tangent to each other? (See Figure 3.44.)

2. Define the *reflection* in a line m to be the transformation R_m of the plane which leaves each point of m fixed and transforms a point A not on m as follows. Let M be the foot of the perpendicular from A to m. Then, by definition, R_m(A) is the unique point A' such that A' * M * A and A'M $\cong$ MA. (See Figure 3.45.) This definition uses the result from Chapter 4 that the perpendicular from A to m is unique, so that the *foot* M is uniquely determined as the intersection with m. Prove that R_m is a *motion*, i.e., that AB $\cong$ A'B' for any segment AB. Prove also that AB $\cong$ CD $\Rightarrow$ A'B' $\cong$ C'D', and that $\angle$A $\cong$ $\angle$B $\Rightarrow$ $\angle$A' $\cong$ $\angle$B'. (Chapter 9 will be devoted to a thorough study of motions; the reflections generate the group of all such transformations.) (Hint: The proof breaks into the cases (i) A or B lies on m, (ii) A and B lie on opposite sides of m, and (iii) A and B

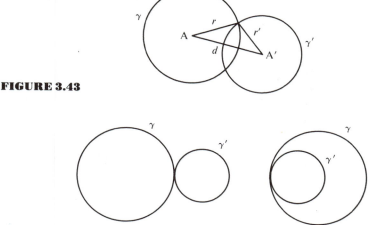

FIGURE 3.43

FIGURE 3.44

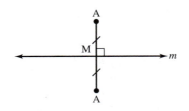

FIGURE 3.45

lie on the same side of *m*. In (ii), let M, N be the midpoints of AA′, BB′ and let C be the point at which AB meets *m*; prove that A′ * C * B′ by showing that ∢A′CM ≅ B′CN and apply Axiom C-3. In (iii), let C be the point at which AB′ meets *m*, and use B = (B′)′ and the first two cases to show that △ABC ≅ △A′B′C. Take care not to use results that are valid only in Euclidean geometry.)

Note. In elliptic geometry the perpendicular from A to *m* is unique except for one point P called the *pole* of *m* (see Figure 3.24, where *m* is the equator and P is the north pole); the definition of reflection is modified in elliptic geometry so that $R_m(P) = P$. Can you see that R_m is then the same as the 180° rotation about P? Recall that antipodal points are identified.

3. Consider the following statements on congruence:
    1. Given triangle △ABC and segment DE such that AB ≅ DE. Then on a given side of $\overleftrightarrow{DE}$ there is a unique point F such that AC ≅ DF and BC ≅ EF.
    2. Given triangles △ADC and △A′D′C′ and given A * B * C and A′ * B′ * C′. If AB ≅ A′B′, BC ≅ B′C′, AD ≅ A′D′, and BD ≅ B′D′, then CD ≅ C′D′ ("rigidity of a triangle with a tail" — see Figure 3.46).

Prove these statements. Also, prove a statement 2a obtained from statement 2 by substituting CD ≅ C′D′ for BD ≅ B′D′ in the hypothesis and making BD ≅ B′D′ the conclusion.

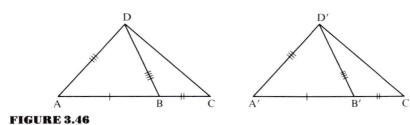

FIGURE 3.46

FIGURE 3.47

In Borsuk and Szmielew (1960), statements 1 and 2 are taken as axioms, in place of our Axioms C-4, C-5, and C-6. The advantage of this change is that these new congruence axioms refer only to congruence of segments. Congruence of angles, ∢ABC ≅ ∢A'B'C', can then be *defined* by specifying that A and C (respectively, A' and C') can be chosen on the sides of ∢B (respectively, ∢B') so that AB ≅ A'B', BC ≅ B'C', and AC ≅ A'C'. With this definition, keeping the same incidence and betweenness axioms as before, show that C-4, C-5, and C-6 can be proved from C-1, C-2, C-3, and statements 1 and 2. (Hint: First prove statement 2a by an RAA argument. Then show that if ∢ABC ≅ ∢A'B'C', and that if we had chosen other points D, E, D', and E' on the sides of ∢B and ∢B' such that DB ≅ D'B' and EB ≅ E'B', then DE ≅ D'E'. See Figure 3.47.)

4. Let AB be a diameter of circle γ with center O. The intersection σ of γ with one of the half-planes determined by $\overleftrightarrow{AB}$ is called an *open semicircle of* γ with endpoints A, B; adding the points A, B gives the *semicircle* $\bar{\sigma}$. Define a betweenness relation # on σ as follows: P # Q # R means that P, Q, and R are distinct points on σ and $\overrightarrow{OP} * \overrightarrow{OQ} * \overrightarrow{OR}$ (see Exercise 15). Specify also that A # P # B for any P on σ.

   (a) Let M be the point on σ such that $\overleftrightarrow{MO} \perp \overleftrightarrow{AB}$ (see Figure 3.48). Let AMB = AM ∪ MB. For any point P on σ, prove that ray $\overrightarrow{OP}$ intersects AMB in a point P' and that the mapping P ↦ P' is one-to-one from $\bar{\sigma}$ onto AMB.

   (b) Define P' # Q' # R' to mean P # Q # R. If P', Q', and R' all lie on segment AM or all lie on MB, prove that P'' # Q' # R' ⟹ P' * Q' * R'.

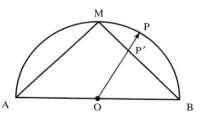

FIGURE 3.48

(c) Prove that Dedekind's axiom holds for AMB and hence for $\bar{\sigma}$ (use Exercise 7).

PROJECTS

1. Report on T. L. Heath's (1956) proof for the circular continuity principle.
2. Incidence, points, and lines in the real plane $\mathbb{R}^2$ were given in Major Exercise 9, Chapter 2. Distance is given by the usual Pythagorean formula

$$d(\text{AB}) = \sqrt{(a_1 - b_1)^2 + (a_2 - b_2)^2}$$

   where $\text{A} = (a_1, a_2)$, $\text{B} = (b_1, b_2)$. Define $\text{A} * \text{B} * \text{C}$ to mean $d(\text{AC}) = d(\text{AB}) + d(\text{BC})$, and define $\text{AB} \cong \text{CD}$ to mean $d(\text{AB}) = d(\text{CD})$. Define $\sphericalangle \text{ABC} \cong \sphericalangle \text{DEF}$ if A, C, D, and F can be chosen on the sides of these angles so that $\text{AB} \cong \text{ED}$, $\text{CB} \cong \text{FE}$, and $\text{AC} \cong \text{DF}$. With these interpretations, verify all the axioms for Euclidean geometry (see Moise, 1990, Chapter 26, or Borsuk and Szmielew, 1960, Chapter 4).
3. Suppose in Project 2 the field $\mathbb{R}$ of real numbers is replaced by an arbitrary *Euclidean field F* (an ordered field in which every positive number has a square root). Show that all the axioms for Euclidean geometry except Dedekind's and Archimedes' axioms are satisfied; show also that the circular continuity principle is satisfied.
4. In Euclidean geometry, Hilbert showed how to construct perpendiculars using ruler (marked straightedge) alone. His construction uses the theorem that the altitudes of a triangle are concurrent. Report on Hilbert's results. (Refer to D. Hilbert, 1987, p. 100.)

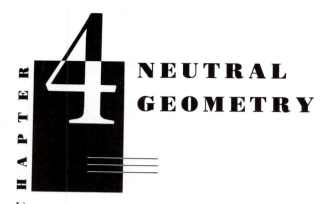

NEUTRAL GEOMETRY

CHAPTER 4

> If only it could be proved . . . that "there is a
> Triangle whose angles are together not *less* than two
> right angles"! But alas, *that* is an *ignis fatuus* that has
> never yet been caught!
>
> C. L. DODGSON (LEWIS CARROLL)

GEOMETRY WITHOUT THE PARALLELL AXIOM

In the exercises of the previous chapter you gained experience in proving some elementary results from Hilbert's axioms. Many of these results were taken for granted by Euclid. You can see that filling in the gaps and rigorously proving every detail is a long task. In any case, we must show that Euclid's postulates are consequences of Hilbert's. We have seen that Euclid's first postulate is the same as Hilbert's Axiom I-1. In our new language, Euclid's second postulate says the following: given segments AB and CD, there exists a point E such that A * B * E and CD ≅ BE. This follows immediately from Hilbert's Axiom C-1 applied to the ray emanating from B opposite to $\overrightarrow{BA}$ (see Figure 4.1).

The third postulate of Euclid becomes a definition in Hilbert's system. The *circle* with *center* O and *radius* OA is defined as the set of all points P such that OP is congruent to OA. Axiom C-1 then guarantees that on every ray emanating from O there exists such a point P.

The fourth postulate of Euclid — all right angles are congruent — becomes a theorem in Hilbert's system, as was shown in Proposition 3.23.

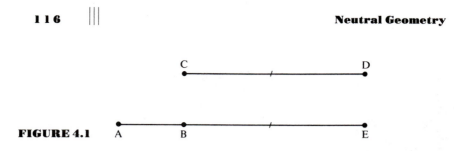

FIGURE 4.1   A   B   E

Euclid's parallel postulate is discussed later in this chapter. In this chapter we shall be interested in neutral geometry — by definition, all those geometric theorems that can be proved using only the axioms of incidence, betweenness, congruence, and continuity and without using the axiom of parallelism. Every result proved previously is a theorem in neutral geometry. You should review all the statements in the theorems, propositions, and exercises of Chapter 3 because they will be used throughout the book. Our proofs will be less formal henceforth.

What is the purpose of studying neutral geometry? We are not interested in studying it for its own sake. Rather, we are trying to clarify the role of the parallel postulate by seeing which theorems in the geometry do not depend on it, i.e., which theorems follow from the other axioms alone without ever using the parallel postulate in proofs. This will enable us to avoid many pitfalls and to see much more clearly the logical structure of our system. Certain questions that can be answered in Euclidean geometry (e.g., whether there is a unique parallel through a given point) may not be answerable in neutral geometry because its axioms do not give us enough information.

ALTERNATE INTERIOR ANGLE THEOREM

The next theorem requires a definition: let t be a transversal to lines l and l', with t meeting l at B and l' at B'. Choose points A and C on l such that A * B * C; choose points A' and C' on l' such that A and A' are on the same side of t and such that A' * B' * C'. Then the following four angles are called *interior:* ∢A'B'B, ∢ABB', ∢C'B'B, ∢CBB'. The two pairs (∢ABB', ∢C'B'B) and (∢A'B'B, ∢CBB') are called pairs of *alternate interior angles* (see Figure 4.2).

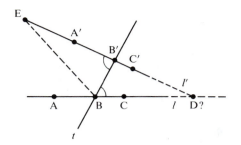

FIGURE 4.2

THEOREM 4.1 (Alternate Interior Angle Theorem). If two lines cut by a transversal have a pair of congruent alternate interior angles, then the two lines are parallel.

Proof:
Given ⊀A′B′B ≅ ⊀CBB′. Assume on the contrary *l* and *l′* meet at a point D. Say D is on the same side of *t* as C and C′. There is a point E on B⃗′A⃗′ such that B′E ≅ BD (Axiom C-1). Segment BB′ is congruent to itself, so that △B′BD ≅ △BB′E (SAS). In particular, ⊀DB′B ≅ ⊀EBB′. Since ⊀DB′B is the supplement of ⊀EB′B, ⊀EBB′ must be the supplement of ⊀DBB′ (Proposition 3.14 and Axiom C-4). This means that E lies on *l*, and hence *l* and *l′* have the two points E and D in common, which contradicts Proposition 2.1 of incidence geometry. Therefore, *l* ∥ *l′*. ■

This theorem has two very important corollaries.

COROLLARY 1. Two lines perpendicular to the same line are parallel. Hence, the perpendicular dropped from a point P not on line *l* to *l* is *unique* (and the point at which the perpendicular intersects *l* is called its *foot*).

Proof:
If *l* and *l′* are both perpendicular to *t*, the alternate interior angles are right angles and hence are congruent (Proposition 3.23). ■

COROLLARY 2. If *l* is any line and P is any point not on *l*, there exists at least one line *m* through P parallel to *l* (see Figure 4.3).

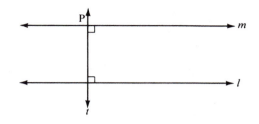

FIGURE 4.3

Proof:
There is a line *t* through P perpendicular to *l*, and again there is a unique line *m* through P perpendicular to *t* (Proposition 3.16). Since *l* and *m* are both perpendicular to *t*, Corollary 1 tells us that *l* ∥ *m*. (This construction will be used repeatedly.) ■

To repeat, there always exists a line *m* through P parallel to *l* — this has been proved in neutral geometry. But we don't know that *m* is *unique*. Although Hilbert's parallel postulate says that *m* is indeed unique, we are not assuming that postulate. We must keep our minds open to the strange possibility that there may be other lines through P parallel to *l*.

Warning. You are accustomed in Euclidean geometry to use the *converse* of Theorem 4.1, which states, "If two lines are parallel, then alternate interior angles cut by a transversal are congruent." We haven't proved this converse, so don't use it! (It turns out to be logically equivalent to the parallel postulate — see Exercise 5.)

EXTERIOR ANGLE THEOREM

Before we continue our list of theorems, we must first make another definition: an angle supplementary to an angle of a triangle is called an *exterior angle* of the triangle; the two angles of the triangle not adjacent to this exterior angle are called the *remote interior angles*. The following theorem is a consequence of Theorem 4.1:

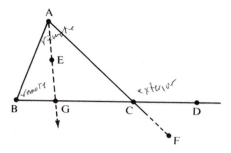

FIGURE 4.4

THEOREM 4.2 (Exterior Angle Theorem). An exterior angle of a triangle is greater than either remote interior angle (see Figure 4.4).

To prove ∡ACD is greater than ∡B and ∡A:

Proof:
Consider the remote interior angle ∡BAC. If ∡BAC ≅ ∡ACD, then $\overleftrightarrow{AB}$ is parallel to $\overleftrightarrow{CD}$ (Theorem 4.1), which contradicts the hypothesis that these lines meet at B. Suppose ∡BAC were greater than ∡ACD (RAA hypothesis). Then there is a ray $\overrightarrow{AE}$ between $\overrightarrow{AB}$ and $\overrightarrow{AC}$ such that ∡ACD ≅ ∡CAE (by definition). This ray $\overrightarrow{AE}$ intersects BC in a point G (crossbar theorem, Chapter 3). But according to Theorem 4.1, lines $\overleftrightarrow{AE}$ and $\overleftrightarrow{CD}$ are parallel. Thus, ∡BAC cannot be greater than ∡ACD (RAA conclusion). Since ∡BAC is also not congruent to ∡ACD, ∡BAC must be less than ∡ACD (Proposition 3.21(a)).

For remote angle ∡ABC, use the same argument applied to exterior angle ∡BCF, which is congruent to ∡ACD by the vertical angle theorem (Proposition 3.15(a)). ∎

The exterior angle theorem will play a very important role in what follows. It was the 16th proposition in Euclid's *Elements*. Euclid's proof had a gap due to reasoning from a diagram. He considered the line $\overleftrightarrow{BM}$ joining B to the midpoint of AC and he constructed point B' such that B * M * B' and BM ≅ MB' (Axiom C-1). He then assumed from the diagram that B' lay in the interior of ∡ACD (see Figure 4.5). Since ∡B'CA ≅ ∡A (SAS), Euclid concluded correctly that ∡ACD > ∡A.

The gap in Euclid's argument can easily be filled with the tools we have developed. Since segment BB' intersects AC at M, B and B' are

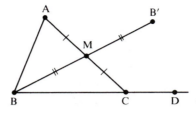

FIGURE 4.5

on opposite sides of $\overleftrightarrow{AC}$ (by definition). Since BD meets $\overleftrightarrow{AC}$ at C, B and D are also on opposite sides of $\overleftrightarrow{AC}$. Hence, B' and D are on the same side of $\overleftrightarrow{AC}$ (Axiom B-4). Next, B' and M are on the same side of $\overleftrightarrow{CD}$, since segment MB' does not contain the point B at which $\overleftrightarrow{MB'}$ meets $\overleftrightarrow{CD}$ (by construction of B' and Axioms B-1 and B-3). Also, A and M are on the same side of $\overleftrightarrow{CD}$ because segment AM does not contain the point C at which $\overleftrightarrow{AM}$ meets $\overleftrightarrow{CD}$ (by definition of midpoint and Axiom B-3). So again, Separation Axiom B-4 ensures that A and B' are on the same side of $\overleftrightarrow{CD}$. By definition of "interior" (in Chapter 3, preceding Proposition 3.7), we have shown that B' lies in the interior of ∢ACD.

Note on Elliptic Geometry. Figure 3.24 shows a triangle on the sphere with both an exterior angle and a remote interior angle that are right angles, so the exterior angle theorem doesn't hold. Our proof of it was based on the alternate interior angle theorem, which can't hold in elliptic geometry because there are no parallels. The proof we gave of Theorem 4.1 breaks down in elliptic geometry because Axiom B-4, which asserts that a line separates the plane into two sides, doesn't hold; we knew points E and D in that proof were distinct because they lay on opposite sides of line *t*. Or, thinking in terms of spherical geometry, where a great circle does separate the sphere into two hemispheres, if points E and D are distinct, there is no contradiction because great circles do meet in two antipodal points.

Euclid's proof of Theorem 4.2 breaks down on the sphere because "lines" are circles and if segment BM is long enough, the reflected point B' might lie on it (e.g., if BM is a semicircle, B' = B).

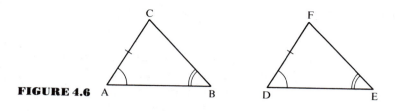

FIGURE 4.6

As a consequence of the exterior angle theorem (and our previous results), you can now prove as exercises the following familiar propositions.

PROPOSITION 4.1 (SAA Congruence Criterion). Given $AC \cong DF$, $\angle A \cong \angle D$, and $\angle B \cong \angle E$. Then $\triangle ABC \cong \triangle DEF$ (Figure 4.6).

PROPOSITION 4.2. Two right triangles are congruent if the hypotenuse and a leg of one are congruent respectively to the hypotenuse and a leg of the other (Figure 4.7).

PROPOSITION 4.3 (Midpoints). Every segment has a unique midpoint.

PROPOSITION 4.4 (Bisectors). (a) Every angle has a unique bisector. (b) Every segment has a unique perpendicular bisector.

PROPOSITION 4.5. In a triangle $\triangle ABC$, the greater angle lies opposite the greater side and the greater side lies opposite the greater angle, i.e., $AB > BC$ if and only if $\angle C > \angle A$.

PROPOSITION 4.6. Given $\triangle ABC$ and $\triangle A'B'C'$, if $AB \cong A'B'$ and $BC \cong B'C'$, then $\angle B < \angle B'$ if and only if $AC < A'C'$.

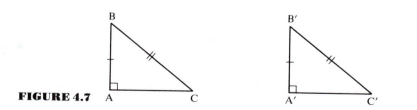

FIGURE 4.7

MEASURE OF ANGLES AND SEGMENTS

Thus far in our treatment of geometry we have refrained from using numbers that measure the sizes of angles and segments — this was in keeping with the spirit of Euclid. From now on, however, we will not be so austere. The next theorem (Theorem 4.3) asserts the possibility of measurement and lists its properties. The proof requires the axioms of continuity for the first time (in keeping with the elementary level of this book, the interested reader is referred to Borsuk and Szmielew, 1960, Chapter 3, Sections 9 and 10). In some popular treatments of geometry this theorem is taken as an axiom (ruler-and-protractor postulates — see Moise, 1990). The familiar notation $(\sphericalangle A)°$ will be used for the number of degrees in $\sphericalangle A$, and the length of segment $\underline{AB}$ (with respect to some unit of measurement) will be denoted by $\overline{AB}$.

THEOREM 4.3. A. There is a unique way of assigning a degree measure to each angle such that the following properties hold (refer to Figure 4.8):

(0) $(\sphericalangle A)°$ is a real number such that $0 < (\sphericalangle A)° < 180°$
(1) $(\sphericalangle A)° = 90°$ if and only if $\sphericalangle A$ is a right angle.
(2) $(\sphericalangle A)° = (\sphericalangle B)°$ if and only if $\sphericalangle A \cong \sphericalangle B$.
(3) If $\overrightarrow{AC}$ is interior to $\sphericalangle DAB$, then $(\sphericalangle DAB)° = (\sphericalangle DAC)° + (\sphericalangle CAB)°$.
(4) For every real number x between 0 and 180, there exists an angle $\sphericalangle A$ such that $(\sphericalangle A)° = x°$.
(5) If $\sphericalangle B$ is supplementary to $\sphericalangle A$, then $(\sphericalangle A)° + (\sphericalangle B)° = 180°$.
(6) $(\sphericalangle A)° > (\sphericalangle B)°$ if and only if $\sphericalangle A > \sphericalangle B$.

B. Given a segment OI, called a *unit segment*. Then there is a unique way of assigning a length $\overline{AB}$ to each segment AB such that the

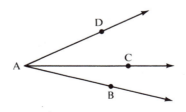

FIGURE 4.8

following properties hold:

(7) $\overline{AB}$ is a positive real number and $\overline{OI} = 1$.

(8) $\overline{AB} = \overline{CD}$ if and only if $AB \cong CD$.

(9) $A * B * C$ if and only if $\overline{AC} = \overline{AB} + \overline{BC}$.

(10) $\overline{AB} < \overline{CD}$ if and only if $AB < CD$.

(11) For every positive real number x, there exists a segment AB such that $\overline{AB} = x$.

Note. So as not to mystify you, here is the method for assigning lengths. We start with a segment OI whose length will be 1. Then any segment obtained by laying off n copies of OI will have length n. By Archimedes' axiom, every other segment AB will have its endpoint B between two points B_{n-1} and B_n such that $\overline{AB_{n-1}} = n - 1$ and $\overline{AB_n} = n$; then $\overline{AB}$ will have to equal $\overline{AB_{n-1}} + \overline{B_{n-1}B}$ by condition (9) of Theorem 4.3, so we may assume $n = 1$ and $B_{n-1} = A$. If B is the midpoint $B_{1/2}$ of AB_1, we set $\overline{AB_{1/2}} = \frac{1}{2}$; otherwise B lies either in $AB_{1/2}$ or in $B_{1/2}B_1$, say, in $AB_{1/2}$. If then B is the midpoint $B_{1/4}$ of $AB_{1/2}$, we set $\overline{AB_{1/4}} = \frac{1}{4}$; otherwise B lies in $AB_{1/4}$, say, and we continue the process. Eventually B will either be obtained as the midpoint of some segment whose length has been determined, in which case $\overline{AB}$ will be determined to some dyadic rational number $a/2^n$; or the process will continue indefinitely, in which case $\overline{AB}$ will be the limit of an infinite sequence of dyadic rational numbers; i.e., $\overline{AB}$ will be determined as an infinite decimal with respect to the base 2.

The axioms of continuity are not needed if one merely wants to define addition for congruence classes of segments and then prove the triangle inequality (Corollary 2 to Theorem 4.3; see Borsuk and Szmielew, 1960, pp. 103–108, for a definition of this operation). It is in order to prove Theorem 4.4, Major Exercise 8, and the parallel projection theorem that we need the measurement of angles and segments by real numbers, and for such measurement Archimedes' axiom is required. However, parts 4 and 11 of Theorem 4.3, the proofs for which require Dedekind's axiom, are never used in proofs in this book. See Appendix B for coordinatization without continuity axioms.

Using degree notation, $\sphericalangle A$ is defined as *acute* if $(\sphericalangle A)° < 90°$, and *obtuse* if $(\sphericalangle A)° > 90°$. Combining Theorems 4.2 and 4.3 gives

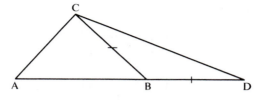

FIGURE 4.9   A   B   D

the following corollary, which is essential for proving the Saccheri-Legendre theorem.

COROLLARY 1.   The sum of the degree measures of any *two* angles of a triangle is less than 180°.

The only immediate application of segment measurement that we will make is in the proof of the next corollary, the famous "triangle inequality."

COROLLARY 2 (Triangle Inequality).   If A, B, and C are three noncollinear points, then $\overline{AC} < \overline{AB} + \overline{BC}$.

Proof:
(1) There is a unique point D such that A * B * D and BD ≅ BC (Axiom C-1 applied to the ray opposite to $\overrightarrow{BA}$). (See Figure 4.9.)
(2) Then ∢BCD ≅ ∢BDC (Proposition 3.10: base angles of an isosceles triangle).
(3) $\overline{AD} = \overline{AB} + \overline{BD}$ (Theorem 4.3(9)) and $\overline{BD} = \overline{BC}$ (step 1 and Theorem 4.3(8)); substituting gives $\overline{AD} = \overline{AB} + \overline{BC}$.
(4) $\overrightarrow{CB}$ is between $\overrightarrow{CA}$ and $\overrightarrow{CD}$ (Proposition 3.7); hence, ∢ACD > ∢BCD (by definition).
(5) ∢ACD > ∢ADC (steps 2 and 4; Proposition 3.21(c)).
(6) AD > AC (Proposition 4.5).
(7) Hence, $\overline{AB} + \overline{BC} > \overline{AC}$ (Theorem 4.3(10); steps 3 and 6). ∎

SACCHERI-LEGENDRE THEOREM

The following very important theorem also requires an axiom of continuity (Archimedes' axiom) for its proof.

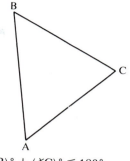

FIGURE 4.10 $(\angle A)° + (\angle B)° + (\angle C)° \leqq 180°$.

THEOREM 4.4 (Saccheri-Legendre). The sum of the degree measures of the three angles in any triangle is *less than or equal to* 180°.

This result may strike you as peculiar, since you are accustomed to the notion of an *exact* sum of 180°. Nevertheless, this exactness cannot be proved in neutral geometry! Saccheri tried, but the best he could conclude was "less than or equal." Max Dehn showed in 1900 that there is no way to prove this theorem without Archimedes' axiom.[1] The idea of the proof is as follows:

Assume, on the contrary, that the angle sum of △ABC is greater than 180°, say 180° + p°, where p is a positive number. It is possible (by a trick you will find in Exercise 15) to replace △ABC with another triangle that has the same angle sum as △ABC but in which one angle has at most half the number of degrees as $(\angle A)°$ We can repeat this trick to get another triangle that has the same angle sum 180° + p° but in which one angle has at most one-fourth the number of degrees as $(\angle A)°$. The Archimedean property of real numbers guarantees that if we repeat this construction enough times, we will eventually obtain a triangle that has angle sum 180° + p° but in which one angle has degree measure at most p°. The sum of the degree measures of the other *two* angles will be greater than or equal to 180°, contradicting Corollary 1 to Theorem 4.3. This proves the theorem.

You should prove the following consequence of the Saccheri-Legendre theorem as an exercise.

---

[1] See the heuristic argument in Project 1. The full significance of Archimedes' axiom was first grasped in the 1880s by M. Pasch and O. Stolz. G. Veronese and T. Levi-Civita developed the first non-Archimedean geometry. Also see Appendix B.

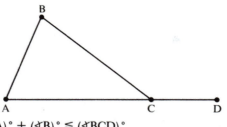

FIGURE 4.11 $(\angle A)° + (\angle B)° \leqq (\angle BCD)°$.

COROLLARY 1. The sum of the degree measures of two angles in a triangle is less than or equal to the degree measure of their remote exterior angle (see Figure 4.11).

It is natural to generalize the Saccheri-Legendre theorem to polygons other than triangles. For example, let us prove that the angle sum of a quadrilateral ABCD is at most 360°. Break □ABCD into two triangles, △ABC and △ADC, by the diagonal AC (see Figure 4.12). By the Saccheri-Legendre theorem,

$$(\angle B)° + (\angle BAC)° + (\angle ACB)° \leqq 180°$$
and $\qquad (\angle D)° + (\angle DAC)° + (\angle ACD)° \leqq 180°.$

Theorem 4.3(3) gives us the equations

$$(\angle BAC)° + (\angle DAC)° = (\angle BAD)°$$
and $\qquad (\angle ACB)° + (\angle ACD)° = (\angle BCD)°$

Using these equations, we add the two inequalities to obtain the desired inequality

$$(\angle B)° + (\angle D)° + (\angle BAD)° + (\angle BCD)° \leqq 360°$$

Unfortunately, there is a gap in this simple argument! To get the equations used above, we assumed by looking at the diagram (Figure

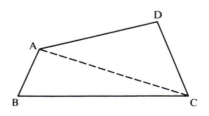

FIGURE 4.12

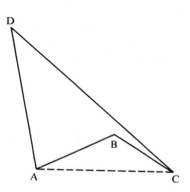

FIGURE 4.13

4.12) that C was interior to ∢BAD and that A was interior to ∢BCD. But what if the quadrilateral looked like Figure 4.13? In this case the equations would not hold. To prevent such a case, we must add a hypothesis; we must assume that the quadrilateral is "convex."

DEFINITION. Quadrilateral □ABCD is called *convex* if it has a pair of opposite sides, e.g., AB and C⃡D, such that CD is contained in one of the half-planes bounded by A⃡B and AB is contained in one of the half-planes bounded by C⃡D.[2]

The assumption made above is now justified by starting with a convex quadrilateral. Thus, we have proved the following corollary:

COROLLARY 2. The sum of the degree measures of the angles in any *convex* quadrilateral is at most 360°.

Note. The Saccheri-Legendre theorem is false in elliptic geometry (see Figure 3.24.). In fact, it can be proved in elliptic geometry that the angle sum of a triangle is always greater than 180° (see Kay, 1969). Since a triangle can have two or three right angles, a *hypotenuse,*

---

[2] It can be proved that this condition also holds for the other pair of opposite sides, AD and BC—see Exercise 23 in this chapter. The use of the word "convex" in this definition does not agree with its use in Exercise 19, Chapter 3; a convex quadrilateral is obviously not a "convex set" as defined in that exercise. However, we can define the *interior* of a convex quadrilateral □ABCD as follows: each side of □ABCD determines a half-plane containing the opposite side; the interior of □ABCD is then the intersection of the four half-planes so determined. You can then prove that the interior of a convex quadrilateral is a convex set (which is one of the problems in Exercise 25).

defined as a side opposite a right angle, need not be unique, and a *leg,* defined as a side of a right triangle not opposite a right angle, need not exist (and if opposite an obtuse angle, a leg could be longer than a hypotenuse).

EQUIVALENCE OF PARALLEL POSTULATES

We shall now prove the equivalence of Euclid's fifth postulate and Hilbert's parallel postulate. Note, however, that we are not proving either or both of the postulates; we are only proving that we *can* prove one *if* we first assume the other. We shall first state Euclid V (all the terms in the statement have now been defined carefully).

EUCLID'S POSTULATE V. If two lines are intersected by a transversal in such a way that the sum of the degree measures of the two interior angles on one side of the transversal is less than 180°, then the two lines meet on that side of the transversal.

THEOREM 4.5. Euclid's fifth postulate $\Longleftrightarrow$ Hilbert's parallel postulate.

Proof:
First, assume Hilbert's postulate. The situation of Euclid V is shown in Figure 4.14. $(\sphericalangle 1)° + (\sphericalangle 2)° < 180°$ (hypothesis) and $(\sphericalangle 1)° + (\sphericalangle 3)° = 180°$ (supplementary angles, Theorem 4.3(5)). Hence, $(\sphericalangle 2)° < 180° - (\sphericalangle 1)° = (\sphericalangle 3)°$. There is a unique ray $\overrightarrow{B'C'}$ such that $\sphericalangle 3$ and $\sphericalangle C'B'B$ are congruent alternate interior angles (Axiom C-4). By Theorem 4.1, $\overleftrightarrow{B'C'}$ is parallel to *l*. Since

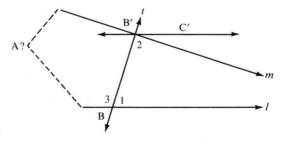

FIGURE 4.14

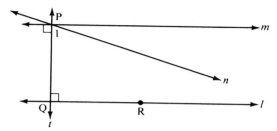

FIGURE 4.15

$m \neq \overleftrightarrow{B'C'}$, m meets l (Hilbert's postulate). To conclude, we must prove that m meets l on the same side of t as C'. Assume, on the contrary, that they meet at a point A on the opposite side. Then $\sphericalangle 2$ is an exterior angle of $\triangle ABB'$. Yet it is smaller than the remote interior $\sphericalangle 3$. This contradiction of Theorem 4.2 proves Euclid V (RAA).

Conversely, assume Euclid V and refer to Figure 4.15, the situation of Hilbert's postulate. Let t be the perpendicular to l through P, and m the perpendicular to t through P. We know that $m \parallel l$ (Corollary 1 to Theorem 4.1). Let n be any other line through P. We must show that n meets l. Let $\sphericalangle 1$ be the acute angle n makes with t (which angle exists because $n \neq m$). Then $(\sphericalangle 1)° + (\sphericalangle PQR)° < 90° + 90° = 180°$. Thus, the hypothesis of Euclid V is satisfied. Hence, n meets l, proving Hilbert's postulate. ■

Since Hilbert's parallel postulate and Euclid V are logically equivalent in the context of neutral geometry, Theorem 4.5 allows us to use them interchangeably. You will prove as exercises that the following statements are also logically equivalent to the parallel postulate.[3]

PROPOSITION 4.7. Hilbert's parallel postulate ⟺ if a line intersects one of two parallel lines, then it also intersects the other.

PROPOSITION 4.8. Hilbert's parallel postulate ⟺ converse to Theorem 4.1 (alternate interior angles).

PROPOSITION 4.9. Hilbert's parallel postulate ⟺ if t is a transversal to l and m, $l \parallel m$, and $t \perp l$, then $t \perp m$.

---

[3] Transitivity of parallelism is also logically equivalent to the parallel postulate.

PROPOSITION 4.10. Hilbert's parallel postulate ⟺ if $k \parallel l$, $m \perp k$, and $n \perp l$, then either $m = n$ or $m \parallel n$.

The next proposition is another statement logically equivalent to Hilbert's parallel postulate, but at this point we can only prove the implication in one direction (the other implication is proved in Chapter 5; see Exercise 14).

PROPOSITION 4.11. Hilbert's parallel postulate ⟹ the angle sum of every triangle is 180°.

ANGLE SUM OF A TRIANGLE

We define the *angle sum* of triangle △ABC as $(\sphericalangle A)° + (\sphericalangle B)° + (\sphericalangle C)°$, which is a certain number of degrees $\leqq 180°$ (by the Saccheri-Legendre theorem). We define the *defect* δABC to be 180° minus the angle sum. In Euclidean geometry we are accustomed to having no "defective" triangles, i.e., we are accustomed to having the defect equal zero (Proposition 4.11).

The main purpose of this section is to show that if *one* defective triangle exists, then *all* triangles are defective. Or, put in the contrapositive form, if one triangle has angle sum 180°, then so do all others. We are not asserting that one such triangle does exist, nor are we asserting the contrary; we are only examining the hypothesis that one might exist.

THEOREM 4.6. Let △ABC be any triangle and D a point between A and B (Figure 4.16). Then $\delta ABC = \delta ACD + \delta BCD$ *(additivity of the defect)*.

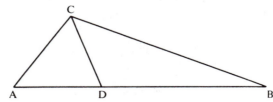

FIGURE 4.16 A   D   B

Proof:

Since $\overrightarrow{CD}$ is interior to $\sphericalangle ACB$ (Proposition 3.7), $(\sphericalangle ACB)° = (\sphericalangle ACD)° + (\sphericalangle BCD)°$ (by Theorem 4.3(3)). Since $\sphericalangle ADC$ and $\sphericalangle BDC$ are supplementary angles, $180° = (\sphericalangle ADC)° + (\sphericalangle BDC)°$ (by Theorem 4.3(5)). To obtain the additivity of the defect, all we have to do is write down the definition of the defect (180° minus the angle sum) for each of the three triangles under consideration and substitute the two equations above (Exercise 1). ■

COROLLARY. Under the same hypothesis, the angle sum of △ABC is equal to 180° if and only if the angle sums of both △ACD and △BCD are equal to 180°.

Proof:

If △ACD and △BCD both have defect zero, then defect of △ABC = 0 + 0 = 0 (Theorem 4.6). Conversely, if △ABC has defect zero, then, by Theorem 4.6, $\delta ACD + \delta BCD = 0$. But the defect of a triangle can never be negative (Saccheri-Legendre theorem). Hence, △ACD and △BCD each have defect zero (the sum of two nonnegative numbers equals zero only when each equals zero). ■

Next, recall that by definition a *rectangle* is a quadrilateral whose four angles are right angles. Hence, the angle sum of a rectangle is 360°. Of course, we don't yet know whether rectangles exist in neutral geometry. (Try to construct one without using the parallel postulate or any statement logically equivalent to it—see Exercise 19.)

The next theorem is the result we seek. Its proof will be given in five steps.

THEOREM 4.7. If a triangle exists whose angle sum is 180°, then a rectangle exists. If a rectangle exists, then every triangle has angle sum equal to 180°.

Proof:

(1) Construct a *right* triangle having angle sum 180°.

Let △ABC be the given triangle with defect zero (hypothesis). Assume it is not a right triangle; otherwise we are done. At least two of the angles in this triangle are acute, since the angle

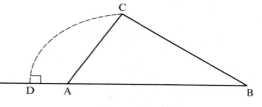

FIGURE 4.17

sum of two angles in a triangle must be less than 180° (corollary to Theorem 4.3); e.g., assume ∢A and ∢B are acute. Let CD be the altitude from vertex C (which exists, by Proposition 3.16). We claim that D lies between A and B. Assume the contrary, that D * A * B (see Figure 4.17). Then remote interior angle ∢CDA is greater than exterior angle ∢CAB, contradicting Theorem 4.2. Similarly, if A * B * D, we get a contradiction. Thus, A * D * B (Axiom B-3); see Figure 4.18. It now follows from the corollary to Theorem 4.6 that each of the right triangles △ADC and △BDC has defect zero.

(2) From a right triangle of defect zero construct a rectangle.

Let △CDB be a right triangle of defect zero with ∢D a right angle. By Axiom C-4, there is a unique ray $\overrightarrow{CX}$ on the opposite side of $\overleftrightarrow{CB}$ from D such that ∢DBC ≅ ∢BCX. By Axiom C-1, there is a unique point E on $\overrightarrow{CX}$ such that CE ≅ BD (Figure 4.19). Then △CDB ≅ △BEC (SAS). Hence, △BEC is also a right triangle of defect zero with right angle at E. Also, since (∢DBC)° + (∢BCD)° = 90° by our hypothesis, we obtain by substitution (∢ECB)° + (∢BCD)° = 90° and (∢DBC)° + (∢EBC)° = 90° Moreover, B is an interior point of ∢ECD, since the alternate interior angle theorem implies $\overleftrightarrow{CE} \parallel \overleftrightarrow{DB}$ and $\overleftrightarrow{CD} \parallel \overleftrightarrow{BE}$ and C is interior to ∢EBD (for the same reason).

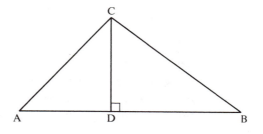

FIGURE 4.18   A   D   B

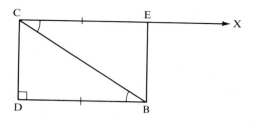

FIGURE 4.19

Thus, we can apply Theorem 4.3(3) to conclude that $(\angle ECD)^\circ = 90^\circ = (\angle EBD)^\circ$. This proves that $\square CDBE$ is a rectangle.

(3) From one rectangle, construct "arbitrarily large" rectangles.

More precisely, given any right triangle $\triangle D'E'C'$, construct a rectangle $\square AFBC$ such that $AC > D'C'$ and $BC > E'C'$. This can be done using Archimedes' axiom. We simply "lay off" enough copies of the rectangle we have to achieve the result (see Figures 4.20 and 4.21; you can make this "laying off" precise as an exercise).

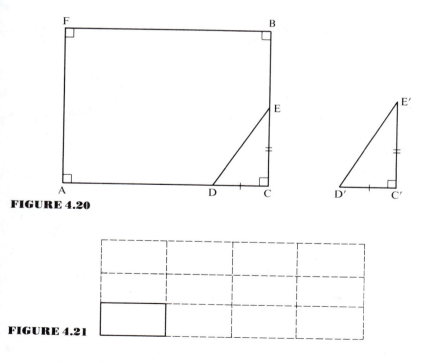

FIGURE 4.20

FIGURE 4.21

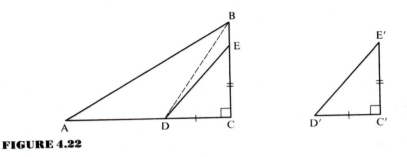

FIGURE 4.22

(4) Prove that all *right* triangles have defect zero.

This is achieved by "embedding" an arbitrary right triangle $\triangle D'C'E'$ in a rectangle, as in step 3, and then showing successively (by twice applying the corollary to Theorem 4.6) that $\triangle ACB$, $\triangle DCB$, and $\triangle DCE$ each have defect zero (see Figure 4.22).

(5) If every *right* triangle has defect zero, then *every* triangle has defect zero.

As in step 1, drop an altitude to decompose an arbitrary triangle into two right triangles (Figure 4.18) and apply the corollary to Theorem 4.6. ■

Historians credit Theorem 4.7 to Saccheri and Legendre, but we will not name it after them, so as to avoid confusion with Theorem 4.4.

COROLLARY. If there exists a triangle with positive defect, then all triangles have positive defect.

REVIEW EXERCISE

Which of the following statements are correct?

(1) If two triangles have the same defect, they are congruent.
(2) Euclid's fourth postulate is a theorem in neutral geometry.

(3) Theorem 4.5 shows that Euclid's fifth postulate is a theorem in neutral geometry.

(4) The Saccheri-Legendre theorem tells us that some triangles exist that have angle sum less than 180° and some triangles exist that have angle sum equal to 180°.

(5) The alternative interior angle theorem states that if parallel lines are cut by a transversal, then alternate interior angles are congruent to each other.

(6) It is impossible to prove in neutral geometry that quadrilaterals exist.

(7) The Saccheri-Legendre theorem is false in Euclidean geometry because in Euclidean geometry the angle sum of any triangle is never less than 180°.

(8) According to our definition of "angle," the degree measure of an angle cannot equal 180°.

(9) The notion of one ray being "between" two others is undefined.

(10) It is impossible to prove in neutral geometry that parallel lines exist.

(11) The definition of "remote interior angle" given on p. 118 is incomplete because it used the word "adjacent," which has never been defined.

(12) An exterior angle of a triangle is any angle that is not in the interior of the triangle.

(13) The SSS criterion for congruence of triangles is a theorem in neutral geometry.

(14) The alternate interior angle theorem implies, as a special case, that if a transversal is perpendicular to one of two parallel lines, then it is also perpendicular to the other.

(15) Another way of stating the Saccheri-Legendre theorem is to say that the defect of a triangle cannot be negative.

(16) The ASA criterion for congruence of triangles is one of the axioms for neutral geometry.

(17) The proof of Theorem 4.7 depends on Archimedes' axiom.

(18) If $\triangle ABC$ is any triangle and C is any of its vertices, and if a perpendicular is dropped from C to $\overleftrightarrow{AB}$, then that perpendicular will intersect $\overleftrightarrow{AB}$ in a point between A and B.

(19) It is a theorem in neutral geometry that given any point P and any line l, there is at most one line through P perpendicular to l.

(20) It is a theorem in neutral geometry that vertical angles are congruent to each other.

(21) The proof of Theorem 4.2 (on exterior angles) uses Theorem 4.1 (on alternate interior angles).

(22) The gap in Euclid's attempt to prove Theorem 4.2 can be filled using our axioms of betweenness.

EXERCISES

The following are exercises in neutral geometry, unless otherwise stated. This means that in your proofs you are allowed to use only those results that have been given previously (including results from previous exercises). You are not allowed to use the parallel postulate or other results from Euclidean geometry that depend on it.

1. (a) Finish the last step in the proof of Theorem 4.6. (b) Prove that congruent triangles have the same defect. (c) Prove the corollary to Theorem 4.7. (d) Prove Corollary 1 to Theorem 4.3.
2. The Pythagorean theorem cannot be proved in neutral geometry (as you will show in Exercise 11(d), Chapter 6). Explain why the Euclidean proof suggested by Figure 1.15 of Chapter 1 is not valid in neutral geometry.
3. State the converse to Euclid's fifth postulate. Prove this converse as a theorem in neutral geometry.
4. Prove Proposition 4.7. Deduce as a corollary that transitivity of parallelism is equivalent to Hilbert's parallel postulate.
5. Prove Proposition 4.8. (Hint: Assume the converse to Theorem 4.1. Let *m* be the parallel to *l* through P constructed in the proof of Corollary 2 to Theorem 4.1 and let *n* be any parallel to *l* through P. Use the congruence of alternate interior angles and the uniqueness of perpendiculars to prove *m* = *n*. Assuming next the parallel postulate, use Axiom C-4 and an RAA argument to establish the converse to Theorem 4.1.)
6. Prove Proposition 4.9.
7. Prove Proposition 4.10.
8. Prove Proposition 4.11. (Hint: See Figure 4.23.)
9. The following purports to be a proof in neutral geometry of the SAA criterion for congruence. Find the flaw (see Figure 4.6).

   Given $AC \cong DF$, $\sphericalangle A \cong \sphericalangle D$, $\sphericalangle B \cong \sphericalangle E$. Then $\sphericalangle C \cong \sphericalangle F$, since $(\sphericalangle C)° = 180° - (\sphericalangle A)° - (\sphericalangle B)° = 180° - (\sphericalangle D)° - (\sphericalangle E)° = (\sphericalangle F)°$

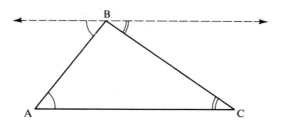

FIGURE 4.23

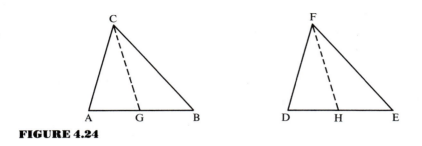

FIGURE 4.24

(Theorem 4.3(2)). Hence, △ABC ≅ △DEF by ASA (Proposition 3.17).

10. Here is a correct proof of the SAA criterion. Justify each step. (1) Assume side AB is not congruent to side DE. (2) Then AB < DE or DE < AB. (3) If DE < AB, then there is a point G between A and B such that AG ≅ DE (see Figure 4.24). (4) Then △CAG ≅ △FDE. (5) Hence, ∢AGC ≅ ∢E. (6) It follows that ∢AGC ≅ ∢B. (7) This contradicts a certain theorem (which?). (8) Therefore, DE is not les than AB. (9) By a similar argument involving a point H between D and E, AB is not less than DE. (10) Hence, AB ≅ DE. (11) Therefore, △ABC ≅ △DEF.

11. Prove Proposition 4.2. (Hint: See Figure 4.7. On the ray opposite to $\overrightarrow{AC}$, lay off segment AD congruent to A′C′. First prove △DAB ≅ △C′A′B′, and then use isosceles triangles and the SAA criterion to conclude.)

12. Here is a proof that segment AB has a midpoint. Justify each step (see Figure 4.25).
(1) Let C be any point not on $\overleftrightarrow{AB}$. (2) There is a unique ray $\overrightarrow{BX}$ on the opposite side of $\overleftrightarrow{AB}$ from C such that ∢CAB ≅ ∢ABX. (3) There is a

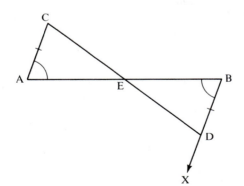

FIGURE 4.25

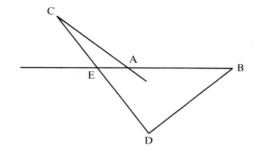

FIGURE 4.26

unique point D on $\overrightarrow{BX}$ such that AC $\cong$ BD. (4) D is on the opposite side of $\overleftrightarrow{AB}$ from C. (5) Let E be the point at which segment CD intersects $\overleftrightarrow{AB}$. (6) Assume E is not between A and B. (7) Then either E = A, or E = B, or E * A * B, or A * B * E. (8) $\overleftrightarrow{AC}$ is parallel to $\overleftrightarrow{BD}$. (9) Hence, E $\neq$ A and E $\neq$ B. (10) Assume E * A * B (Figure 4.26). (11) Since $\overleftrightarrow{CA}$ intersects side EB of $\triangle$EBD at a point between E and B, it must also intersect either ED or BD. (12) Yet this is impossible. (13) Hence, A is not between E and B. (14) Similarly, B is not between A and E. (15) Thus, A * E * B (see Figure 4.25). (16) Then $\sphericalangle$AEC $\cong$ $\sphericalangle$BED. (17) $\triangle$EAC $\cong$ $\triangle$EBD. (18) Therefore, E is a midpoint of AB.

13. (a) Prove that segment AB has only one midpoint. (Hint: Assume the contrary and use Propositions 3.3 and 3.13 to derive a contradiction, or else put another possible midpoint E' into Figure 4.25 and derive a contradiction from congruent triangles.)

    (b) Prove Proposition 4.4 on bisectors. (Hint: Use midpoints.)

14. Prove Corollary 1 to the Saccheri-Legendre theorem.

15. Prove the following result, needed to demonstrate the Saccheri-Legendre theorem (see Figure 4.27). Let D be the midpoint of BC and E the unique point on $\overleftrightarrow{AD}$ such that A * D * E and AD $\cong$ DE. Then $\triangle$AEC has the same angle sum as $\triangle$ABC, and either ($\sphericalangle$EAC)$^\circ$ or ($\sphericalangle$AEC)$^\circ$ is $\leq \frac{1}{2}$ ($\sphericalangle$BAC)$^\circ$. (Hint: First show that $\triangle$BDA $\cong$ $\triangle$CDE, then that ($\sphericalangle$EAC)$^\circ$ + ($\sphericalangle$AEC)$^\circ$ = ($\sphericalangle$BAC)$^\circ$.)

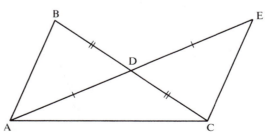

FIGURE 4.27

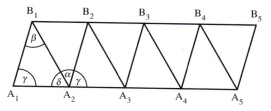

FIGURE 4.28

16. Here is another proof of Theorem 4.4 due to Legendre. Justify the unjustified steps: (1) Let $A_1A_2B_1$ be the given triangle, lay off n copies of segment A_1A_2, and construct a row of triangles $A_jA_{j+1}B_j$, $j = 1, \ldots, n$, congruent to $A_1A_2B_1$ as shown in Figure 4.28. (2) The $B_jA_{j+1}B_{j+1}$, $j = 1, \ldots, n$, are also congruent triangles, the last by construction of B_{n+1}. (3) With angles labeled as in Figure 4.28, $\alpha + \gamma + \delta = 180°$ and $\beta + \gamma + \delta$ equals the angle sum of $A_1A_2B_1$. (4) Assume on the contrary that $\beta > \alpha$. (5) Then $A_1A_2 > B_1B_2$, by Proposition 4.6. (6) Also $\overline{A_1B_1} + n \cdot \overline{B_1B_2} + \overline{B_{n+1}A_{n+1}} > n \cdot \overline{A_1A_2}$, by repeated application of the triangle inequality. (7) $A_1B_1 \cong B_{n+1}A_{n+1}$. (8) $2\overline{A_1B_1} > n(\overline{A_1A_2} - \overline{B_1B_2})$. (9) Since n was arbitrary, this contradicts Archimedes' axiom. (10) Hence the triangle has angle sum $\leqq 180°$.

17. Prove the following theorems:
    (a) Let γ be a circle with center O, and let A and B be two points on γ. The segment AB is called a *chord* of γ; let M be its midpoint. If $O \neq M$, then $\overleftrightarrow{OM}$ is perpendicular to $\overleftrightarrow{AB}$. (Hint: Corresponding angles of congruent triangles are congruent.)
    (b) Let AB be a chord of the circle γ having center O. Prove that the perpendicular bisector of AB passes through the center O of γ.

18. Prove the theorem of Thales *in Euclidean geometry* that an angle inscribed in a semicircle is a right angle. Prove in *neutral geometry* that this statement implies the existence of a right triangle with zero defect (see Figure 4.29).

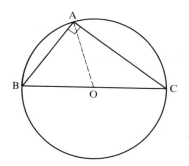

FIGURE 4.29

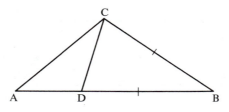

FIGURE 4.30

19. Find the flaw in the following argument purporting to construct a rectangle. Let A and B be any two points. There is a line l through A perpendicular to $\overleftrightarrow{AB}$ (Proposition 3.16) and, similarly, there is a line m through B perpendicular to $\overleftrightarrow{AB}$. Take any point C on m other than B. There is a line through C perpendicular to l—let it intersect l at D. Then $\square ABCD$ is a rectangle.

20. The sphere, with "lines" interpreted as great circles, is not a model of neutral geometry. Here is a proposed construction of a rectangle on a sphere. Let α, β be two circles of longitude and let them intersect the equator at A and D. Let γ be a circle of latitude in the northern hemisphere, intersecting α and β at two other points, B and C. Since circles of latitude are perpendicular to circles of longitude, the quadrilateral with vertices ABCD and sides the arcs of α, γ, and β and the equator traversed in going from A north to B east to C south to D west to A should be a rectangle. Explain why this construction doesn't work.

21. Prove Proposition 4.5. (Hint: If $AB > BC$, then let D be the point between A and B such that $BD \cong BC$ (Figure 4.30). Use isosceles triangle $\triangle CBD$ and exterior angle $\sphericalangle BDC$ to show that $\sphericalangle ACB > \sphericalangle A$. Use this result and trichotomy of ordering to prove the converse.)

22. Prove Proposition 4.6. (Hint: Given $\sphericalangle B < \sphericalangle B'$. Use the hypothesis of Proposition 4.6 to reduce to the case $A = A'$, $B = B'$, and C interior to $\sphericalangle ABC'$, so that you must show $AC < AC'$ (see Figure 4.31). This is easy in case $C = D$, where point D is obtained from the crossbar theorem. In case $C \neq D$, Proposition 4.5 reduces the problem to showing that $\sphericalangle AC'C < \sphericalangle ACC'$. In case $B * D * C$ (as in Figure 4.31), you

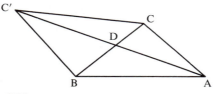

FIGURE 4.31 $BC \cong BC'$.

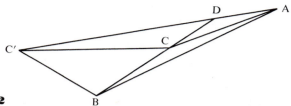

FIGURE 4.32

can prove this inequality using the congruence ∢BCC′ ≅ ∢BC′C. In case B ∗ C ∗ D (Figure 4.32), apply the congruence ∢BCC′ ≅ ∢BC′C and Theorem 4.2 to exterior angle ∢BCC′ of △DCC′ and exterior angle ∢DCC′ of △BCC′. (The converse implication in Proposition 4.6 follows from the direct implication, just shown, if you apply trichotomy.)

23. For the purpose of this exercise, call segments AB and CD *semiparallel* if segment AB does not intersect line $\overleftrightarrow{CD}$ and segment CD does not intersect line $\overleftrightarrow{AB}$. Obviously, if $\overleftrightarrow{AB} \parallel \overleftrightarrow{CD}$, then AB and CD are semiparallel, but the converse need not hold (see Figure 4.33). We have defined a quadrilateral to be convex if one pair of opposite sides is semiparallel. Prove that the other pair of opposite sides is also semiparallel. (Hint: Suppose AB is semiparallel to CD and assume, on the contrary, that AD meets $\overleftrightarrow{BC}$ in a point E. Use the definition of quadrilateral (Exercise 3, Chapter 1) to show either that E ∗ B ∗ C or B ∗ C ∗ E; in either case, use Pasch's theorem to derive a contradiction.)

24. Prove that the diagonals of a convex quadrilateral intersect. (Hint: Apply the crossbar theorem.)

25. Prove that the intersection of convex sets (defined in Exercise 19, Chapter 3) is again a convex set. Use this result to prove that the interior of a convex quadrilateral is a convex set and that the point at which the diagonals intersect lies in the interior.

26. The *convex* hull of a set of points S is the intersection of all the convex sets containing S; i.e., it is the smallest convex set containing S. Prove that the convex hull of three noncollinear points A, B, and C consists of the sides and interior of △ABC.

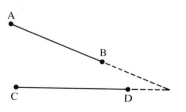

FIGURE 4.33

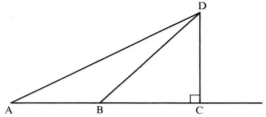

FIGURE 4.34

27. Given A * B * C and $\overleftrightarrow{DC} \perp \overleftrightarrow{AC}$. Prove that AD > BD > CD (Figure 4.34; use Proposition 4.5).
28. Given any triangle △DAC and any point B between A and C. Prove that either DB < DA or DB < DC. (Hint: Drop a perpendicular from D to $\overleftrightarrow{AC}$ and use the previous exercise.)
29. Prove that the interior of a circle is a convex set. (Hint: Use the previous exercise.)
30. Prove that if D is an exterior point of △ABC, then there is a line $\overleftrightarrow{DE}$ through D that is contained in the exterior of △ABC (see Figure 4.35).
31. Suppose that line *l* meets circle γ in two points C and D. Prove that:
    (a) Point P on *l* lies inside γ if and only if C * P * D.
    (b) If points A and B are inside γ and on opposite sides of *l*, then the point E at which AB meets *l* is between C and D.
32. In Figure 4.36, the pairs of angles (⦣A′B′B″, ⦣ABB″) and (⦣C′B′B″, ⦣CBB″) are called pairs of *corresponding angles* cut off on *l* and *l*′ by transversal *t*. Prove that corresponding angles are congruent if and only if alternate interior angles are congruent.
33. *Prove* that there exists a triangle which is not isosceles.

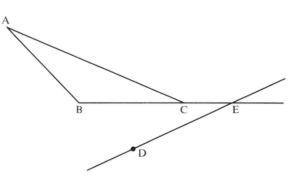

FIGURE 4.35

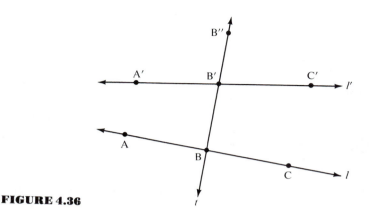

FIGURE 4.36

MAJOR EXERCISES

1. THEOREM.   If line *l* passes through a point A inside circle γ then *l* intersects γ in two points.

Here is the idea of the proof; fill in the details using the circular continuity principle (instead of the stronger axiom of Dedekind) and Exercise 27 (see Figure 4.37). Let O be the center of γ. Point B is taken to be the foot of the perpendicular from O to *l*, point C is taken such that B is the midpoint of OC, and γ' is the circle centered at C having the same radius as γ. Prove that γ' intersects $\overleftrightarrow{OC}$ in a point E' inside γ and a point E outside γ, so that γ' intersects γ in two points P, P', and that these points lie on the

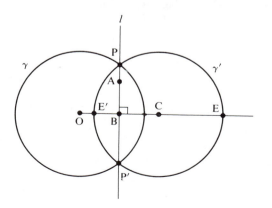

FIGURE 4.37

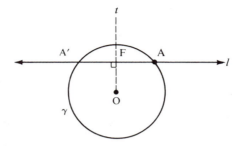

FIGURE 4.38

original line *l*. (We located the intersections of γ with *l* by intersecting γ with its *reflection* γ' across *l*—see p. 111.)

4. Apply the previous exercise to prove that the circular continuity principle implies the elementary continuity principle. (Hint: Use Exercise 27.)

3. Let line *l* intersect circle γ at point A. If $l \perp \overleftrightarrow{OA}$, where O is the center of γ, we call *l* *tangent* to γ at A; otherwise *l* is called *secant* to γ.

   (a) Suppose *l* is secant to γ. Prove that the foot F of the perpendicular *t* from O to *l* lies inside γ and that the reflection A' of A across *t* is a second point at which *l* meets γ. (See Figure 4.38.)

   (b) Suppose now *l* is tangent to γ. Prove that every point B $\neq$ A lying on *l* is outside γ, and hence A is the unique point at which *l* meets γ.

   (c) Let point P lie outside γ. Proposition 7.3, Chapter 7, applies the circular continuity principle to construct a line through P tangent to γ. Explain why that construction is valid only in Euclidean geometry. Prove that the tangent line exists in neutral geometry. (Hint: Let Q $\neq$ P be any point on the perpendicular to $\overleftrightarrow{OP}$ through P. Prove that $\overrightarrow{PQ}$ does not intersect γ whereas $\overrightarrow{PO}$ does. Apply Dedekind's axiom to ray $\overrightarrow{OQ}$. See Figure 4.39.) Once one tangent *l* through P is obtained, prove that the reflection of *l* across $\overleftrightarrow{OP}$ is a second one.

4. *Converse to the triangle inequality.* If *a, b,* and *c* are lengths of segments such that the sum of any two is greater than the third, then there exists a

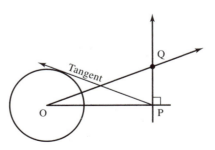

FIGURE 4.39

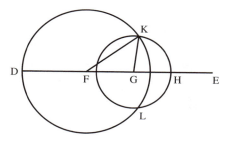

FIGURE 4.40

triangle whose sides have those lengths (Euclid's Proposition 22). Use the circular continuity principle to fill the gap in Euclid's proof and justify the steps: Assume $a \geq b \geq c$. Take any point D and any ray emanating from D. Starting from D, lay off successively on that ray points F, G, H so that $a = \overline{DF}, b = \overline{FG}, c = \overline{GH}$. Then the circle with center F and radius a meets the circle with center G and radius c at a point K, and $\triangle FGK$ is the triangle called for in the proposition. (See Figure 4.40.)

5. Prove that the converse to the triangle inequality implies the circular continuity principle (assuming the incidence, betweenness, and congruence axioms).

6. Prove: If b and c are lengths of segments, then there exists a right triangle with hypotenuse c and leg b if and only if $b < c$. (Hint for the "if" part: Take any point C and any perpendicular lines through C. There exists a point A on one line such that $\overline{AC} = b$. If α is the circle centered at A of radius c, point C lies inside α, and hence α intersects the other line in some point B. Then $\triangle ABC$ is the requisite right triangle.)

7. Show how the previous exercise furnishes a solution to Major Exercise 3(c) that avoids the use of Dedekind's axiom. (Hint: Let $c = \overline{OP}$ and $b = $ radius of γ and lay off $\sphericalangle A$ at O with $\overrightarrow{OP}$ as one side.)

8. Here is an Archimedean proof in neutral geometry of the "important corollary" to Aristotle's axiom, Chapter 3. We must show that given any positive real number a there is a point R on line l such that $(\sphericalangle QRP)° < a°$ (intuitively, by taking R sufficiently far out we can get as small an angle as we please). The idea is to construct a sequence of angles $\sphericalangle QR_1P$, $\sphericalangle QR_2P, \ldots$ each one of which is at most half the size of its predecessor. Justify the following steps (Figure 4.41):

FIGURE 4.41

There exists a point R_1 on l such that $PQ \cong QR_1$ (why?), so that $\triangle PQR_1$ is isosceles. It follows that $(\sphericalangle QR_1P)° \leq 45°$ (why?). Next, there exists a point R_2 such that $Q * R_1 * R_2$ and $PR_1 \cong R_1R_2$, so that $\triangle PR_1R_2$ is isosceles. It follows that $(\sphericalangle QR_2P)° \leq 22\frac{1}{2}°$ (to justify this step, use Corollary 1 to the Saccheri-Legendre theorem). Continuing in this way, we get angles successively less than or equal to $11\frac{1}{4}°$, $5\frac{5}{8}°$, etc. so that by the Archimedean property of real numbers, we eventually get an angle $\sphericalangle QR_nP$ with $\sphericalangle(QR_nP)° < a°$.

PROJECTS

1. Here is a heuristic argument showing that Archimedes' axiom is necessary to prove the Saccheri-Legendre theorem. It is known that on a sphere, the angle sum of every triangle is greater than 180° (see Kay, 1969); that doesn't contradict the Saccheri-Legendre theorem, because a sphere is not a model of neutral geometry. Fix a point O on a sphere. Consider the set N of all points on the sphere whose distance from O is *infinitesimal*. Interpret "line" to be the arc in N of any great circle. Give "between" its natural interpretation on an arc, and interpret "congruence" as in spherical geometry. Then N becomes a model of our I, B, and C axioms in which Archimedes' axiom and the Saccheri-Legendre theorem do not hold. Similarly, if we fix a point O in a Euclidean plane and take N to be its infinitesimal neighborhood, the angle sum of every triangle is 180°, yet Euclid V does not hold in N (because the point at which the lines are supposed to meet is too far away); thus the converse to Proposition 4.11 cannot be proved from our I, B, and C axioms alone (Aristotle's axiom is needed; see Chapter 5).

   For a rigorous elaboration of this argument, see Hessenberg and Diller (1967) (if you can read German; if you can't, then report on Chapter 32 of Moise, 1990, which constructs a Euclidean ordered field that is not Archimedean).

2. Report on the proof of Theorem 4.3 given in Borsuk and Szmielew, Chapter 3, Sections 9 and 10. The key to the proof is that every Dedekind cut on the ordered set of dyadic rational numbers (see Exercise 18, Chapter 3) determines a unique real number.

3. Our proof of Theorem 4.7 used Archimedes' axiom again. Report on the proof in Martin (1982), Chapter 22, that avoids using this axiom.

4. Given a sphere of radius r, let ϵ be any positive real number $\leq \frac{1}{2}\pi r$ and let N_ϵ be the set of all points on the sphere whose spherical distance from a fixed point O on the sphere is less than ϵ. Interpret "line," "between," and "congruent" as they were interpreted for N in Project 1. Then N_ϵ is not a model of our I, B, and C axioms. Tell which axioms hold and which ones fail. For those that fail, explain heuristically why they hold in N.

HISTORY OF THE PARALLEL POSTULATE

Like the goblin "Puck," [the feat of proving Euclid V] has led me "up and down, up and down," through many a wakeful night: but always, just as I thought I had it, some unforeseen fallacy was sure to trip me up, and the tricksy sprite would "leap out, laughing ho, ho, ho!"

C. L. DODGSON (LEWIS CARROLL)

Let us summarize what we have done so far. We have discovered certain gaps in Euclid's definitions and postulates for plane geometry. We filled in these gaps and firmed up the foundations for this geometry by presenting (a modified version of) Hilbert's definitions and axioms. We then built a structure of theorems on these foundations. However, the structure thus far erected does not rest on the parallel postulate, and we called that structure "neutral geometry." One reason we postponed building on the parallel postulate is that we have less confidence in it than in the other axioms.

You may feel that to deny the Euclidean parallel postulate would go against common sense. Albert Einstein once said that "common sense is, as a matter of fact, nothing more than layers of preconceived notions stored in our memories and emotions for the most part before age eighteen."

That Euclid himself did not quite trust this postulate is shown by the fact that he postponed using it in a proof for as long as possible — until his 29th proposition. In this chapter, we will examine a few illuminating attempts to prove Euclid's parallel postulate (many other attempts are presented in Bonola, 1955; Gray, 1989; and Rosenfeld, 1988). It should be emphasized that most of these attempts were made by outstanding mathematicians, not incompetents. And even though each attempt was flawed, the effort was usually not wasted; for, assuming that all but one step can be justified, when we detect the flawed step, we find another statement which to our surprise is equivalent[1] to the parallel postulate. You have the opportunity to do this enjoyable detective work in Exercises 8 through 13.

PROCLUS

Proclus (A.D. 410–485), whose commentary is one of the main sources of information on Greek geometry, criticized the parallel postulate as follows: "This ought even to be struck out of the Postulates altogether; for it is a theorem involving many difficulties, which Ptolemy, in a certain book, set himself to solve. . . . The statement that since [the two lines] converge more and more as they are produced, they will sometime meet is plausible but not necessary." Proclus offers the example of a hyperbola that approaches its asymptotes as closely as you like without ever meeting them (see Figure 5.1). This example shows that the opposite of Euclid's conclusions can at least be imagined.[2] Proclus says: "It is then clear from this that we must seek a proof of the present theorem, and that it is alien to the special character of postulates."

For over two thousand years some of the best mathematicians tried to prove Euclid's fifth postulate. What does it mean, according to our terminology, to have a proof? It should not be necessary to *assume* the parallel postulate as an axiom; we should be able to prove it *from the other axioms*. If we were able to prove Euclid V in this way, it would become a theorem in neutral geometry and neutral geometry would encompass all of Euclidean geometry.

The first known attempted proof was by Ptolemy. Without going through the details of his argument (see Heath, 1956, pp. 204–206), we might say that he assumed Hilbert's parallel postulate without

---

[1] Actually, the flawed argument only proves that the unjustified statement *implies* the parallel postulate; the converse requires further argument. I do not present any attempts that are uninformative.

[2] Students always object to Figure 5.1 on the grounds that the hyperbola is not "straight." We agreed not to use this word because we don't have a precise definition. A precise definition can be given in differential geometry. See Appendix A.

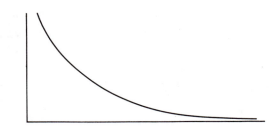

FIGURE 5.1

realizing it. We have seen in Chapter 4 that Hilbert's parallel postulate is logically equivalent to Euclid V (Theorem 4.5), so that Ptolemy was assuming what he was trying to prove; i.e., his reasoning was essentially circular.

Proclus attempted to prove the parallel postulate as follows (see Figure 5.2): Given two parallel lines *l* and *m*. Suppose line *n* cuts *m* at P. We wish to show *n* intersects *l* also (see Proposition 4.7). Let Q be the foot of the perpendicular from P to *l* (Corollary 1 to Theorem 4.1). If *n* coincides with $\overleftrightarrow{PQ}$, then it intersects *l* at Q. Otherwise, one ray $\overrightarrow{PY}$ of *n* lies between $\overrightarrow{PQ}$ and a ray $\overrightarrow{PX}$ of *m*. Take X to be the foot of the perpendicular from Y to *m*.

Now, as the point Y recedes endlessly from P on *n*, the segment XY increases indefinitely in size, and so eventually becomes greater than segment PQ. Therefore, Y must cross over to the other side of *l*, so that *n* must meet *l*.

The preceding paragraph is the heart of Proclus' argument; it is a rather sophisticated argument, involving motion and continuity. Moreover, every step in the argument can be shown to be correct — except that the conclusion doesn't follow! (In Exercise 6 you are asked to prove Aristotle's principle that XY increases indefinitely, where "indefinitely" means "without bound." For example, the sequence of numbers $\frac{1}{2}, \frac{3}{4}, \frac{7}{8}, \frac{15}{16}, \frac{31}{32}, \ldots$ increases but not "indefinitely" in the sense of "without bound," because 1 is a bound for these numbers.)

How could one justify the last step? Let us drop a perpendicular YZ from Y to *l*. You might then say that (1) X, Y, and Z are collinear, and (2) XZ $\cong$ PQ. Thus, when XY becomes greater than PQ, XY must also be greater than XZ, so that Y must be on the other side of *l*. Here the conclusion does indeed follow from statements 1 and 2. The trouble is that there is no justification for these statements!

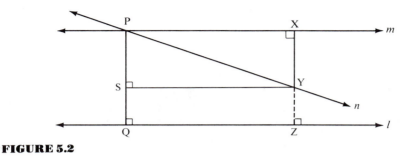

FIGURE 5.2

If this boggles your mind, it may be because Figure 5.2 makes statements 1 and 2 *seem* correct. You recall, however, that we are not allowed to use a diagram to justify a step in a proof. Each step must be proved from stated axioms or previously proven theorems. (We will show later that it is not possible in neutral geometry to prove statements 1 and 2. They can be proved only in Euclidean geometry and only by using the parallel postulate; this makes Proclus' argument circular.)

This analysis of Proclus' faulty argument illustrates how careful you must be in the way you think about parallel lines from now on. You probably visualize parallel lines as railroad tracks, everywhere equidistant from each other, and the ties of the tracks perpendicular to both parallels. This imagery is valid only in Euclidean geometry. Without the parallel postulate, the only thing we can say about two lines that are "parallel" is that, by definition of "parallel," they have no point in common. You can't assume they are equidistant; you can't even assume they have one common perpendicular segment. As Humpty Dumpty remarked: "When I use a word it means what I wish it to mean, neither more nor less."

WALLIS

The next important attempt to prove the parallel postulate was made by the Persian astronomer and mathematician Nasir Eddin al-Tusi (1201 – 1274). But since his attempted proof had several unjustified

John Wallis

assumption:, let us move ahead to John Wallis (1616–1703).[3] Wallis gave up trying to prove the parallel postulate in neutral geometry. Instead, he proposed a new axiom, which he felt was more plausible than the parallel postulate, and then proved the parallel postulate from his new axiom and the other axioms of neutral geometry.

WALLIS' POSTULATE. Given any triangle △ABC and given any segment DE. There exists a triangle △DEF (having DE as one of its sides) that is similar to △ABC (denoted △DEF ~ △ABC). (See Figure 5.3.)

---

[3] Wallis was the leading English mathematician before Isaac Newton. In his treatise *Arithmetica infinitorum* (which Newton studied), Wallis introduced the symbol ∞ for "infinity," developed formulas for certain integrals, and presented his famous infinite product formula

$$\frac{\pi}{2} = \frac{2 \cdot 2 \cdot 4 \cdot 4 \cdot 6 \cdot 6 \cdot 8 \cdots}{1 \cdot 3 \cdot 3 \cdot 5 \cdot 5 \cdot 7 \cdot 7 \cdots}$$

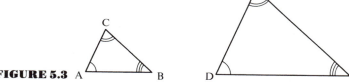

FIGURE 5.3

Similar triangles are triangles whose vertices can be put in one-to-one correspondence so that corresponding angles are congruent. In Euclidean geometry it is proved that corresponding sides of similar triangles are proportional (see Exercise 18); for example, each side of △DEF might be twice as long as the corresponding side of △ABC. Thus, the intuitive meaning of Wallis' postulate is that you can either magnify or shrink a triangle as much as you like, without distortion.

Using Wallis' postulate, the parallel postulate can be proved as follows (see Figure 5.4):

Proof:

Given a point P not on line *l*, construct one parallel *m* to *l* through P as before — by dropping a perpendicular $\overleftrightarrow{PQ}$ to *l* and erecting *m* perpendicular to $\overleftrightarrow{PQ}$. Let *n* be any other line through P. We must show that *n* meets *l*. As before, we consider a ray of *n* emanating from P that is between a ray of *m* and $\overrightarrow{PQ}$; for any point R on this ray, we drop $\overleftrightarrow{RS}$ perpendicular to $\overleftrightarrow{PQ}$ (see Proposition 3.16 and Corollary 1 to Theorem 4.1 for the existence and uniqueness of all our perpendiculars).

We now apply Wallis' postulate to △PSR and segment PQ. It tells us that there is a point T such that △PSR is similar to △PQT.

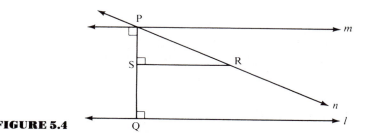

FIGURE 5.4

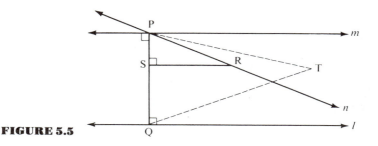

FIGURE 5.5

Assume T lies on the same side of $\overleftrightarrow{PQ}$ as R (Figure 5.5) — if not, reflect across $\overrightarrow{PQ}$.

By definition of similar triangles, $\sphericalangle TPQ \cong \sphericalangle RPS$. But since these angles have the ray $\overrightarrow{PQ} = \overrightarrow{PS}$ as a common side and since T lies on the same side of $\overleftrightarrow{PQ}$ as R, the only way they can be congruent is to be equal (Axiom C-4). Thus, $\overrightarrow{PR} = \overrightarrow{PT}$, so that T lies on n. Similarly, $\sphericalangle PQT \cong \sphericalangle PSR$, a right angle; hence, T lies on l. Thus, n and l meet at T; m is therefore the only line through P parallel to l. ■

There is no longer reason to consider Wallis' postulate any more plausible than Euclid V, because it turns out to be logically equivalent to Euclid V (see Exercise 7(a)).

SACCHERI

We next consider the remarkable work of the logician and Jesuit priest Girolamo Saccheri (1667–1733). Just before he died he published a little book entitled *Euclides ab omni naevo vindicatus (Euclid Freed of Every Flaw)*, which was not really noticed until a century and a half later, when Eugenio Beltrami rediscovered it.

Saccheri's idea was to use a reductio ad absurdum argument. He assumed the negation of the parallel postulate and tried to deduce a contradiction. Specifically, he studied certain quadrilaterals (Figure 5.6) whose base angles are right angles and whose base-adjacent sides

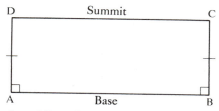

FIGURE 5.6   Saccheri quadrilateral.

are congruent to each other. These quadrilaterals have subsequently become known as *Saccheri quadrilaterals* (though they were studied many centuries earlier by the poet Omar Khayyám and by Nasir Eddin al-Tusi). It is easy to prove in neutral geometry that the summit angles are congruent (Exercise 1), i.e., $\angle C \cong \angle D$.

There are three possible cases:

Case 1: The summit angles are right angles.
Case 2: The summit angles are obtuse.
Case 3: The summit angles are acute.

Wanting to prove the first case, which is the case in Euclidean geometry, Saccheri tried to show that the other two cases led to contradictions. He succeeded in showing that case 2 leads to a contradiction: if the summit angles were obtuse, the angle sum of the quadrilateral would be more than 360°, contradicting Corollary 2 to the Saccheri-Legendre theorem (to verify the hypothesis of Corollary 2, see Exercise 17).

However hard he tried, he could not squeeze a contradiction out of case 3, "the inimical acute angle hypothesis," as he called it. He was able to deduce many strange results,[4] but not a contradiction. Finally, he exclaimed in frustration: "The hypothesis of the acute angle is absolutely false, because [it is] repugnant to the nature of the straight line!" It is as if a man had discovered a rare diamond, but, unable to believe what he saw, announced it was glass. Although he did not recognize it, Saccheri had discovered non-Euclidean geometry.

---

[4] See the translation of Saccheri's treatise by G. B. Halsted (Saccheri, 1970).

CLAIRAUT

Alexis Claude Clairaut (1713–1765) was a leading French geometer. Like Wallis, he did not try to prove the parallel postulate in neutral geometry but replaced it in his 1741 text *Éléments de géometrie* with another axiom.

CLAIRAUT'S AXIOM. Rectangles exist.

One can argue that Euclid V is not obvious because one might have to travel very far indeed to verify that the "physical lines" predicted to meet by that postulate actually do meet. According to Saccheri, it suffices to show the existence of one rectangle, which could be quite "small." Clairaut made that his axiom, arguing that "we observe rectangles all around us in houses, gardens, rooms, walls." So why didn't that settle the matter? Perhaps because the game of trying to prove Euclid V had been going on for so many centuries that it became a challenging obsession for mathematicians. Or did mathematicians finally recognize that geometry was not about "physical space"? After all, if you believe that a rectangle can be drawn on the ground, then you cannot also believe that the earth is spherical, because rectangles do not exist on a sphere. If you think you have drawn a "physical rectangle," you could be mistaken because exact measurements are physically impossible. Or did it finally dawn on mathematicians that any postulate proposed to replace Euclid V — no matter how intuitively appealing — was logically equivalent to Euclid V and therefore nothing was gained *logically* by the replacement?

Let us prove that *Clairaut's axiom is logically equivalent in neutral geometry to the parallel postulate.*

Proof:
If we assume the latter, then the existence of rectangles follows easily from Proposition 4.11 and Theorem 4.7. Conversely, assume Clairaut's axiom. Then by Theorem 4.7, all triangles have angle sum 180°, and by introducing a diagonal, all convex quadrilaterals have angle sum 360°. Return to Proclus' argument as illustrated in Figure 5.2. Let S be the foot of the perpendicular from Y to $\overleftrightarrow{PQ}$. S is on the same side of *m* as Y and Q because $\overleftrightarrow{SY}$ is parallel to *m*

(Corollary 1 to Theorem 4.1). Moreover, □PXYS, which has three right angles, is now known to be a rectangle. You can easily prove (Exercise 4) that opposite sides of a rectangle are congruent, so PS ≅ XY. By Aristotle's axiom (Chapter 3), Y can be chosen on the given ray of n so that XY > PQ. Then PS > PQ and P * Q * S. As above, Y is on the same side of l as S, hence on the opposite side of l from P. Therefore l meets n at some point between P and Y. ∎

LEGENDRE

Legendre did not know of Saccheri's work and rediscovered Saccheri's theorems in neutral geometry that are our Theorems 4.4 and 4.7. Legendre certainly knew of Clairaut's text and rejected Clairaut's axiom. We already discussed, in Chapter 1, one of Legendre's attempts to prove the parallel postulate, whose flaw we ask you to detect in Exercise 8. Legendre published a collection of his many attempts as late as 1833, the year he died. Here is his attempt to prove that the angle sum of every triangle is 180°. (Using our modification of Proclus' argument above, we could then prove Hilbert's parallel postulate.)

Proof (see Figure 5.7):
Suppose, on the contrary, there exists a triangle △ABC having defect $d \neq 0$. By the Saccheri-Legendre Theorem 4.4, $d > 0$. One of the angles of the triangle — say ∢A — must then be acute (in fact, less than 60°). On the opposite side of $\overleftrightarrow{BC}$ from A, let D be the unique point such that ∢DBC ≅ ∢ACB and BD ≅ AC (Axioms

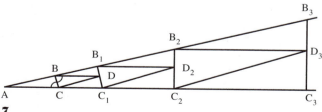

FIGURE 5.7

C-1 and C-4). Then $\triangle ACB \cong \triangle DBC$ (SAS). Also $\overset{\leftrightarrow}{BD} \parallel \overset{\leftrightarrow}{AC}$ and $\overset{\leftrightarrow}{BA} \parallel \overset{\leftrightarrow}{DC}$ (by the alternate interior angle theorem, Theorem 4.1), so that D lies in the interior of the acute $\sphericalangle A$. Hence there is a line l through D such that l intersects side $\overrightarrow{AB}$ in a point $B_1 \neq A$ and side $\overrightarrow{AC}$ in a point $C_1 \neq A$. Because of the parallel lines, we know that $B_1 \neq B$ and $C_1 \neq C$.

Suppose B_1 was on segment AB. Then A and B_1 would be on the same side of $\overset{\leftrightarrow}{BD}$. Since $\overset{\leftrightarrow}{BD} \parallel \overset{\leftrightarrow}{AC}$, A and C_1 are on the same side of $\overset{\leftrightarrow}{BD}$. Thus B_1 and C_1 are on the same side of $\overset{\leftrightarrow}{BD}$ (Axiom B-4). But since D lies in the interior of $\sphericalangle A$, $B_1 * D * C_1$ (Proposition 3.7). This contradiction shows that $A * B * B_1$. Similarly, we have $A * C * C_1$. Since $\triangle ACB \cong \triangle DBC$, the defect of $\triangle DBC$ is also d. Therefore, by the additivity of the defect applied to the four triangles into which $\triangle AB_1C_1$ has been decomposed, the defect of $\triangle AB_1C_1$ is greater than or equal to $2d$.

Repeating this construction for $\triangle AB_1C_1$, we obtain $\triangle AB_2C_2$ with defect greater than or equal to $4d$. Iterating the construction n times, we obtain a triangle with defect greater than or equal to $2^n d$, which can be made as large as we like by taking n sufficiently large. But the defect of a triangle cannot be more than $180°$! This contradiction shows that every triangle $\triangle ABC$ has defect 0. ∎

Can you see the flaw? It is easy, because we have justified every step but one, the sentence beginning with "Hence." That is the assumption you were warned in Chapter 3 not to make. Legendre made the same error as was made many centuries earlier by Simplicius (Byzantine, sixth century), al-Jawhari (Persian, ninth century), Nasir Eddin al-Tusi, and others. He has failed to prove in neutral geometry that the defect of every triangle is zero. Nevertheless, Legendre has succeeded in proving the following theorem in neutral geometry.

THEOREM 5.1. Hypothesis: For any acute $\sphericalangle A$ and any point D in the interior of $\sphericalangle A$, there exists a line through D and not through A which intersects both sides of $\sphericalangle A$. Conclusion: The angle sum of every triangle is $180°$.

You will easily see from the Klein model in Chapter 7 that the hypothesis of Theorem 5.1 fails in non-Euclidean geometry (Figure 7.5). Let us show that *the hypothesis can be proved in Euclidean geometry* (hence this hypothesis is another statement equivalent to Euclid V). Drop a perpendicular from interior point D to one of the sides of ∡A, and let B be the foot of that perpendicular. Since ∡A is acute, $(∡A)° + (∡DBA)° = (∡A)° + 90° < 180°$. So $\overrightarrow{BD}$ meets the other side of ∡A, by Euclid V. ∎

LAMBERT AND TAURINUS

Regarding Euclid V, Johann Heinrich Lambert (1728 – 1777) wrote:

> Undoubtedly, this basic assertion is far less clear and obvious than the others. Not only does it naturally give the impression that it should be proved, but to some extent it makes the reader feel that he is capable

Johann Heinrich Lambert

of giving a proof, or that he should give it. However, to the extent to which I understand this matter, that is just a *first* impression. He who reads Euclid further is bound to be amazed not only at the rigor of his proofs but also at the delightful simplicity of his exposition. This being so, he will marvel all the more at the position of the fifth postulate when he finds out that Euclid proved propositions that could far more easily be left unproved.

Lambert studied quadrilaterals having at least three right angles, which are now named after him (though they were studied seven centuries earlier by the Egyptian scientist ibn-al-Haytham). You showed in Exercise 19, Chapter 4, that Lambert quadrilaterals exist. A Lambert quadrilateral can be "doubled" (by reflecting it across an included side of two right angles) to obtain a Saccheri quadrilateral. Like Saccheri, Lambert disproved the obtuse angle hypothesis and studied the implications of the "inimical" acute angle hypothesis. He observed that it implied that similar triangles must then be congruent, which in turn implied the existence of an absolute unit of length (see Theorem 6.2, Chapter 6). He called this consequence "exquisite" but did not want it to be true, worrying that the absence of similar, proportional figures "would result in countless inconveniences," especially for astronomers (he did not realize that an elegant non-Euclidean trigonometry could be developed).

He also noticed that the defect of a triangle was proportional to its area (see Chapter 10). He recalled that on a sphere in Euclidean space, the angle sum of a triangle formed by arcs of great circles was greater than $180°$, and the excess over $180°$ of the angle sum of the triangle was proportional to the area of the triangle, the constant of proportionality being the square r^2 of the radius of the sphere (see Rosenfeld, 1988, Chapter 1). If r is replaced by ir ($i = \sqrt{-1}$), squaring introduces a minus sign that converts the excess into the defect in that proportionality. Lambert therefore speculated that the acute angle hypothesis described geometry on a "sphere of imaginary radius."[5]

---

[5] In fact, this idea can be explained in terms of a natural embedding of the non-Euclidean plane in *relativistic* three-space (see Chapter 7). Lambert is known for proving the irrationality of π and of e^x and $\tan x$ when x is rational, as well as for important laws he discovered in optics and astronomy. The quote is from B. A. Rosenfeld (1988), p. 100.

Fifty years passed before this brilliant idea was further elaborated in a booklet dated 1826 by F. A. Taurinus, who transformed the formulas of spherical trigonometry into formulas for what he called "log-spherical geometry" by substituting ir for r (his formulas are proved by a different method in Theorem 10.4, Chapter 10). Taurinus vacillated over whether such a geometry actually "existed." He sent a copy of his booklet to C. F. Gauss (see Chapter 6) and later burned the remaining copies in despair when Gauss did not respond.

Lambert cautiously did not submit his *Theory of Parallels* for publication (it was published posthumously in 1786). It contained an erroneous attempt to disprove the acute angle hypothesis. Given line l and distance d, let us call the locus m of all points on a given side of l at perpendicular distance d from l an *equidistant curve*. The flaw in many early attempts to prove the parallel postulate was the tacit assumption that m was a line. Lambert tried to prove this assumption, but he only succeeded in proving that an arc of m could not be a circular arc. Saccheri also erred using differential calculus in his attempt to prove that m was a line.[6]

FARKAS BOLYAI

There were so many attempts to prove Euclid V that by 1763 G. S. Klügel was able to submit a doctoral thesis finding the flaws in 28 different supposed proofs of the parallel postulate, expressing doubt that it could be proved. The French encyclopedist and mathematician J. L. R. d'Alembert called this "the scandal of geometry." Mathematicians were becoming discouraged. The Hungarian Farkas Bolyai wrote to his son János:

> You must not attempt this approach to parallels. I know this way to its very end. I have traversed this bottomless night, which extinguished all light and joy of my life. I entreat you, leave the science of parallels alone. . . . I thought I would sacrifice myself for the sake of the truth. I was ready to become a martyr who would remove the flaw from

---

[6] As an example in elliptic geometry: If l is the equator of the sphere, the equidistant curves are the other circles of latitude.

Farkas Bolyai

geometry and return it purified to mankind. I accomplished monstrous, enormous labors; my creations are far better than those of others and yet I have not achieved complete satisfaction. . . . I turned back when I saw that no man can reach the bottom of the night. I turned back unconsoled, pitying myself and all mankind.

I admit that I expect little from the deviation of your lines. It seems to me that I have been in these regions; that I have traveled past all reefs of this infernal Dead Sea and have always come back with broken mast and torn sail. The ruin of my disposition and my fall date back to this time. I thoughtlessly risked my life and happiness — *aut Caesar aut nihil.*[7]

But the young Bolyai was not deterred by his father's warnings, for he had a completely new idea. He assumed that the negation of

---

[7] The correspondence between Farkas and János Bolyai is from Meschkowski (1964).

Euclid's parallel postulate was not absurd, and in 1823 was able to write to his father:

> It is now my definite plan to publish a work on parallels as soon as I can complete and arrange the material and an opportunity presents itself; at the moment I still do not clearly see my way through, but the path which I have followed gives positive evidence that the goal will be reached, if it is at all possible; I have not quite reached it, but I have discovered such wonderful things that I was amazed, and it would be an everlasting piece of bad fortune if they were lost. When you, my dear Father, see them, you will understand; at present I can say nothing except this: that *out of nothing I have created a strange new universe.* All that I have sent you previously is like a house of cards in comparison with a tower. I am no less convinced that these discoveries will bring me honor than I would be if they were completed.

We will explore this "strange new universe" in the following chapters. A century after János Bolyai wrote this letter, the English physicist J. J. Thomson remarked, somewhat facetiously:

> We have Einstein's space, de Sitter's space, expanding universes, contracting universes, vibrating universes, mysterious universes. In fact, the pure mathematician may create universes just by writing down an equation, and indeed if he is an individualist he can have a universe of his own.

In fact, in 1949 the renowned logician Kurt Gödel found a model of the universe that satisfies Einstein's gravitational equations, one in which it is theoretically possible to travel backward in time![8]

REVIEW EXERCISE

Which of the following statements are correct?

(1) Wallis' postulate implies that there exist two triangles that are similar but not congruent.

---

[8] To date, attempts to refute Gödel's model on either mathematical or philosophical grounds have failed. See "On the paradoxical time-structures of Gödel," by Howard Stein, *Journal of the Philosophy of Science,* v. 37, December 1970, p. 589.

(2) A "Saccheri quadrilateral" is a quadrilateral □ABDC such that ∢CAB and ∢DBA are right angles and AC ≅ BD.

(3) A "Lambert quadrilateral" is a quadrilateral having at least three right angles.

(4) A quadrilateral that is both a Saccheri and a Lambert quadrilateral must be a rectangle.

(5) A hyperbola comes arbitrarily close to its asymptotes without ever intersecting them.

(6) János Bolyai warned his son Farkas not to work on the parallel problem.

(7) Saccheri succeeded in disproving the "inimical" acute angle hypothesis.

(8) In trying to prove Euclid's fifth postulate, Ptolemy tacitly assumed what we have been calling Hilbert's parallel postulate.

(9) It is a theorem in neutral geometry that if $l \parallel m$ and $m \parallel n$, then $l \parallel n$.

(10) It is a theorem in neutral geometry that every segment has a unique midpoint.

(11) It is a theorem in neutral geometry that if a rectangle exists, then the angle sum of any triangle is 180°.

(12) It is a theorem in neutral geometry that if l and m are parallel lines, then alternate interior angles cut out by any transversal to l and m are congruent to each other.

(13) Legendre proved in neutral geometry that for any acute ∢A and any point D in the interior of ∢A, there exists a line through D and not through A which intersects both sides of ∢A.

(14) Clairaut showed that Euclid's fifth postulate could be replaced in the logical presentation of Euclidean geometry by the "more obvious" postulate that rectangles exist, yet mathematicians were not appeased by Clairaut's replacement and they continued to try to prove Euclid V.

EXERCISES

Again, in proofs in Exercises 1–17 you are allowed to use only our previous results from neutral geometry.

1. Let □ABDC be a Saccheri quadrilateral, so that ∢B and ∢A are right angles and CA ≅ DB (Figure 5.8). Prove that ∢C ≅ ∢D. (Hint: Prove △CAB ≅ △DBA, then △CDB ≅ △DCA.) Also prove that Saccheri quadrilaterals exist.

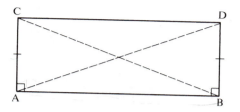

FIGURE 5.8

2. Let □ABDC be a quadrilateral whose base angles ∢A and ∢B are right angles. Prove that if AC < BD, then (∢D)° < (∢C)° (Figure 5.9). (Hint: If AC ≅ BE, with B * E * D, use Exercise 23, Chapter 4, to show that E is interior to ∢ACD, then apply Exercise 1 and the exterior angle theorem.)

3. With the same hypothesis as in Exercise 2, prove the converse, that if (∢D)° < (∢C)°, then AC < BD. (Hint: Assume the contrary, which involves the two cases AC ≅ BD and AC > BD. In each case, derive a contradiction.)

4. The Swiss-German mathematician Lambert considered quadrilaterals with at least three right angles, which are now named after him (Figure 5.10). Prove the following:

   (a) The fourth angle ∢D of a Lambert quadrilateral is never obtuse.

   (b) If ∢D is a right angle, then the opposite sides of □ABCD are congruent (use Exercise 2 and an RAA argument).

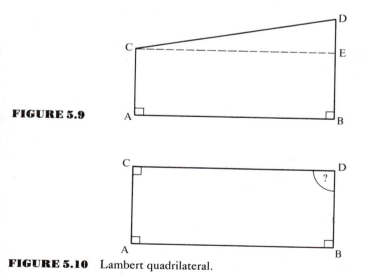

FIGURE 5.9

FIGURE 5.10 Lambert quadrilateral.

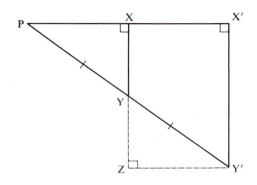

FIGURE 5.11

(c) If ∢D is acute, then each side adjacent to ∢D is greater than its opposite side, i.e., DB > CA and CD > AB (use Exercise 3).

(d) A quadrilateral is both a Lambert and a Saccheri quadrilateral if and only if it is a rectangle.

We can combine statements (a), (b), and (c) of this exercise into the following statement: a side adjacent to the fourth angle of a Lambert quadrilateral is greater than or congruent to its opposite side. As you know, case (b) always holds if the geometry is Euclidean; in the next chapter we will show that case (c) always holds if the geometry is hyperbolic. In elliptic geometry the fourth angle of a Lambert quadrilateral is always obtuse, and a side adjacent to the fourth angle is always smaller than its opposite side.

5. Given a right triangle △PXY with right angle at X, form a new right triangle △PX′Y′ that has acute angle ∢P in common with the given triangle but double the hypotenuse (prove that this can be done); see Figure 5.11. Prove that the side opposite the acute angle is *at least* doubled, whereas the side adjacent to the acute angle is *at most* doubled. (Hint: Extend side XY far enough to drop a perpendicular Y′Z to $\overleftrightarrow{XY}$. Prove that △PXY ≅ △Y′ZY, and apply Exercise 4 to the Lambert quadrilateral □XZY′X′.)

6. Use Exercise 5 to prove Aristotle's axiom (used in Proclus' argument) that as Y recedes endlessly from P, segment XY increases indefinitely (see p. 96). (Hint: Use Archimedes' axiom and the fact that $2^n \to \infty$ as $n \to \infty$.[9]) Does segment PX also increase indefinitely?

7. (a) Prove that Euclid's fifth postulate implies Wallis' postulate (see Figure 5.12). (Hint: Use Axiom C-4 and the fact that in Euclidean

---

[9] Euclid had a version of Archimedes' axiom in Book V, but he presented it as a "definition": If $a < b$, then there is a number n such that $2^n a > b$.

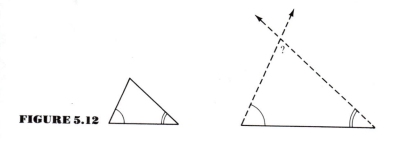

FIGURE 5.12

geometry the angle sum of a triangle is 180° — Proposition 4.11.)
(b) Suppose that in the statement of Wallis' postulate we add the
assumption AB ≅ DE and replace the word "similar" by "congruent." Prove this new statement in neutral geometry.

8. Reread Legendre's attempted proof of the parallel postulate in Chapter
1. Find the flaw, and justify all the steps that are correct. Prove the
flawed statement in Euclidean geometry.

9. Find the unjustified assumption in the following "proof" of the parallel
postulate by Farkas Bolyai (see Figure 5.13). Given P not on line *l*, $\overleftrightarrow{PQ}$
perpendicular to *l* at Q, and line *m* perpendicular to $\overleftrightarrow{PQ}$ at P. Let *n* be any
line through P distinct from *m* and $\overleftrightarrow{PQ}$. We must show that *n* meets *l*. Let
A be any point between P and Q. Let B be the unique point such that

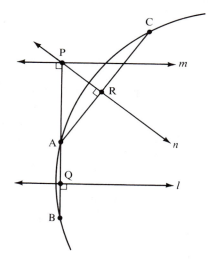

FIGURE 5.13

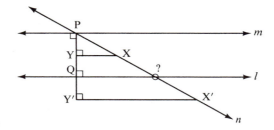

FIGURE 5.14

A * Q * B and AQ ≅ QB. Let R be the foot of the perpendicular from A to *n*. Let C be the unique point such that A * R * C and AR ≅ RC. Then A, B, and C are not collinear (else R = P); hence there is a unique circle γ passing through them. Since *l* is the perpendicular bisector of chord AB of γ and *n* is the perpendicular bisector of chord AC of γ, *l* and *n* meet at the center of γ (Exercise 17(b), Chapter 4).

10. The following attempted proof of the parallel postulate is similar to Proclus' but the flaw is different; detect the flaw with the help of Exercise 5. (See Figure 5.14.) Given P not on line *l*, $\overleftrightarrow{PQ}$ perpendicular to *l* at Q, and line *m* perpendicular to $\overleftrightarrow{PQ}$ at P. Let *n* be any line through P distinct from *m* and $\overleftrightarrow{PQ}$. We must show that *n* meets *l*. Let $\overrightarrow{PX}$ be a ray of *n* between $\overleftrightarrow{PQ}$ and a ray of *m*, and let Y be the foot of the perpendicular from X to $\overleftrightarrow{PQ}$. As X recedes endlessly from P, PY increases indefinitely. Hence, Y eventually reaches a position Y′ on $\overrightarrow{PQ}$ such that PY′ > PQ. Let X′ be the corresponding position reached by X on line *n*. Now X′ and Y′ are on the same side of *l* because $\overleftrightarrow{X'Y'}$ is parallel to *l*. But Y′ and P are on opposite sides of *l*. Hence, X′ and P are on opposite sides of *l*, so that segment PX′ (which is part of *n*) meets *l*.

11. Find the flaw in the following attempted proof of the parallel postulate given by J. D. Gergonne (see Figure 5.15). Given P not on line *l*, $\overleftrightarrow{PQ}$ perpendicular to *l* at Q, line *m* perpendicular to $\overleftrightarrow{PQ}$ at P, and point A ≠ P on *m*. Let $\overrightarrow{PB}$ be the last ray between $\overrightarrow{PA}$ and $\overrightarrow{PQ}$ that intersects *l*, B being the point of intersection. There exists a point C on *l* such that Q * B * C (Axioms B-1 and B-2). It follows that $\overrightarrow{PB}$ is not the last ray between $\overrightarrow{PA}$ and $\overrightarrow{PQ}$ that intersects *l*, and hence all rays between $\overrightarrow{PA}$ and $\overrightarrow{PQ}$ meet *l*. Thus *m* is the only parallel to *l* through P.

12. Legendre made another attempt to prove that the defect of every triangle is zero, as follows. In any triangle △ABC, if we are given the numbers α, β that measure the angles at A and B, respectively, and the number *x* that measures the length of the included side AB, then by ASA the number γ that measures the third angle is uniquely determined, so we can write γ = *f*(α, β, *x*). Now if right angles are measured by the unit 1,

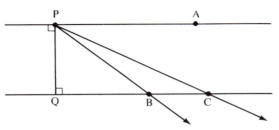

FIGURE 5.15

γ is some number between 0 and 2. But x is not a dimensionless number, since it depends on the arbitrary unit of length (e.g., inch, foot, or meter). Thus the formula for γ cannot actually contain x, and so $\gamma = f(\alpha, \beta)$. (If we knew that the geometry was Euclidean, we would have $f(\alpha, \beta) = 2 - \alpha - \beta$.) Now let D be the midpoint of AB. By Axiom C-4, there is a unique ray $\overrightarrow{DE}$ with E on the same side of $\overleftrightarrow{AB}$ as C such that $\sphericalangle ADE \cong \sphericalangle B$. Then $\overleftrightarrow{DE} \parallel \overleftrightarrow{BC}$, by corresponding angles (Exercise 32, Chapter 4), so, by Pasch's theorem, we can choose label E for a point on AC. Since $\gamma = f(\alpha, \beta)$ for any triangle, $\sphericalangle AED \cong \sphericalangle C$ (i.e., $\triangle ADE$ is similar to $\triangle ABC$). Hence $\square CEDB$ has angle sum 360°, so $\triangle CEB$ and $\triangle BED$ have angle sum 180°. By Theorem 4.7, all triangles have defect zero. What is the flaw? (Hint: See the remarks following Theorem 6.2.)

13. It was stated at the beginning of this chapter that if all steps but one of an attempt to prove the parallel postulate are correct, then the flawed step yields another statement equivalent to Hilbert's parallel postulate. Show that for Proclus' attempt, that statement is: Given parallel lines l, m having a common perpendicular and a point Y not lying on l or m, if X (resp. Z) is the foot of the perpendicular from Y to l (resp. to m), then X, Y, and Z are collinear. (Hint: Use Exercise 4b.)

14. Prove that if the defect of every triangle is 0 then Hilbert's parallel postulate holds (this is the converse to Proposition 4.11). (Hint: See the discussion of Clairaut's axiom.)

15. You will show in Exercise 24 that the following statement can be proved in Euclidean geometry: If points P, Q, R lie on a circle with center O, and if $\sphericalangle PQR$ is acute, then $(\sphericalangle PQR)° = \frac{1}{2}(\sphericalangle POR)°$. In neutral geometry, show that this statement implies the existence of a triangle of defect zero (hence, by Prop. 4.7 and Clairaut's axiom, the geometry is Euclidean).

16. *(Difficult)* Here is Legendre's desperate attempt in the 12th and final edition of his geometry text to prove that the defect d of any triangle $\triangle ABC$ is zero. Label so that $\overline{AC} \geq \overline{AB}$. Then the method of Exercise 15,

Chapter 4 (invented by Legendre to prove Theorem 4.4), gives us $\triangle AEC$ having the same defect d (where $AE = 2AD$, D being the midpoint of BC). From $\overline{AC} \geq \overline{AB}$ it is not difficult to show that $\sphericalangle EAC \leq \sphericalangle EAB$, so that $\sphericalangle EAC \leq \frac{1}{2}\sphericalangle BAC$. If $\overline{AC} \geq \overline{AE}$, set $B_1 = E$ and $C_1 = C$; otherwise, reflect $\triangle AEC$ across the angle bisector of $\sphericalangle EAC$ to obtain $\triangle AB_1C_1$ with defect d, $\sphericalangle B_1AC_1 \leq \frac{1}{2}\sphericalangle BAC$, C_1 on ray $\overrightarrow{AC}$, and $\overline{AC_1} \geq \overline{AB_1}$. Iterate this construction n times to obtain $\triangle AB_nC_n$ with defect d, $\sphericalangle B_nAC_n \leq 2^{-n}\sphericalangle BAC$, and C_n on ray $\overrightarrow{AC}$. Let $n \to \infty$. Since $\sphericalangle B_nAC_n \to 0$, the triangles $\triangle AB_nC_n$ (which all have defect d) converge in the limit to a degenerate triangle on ray $\overrightarrow{AC}$ having angles $0°$, $0°$, and $180°$. Hence $d = 0$. Criticize this argument (Hint: Show that B_n, $C_n \to \infty$.)

17. Prove that a Saccheri quadrilateral is convex. Prove that a Lambert quadrilateral is a parallelogram and that every parallelogram is convex.

The remaining exercises in this chapter are exercises in Euclidean geometry, which means you are allowed to use the parallel postulate and its consequences already established. We will refer to these results in Chapter 7. You are also allowed to use the following result, a proof of which is indicated in the Major Exercises:

PARALLEL PROJECTION THEOREM. Given three parallel lines l, m, and n. Let t and t' be transversals to these parallels, cutting them in points A, B, and C and in points A', B', and C', respectively. Then $\overline{AB}/\overline{BC} = \overline{A'B'}/\overline{B'C'}$. (Figure 5.16.)

18. *Fundamental theorem on similar triangles.* Given $\triangle ABC \sim \triangle A'B'C'$, i.e., given $\sphericalangle A \cong \sphericalangle A'$, $\sphericalangle B \cong \sphericalangle B'$, and $\sphericalangle C \cong \sphericalangle C'$. Then corresponding sides are proportional, i.e., $\overline{AB}/\overline{A'B'} = \overline{AC}/\overline{A'C'} = \overline{BC}/\overline{B'C'}$. (See Figure 5.17.) Prove the theorem. (Hint: Let B'' be the point on $\overrightarrow{AB}$ such that $AB'' \cong A'B'$, and let C'' be the point on $\overrightarrow{AC}$ such that $AC'' \cong A'C'$. Use

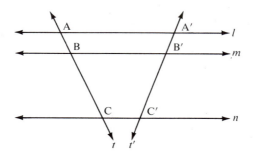

FIGURE 5.16

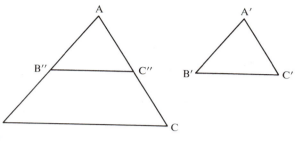

FIGURE 5.17

the hypothesis to show that $\triangle AB''C'' \cong \triangle A'B'C'$ and deduce from corresponding angles that $\overleftrightarrow{B''C''}$ is parallel to $\overleftrightarrow{BC}$. Now apply the parallel projection theorem.)

19. Prove the converse to the fundamental theorem on similar triangles. (Hint: Choose B'' as before. Use Pasch's theorem to show that the parallel to $\overleftrightarrow{BC}$ through B'' cuts AC at a point C''. Then use the hypothesis, Exercise 18, and the SSS criterion to show that $\triangle ABC \sim \triangle A'B'C''$.)

20. *SAS similarity criterion.* If $\angle A \cong \angle A'$ and $\overline{AB}/\overline{A'B'} = \overline{AC}/\overline{A'C'}$, prove that $\triangle ABC \sim \triangle A'B'C'$. (Hint: Same method as in Exercise 19, but using SAS instead of SSS.)

21. Prove the Pythagorean theorem. (Hint: Let CD be the altitude to the hypotenuse; see Figure 5.18. Use the fact that the angle sum of a triangle equals 180° (Exercise 8, Chapter 4) to show that $\triangle ACD \sim \triangle ABC \sim \triangle CBD$. Apply Exercise 18 and a little algebra based on $\overline{AB} = \overline{AD} + \overline{DB}$ to get the result.)

22. The fundamental theorem on similar triangles (Exercise 18) allows the trigonometric functions such as sine and cosine to be defined. Namely, given an acute angle $\angle A$, make it part of a right triangle $\triangle BAC$ with right angle at C, and set

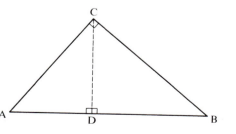

FIGURE 5.18

$$\sin \sphericalangle A = (\overline{BC})/(\overline{AB})$$
$$\cos \sphericalangle A = (\overline{AC})/(\overline{AB}).$$

These definitions are then independent of the choice of the right triangle used. If $\sphericalangle A$ is obtuse and $\sphericalangle A'$ is its supplement, set

$$\sin \sphericalangle A = +\sin \sphericalangle A'$$
$$\cos \sphericalangle A = -\cos \sphericalangle A'.$$

If $\sphericalangle A$ is a right angle, set

$$\sin \sphericalangle A = 1$$
$$\cos \sphericalangle A = 0.$$

Now, given any triangle $\triangle ABC$, if a and b are the lengths of the sides opposite A and B, respectively, prove the law of sines,

$$\frac{a}{b} = \frac{\sin \sphericalangle A}{\sin \sphericalangle B}.$$

(Hint: Drop altitude CD and use the two right triangles $\triangle ADC$ and $\triangle BDC$ to show that $b \sin \sphericalangle A = \overline{CD} = a \sin \sphericalangle B$; see Figure 5.19.) Similarly, prove the law of cosines,

$$c^2 = a^2 + b^2 - 2ab \cos \sphericalangle C,$$

and deduce the converse to the Pythagorean theorem.

23. Given $A * B * C$ and point D not collinear with A, B, and C (Figure 5.20). Prove that

$$\frac{\overline{AB}}{\overline{BC}} = \frac{\overline{AD} \sin \sphericalangle ADB}{\overline{CD} \sin \sphericalangle CDB}$$
$$\frac{\overline{AC}}{\overline{BC}} = \frac{\overline{AD} \sin \sphericalangle ADC}{\overline{BD} \sin \sphericalangle BDC}$$

(Hint: Use the law of sines to compute $\overline{AB}/\overline{AD}$, $\overline{CD}/\overline{BC}$, and $\overline{BD}/\overline{BC}$, and remember that $\sin \sphericalangle ABD = \sin \sphericalangle CBD$.)

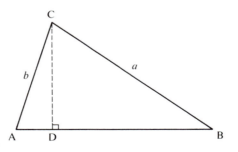

FIGURE 5.19

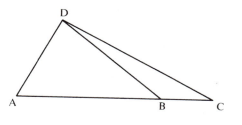

FIGURE 5.20

24. Let γ be a circle with center O, and let P, Q, and R be three points on γ. Prove that if P and R are diametrically opposite, then $\sphericalangle$PQR is a right angle, and if O and Q are on the same side of $\overleftrightarrow{PR}$, then $(\sphericalangle PQR)° = \frac{1}{2}(\sphericalangle POR)°$. (Hint: Again use the fact that the triangular angle sum is 180°. There are four cases to consider, as in Figure 5.21.) State and prove the analogous result when O and Q are on opposite sides of $\overleftrightarrow{PR}$.

25. Prove that if two angles inscribed in a circle subtend the same arc, then they are congruent; see Figure 5.22. (Hint: Apply the previous exercise after carefully defining "subtend the same arc.")

26. Prove that if $\sphericalangle$PQR is a right angle, then Q lies on the circle γ having PR as diameter. (Hint: Use uniqueness of perpendiculars and Exercise 24.)

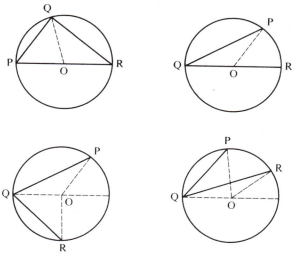

FIGURE 5.21

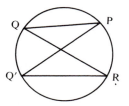

FIGURE 5.22   ⊀PQR ≅ ⊀PQ′R.

MAJOR EXERCISES

These exercises furnish the proof of the parallel projection theorem in Euclidean geometry (text preceding Exercise 18; also see Figure 5.16).

1. Prove the following results about Euclidean parallelograms:
   (a) Opposite sides (and likewise, opposite angles) of a parallelogram are congruent to each other.
   (b) A parallelogram is a rectangle iff its diagonals are congruent, and in that case the diagonals bisect each other.
   (c) A parallelogram has a circumscribed circle iff it is a rectangle. (Hint for the "only if" part: Opposite angles must subtend semicircles.)
   (d) A rectangle is a square iff its diagonals are perpendicular.
2. Let k, l, m, and n be parallel lines, distinct, except that possibly $l = m$. Let transversals t and t' cut these lines in points A, B, C, and D and in A′, B′, C′, and D′, respectively (Figure 5.23). If AB ≅ CD, prove that A′B′ ≅ C′D′. (Hint: Construct parallels to t through A′ and C′. Apply Major Exercise 1(a) and the congruence of corresponding angles.)
3. Prove that parallel projection preserves betweenness, i.e., in Figure 5.16, if A * B * C, then A′ * B′ * C′. (Hint: Use Axiom B-4.)
4. Prove the parallel projection theorem for the special case in which the ratio of lengths $\overline{AB}/\overline{BC}$ is a rational number p/q. (Hint: Divide AB into p congruent segments and BC into q congruent segments so that all $p + q$ segments will be congruent. Use Major Exercise 2, applying it $p + q$ times.)
5. The case where $\overline{AB}/\overline{BC}$ is an irrational number x is the difficult case. Let $\overline{A'B'}/\overline{B'C'} = x'$. The idea is to show that every rational number p/q less than x is also less than x' (and, by symmetry, vice versa). This will imply

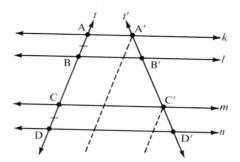

FIGURE 5.23

$x = x'$, since a real number is the least upper bound of all the rational numbers less than it (see any good text on real analysis). To show this, lay off on $\overrightarrow{BA}$ a segment BD of length $p\overline{CB}/q$, and let D' be the parallel projection of D onto t'. From $p/q < x$, deduce B * D * A. Now apply Major Exercises 3 and 4 to show that $p/q < x'$.[10]

6. Given a segment AB of length a with respect to some unit segment OI (see Theorem 4.3). Using straightedge and compass only, show how to construct a segment of length $\sqrt{a}$. (Hint: Extend AB to a segment AC of length $a + 1$; erect a perpendicular through B and let D be one of its intersections with the circle having AC as diameter; apply the theory of similar triangles to show that $\overline{BD} = \sqrt{a}$. Review the construction in Major Exercise 1, Chapter 1.)

7. Prove that given any line l, two points A and B not on l are on the same side S of l if and only if they lie on a circle contained in S. (Hint: If A and B are on opposite sides of l, apply Major Exercise 1, Chapter 4. If they are on the same side S, let M be the midpoint and m the perpendicular bisector of AB. Any circle through A and B has its center on m. If $\overleftrightarrow{AB} \parallel l$, take any point P between M and the point where m meets l, and use the circle through A, B and P (see Exercise 12, Chapter 6). Otherwise, if A is closer to l than B, let the perpendicular from A to l meet m at O. Show that the circle centered at O with radius OA $\cong$ OB lies in S. Be sure to indicate where the hypothesis that *the geometry is Euclidean* is used; see Exercise P-20, Chapter 7.)

---

[10] This clever method of proof was essentially discovered by the ancient Greek mathematician Eudoxus—see E. C. Zeeman, "Research, Ancient and Modern," *Bulletin of the Institute of Mathematics and Its Applications,* 10 (1974): 272–281, Warwick University, England.

PROJECTS

1. Since there is no algebra in Euclid's *Elements,* it was quite a feat for Eudoxus to have discovered a purely geometric treatment of proportions and similar triangles. Report on this, using Heath (1956) and Moise (1990, Chapter 20) as references.

2. Eudoxus was also the founder of theoretical astronomy in antiquity (his work later refined by Ptolemy). In his model, the universe was bounded by "the celestial sphere," so that the physical interpretation of Euclid's second and third postulates was false! Even Kepler and Galileo believed in an outer limit to the world. It was René Descartes (1596 – 1650) who promoted the idea that we live in infinite, unbounded Euclidean space. Report on these issues, using Torretti (1978) as one reference.

3. Report on several other attempts to prove the Euclidean parallel postulate in neutral geometry, using Rosenfeld (1988) as a reference.

4. We remarked after Theorem 4.3 that the definition of π as the ratio of the circumference of any circle to its diameter could only be justified in Euclidean geometry, not neutral geometry. Report on the justification in Moise (1990), section 21.2, which uses the theory of similar triangles.

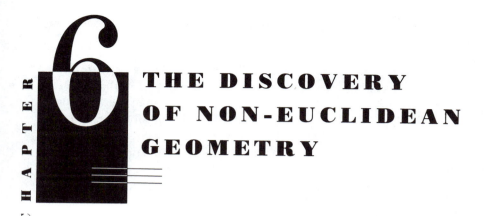

6

THE DISCOVERY
OF NON-EUCLIDEAN
GEOMETRY

Out of nothing I have created a strange new universe.
JÁNOS BOLYAI

JÁNOS BOLYAI

It is remarkable that sometimes when the time is right for a new idea to come forth, the idea occurs to several people more or less simultaneously. Thus it was in the eighteenth century with the discovery of the calculus by Newton in England and Leibniz in Germany, and in the nineteenth century with the discovery of non-Euclidean geometry. When János Bolyai (1802–1860) announced privately his discoveries in non-Euclidean geometry, his father Farkas admonished him:

> It seems to me advisable, if you have actually succeeded in obtaining a solution of the problem, that, for a two-fold reason, its publication be hastened: first, because ideas easily pass from one man to another who, in that case, can publish them; secondly, because it seems to be true that many things have, as it were, an epoch in which they are discovered in several places simultaneously, just as the violets appear on all sides in springtime.[1]

[1] Quoted in Meschkowski (1964). The title of J. Bolyai's appendix is "The Science of Absolute Space with a Demonstration of the Independence of the Truth or Falsity of Euclid's Parallel Postulate (Which Cannot Be Decided a Priori) and, in Addition, the Quadrature of the Circle in Case of Its Falsity."

János Bolyai did publish his discoveries, as a 26-page appendix to a book by his father surveying attempts to prove Euclid V (the *Tentamen*, 1831). Farkas sent a copy to his friend, the German mathematician Carl Friedrich Gauss (1777–1855), undisputedly the foremost mathematician of his time. Farkas Bolyai had become close friends with Gauss 35 years earlier, when they were both students in Göttingen. After Farkas returned to Hungary, they maintained an intimate correspondence,[2] and when Farkas sent Gauss his own attempt to prove the parallel postulate, Gauss tactfully pointed out the fatal flaw.

János was 13 years old when he mastered the differential and integral calculus. His father wrote to Gauss begging him to take the young prodigy into his household as an apprentice mathematician. Gauss never replied to this request (perhaps because he was having enough trouble with his own son Eugene, who had run away from home). Fifteen years later, when Farkas mailed the *Tentamen* to Gauss, he certainly must have felt that his son had vindicated his belief in him, and János must have expected Gauss to publicize his achievement. One can therefore imagine the disappointment János must have felt when he read the following letter to his father from Gauss:

> If I begin with the statement that I dare not praise such a work, you will of course be startled for a moment: but I cannot do otherwise; to praise it would amount to praising myself; for the entire content of the work, the path which your son has taken, the results to which he is led, coincide almost exactly with my own meditations which have occupied my mind for from thirty to thirty-five years. On this account I find myself surprised to the extreme.
>
> My intention was, in regard to my own work, of which very little up to the present has been published, not to allow it to become known during my lifetime. Most people have not the insight to understand our conclusions and I have encountered only a few who received with any particular interest what I communicated to them. In order to understand these things, one must first have a keen perception of what is needed, and upon this point the majority are quite confused. On the other hand, it was my plan to put all down on paper eventually, so that at least it would not finally perish with me.
>
> So I am greatly surprised to be spared this effort, and am overjoyed

---

[2] For the complete correspondence (in German), see Schmidt and Stäckel (1972).

that it happens to be the son of my old friend who outstrips me in such a remarkable way.[3]

Despite the compliment in Gauss' last sentence, János was bitterly disappointed with the great mathematician's reply; he even imagined that his father had secretly informed Gauss of his results and that Gauss was now trying to appropriate them as his own. A man of fiery temperament, who had fought and won 13 successive duels (unlike Galois, who was killed in a duel at age 20), János fell into deep mental depression and never again published his research. A translation of his immortal "appendix" can be found in R. Bonola's *Non-Euclidean Geometry* (1955). His father did not understand János' discovery and subsequently published another clever attempt to prove Euclid V (Exercise 9, Chapter 5).

In 1851, János wrote:

In my opinion, and as I am persuaded, in the opinion of anyone judging without prejudice, all the reasons brought up by Gauss to explain why he would not publish anything in his life on this subject are powerless and void; for in science, as in common life, it is necessary to clarify things of public interest which are still vague, and to awaken, to strengthen and to promote the lacking or dormant sense for the true and right. Alas, to the great detriment and disadvantage of mankind, only very few people have a sense for mathematics; and for such a reason and pretence Gauss, in order to remain consistent, should have kept a great part of his excellent work to himself. It is a fact that, among mathematicians, and even among celebrated ones, there are, unfortunately, many superficial people, but this should not give a sensible man a reason for writing only superficial and mediocre things and for leaving science lethargically in its inherited state. Such a supposition may be said to be unnatural and sheer folly; therefore I take it rightly amiss that Gauss, instead of acknowledging honestly, definitely and frankly the great worth of the Appendix and the Tentamen, and instead of expressing his great joy and interest and trying to prepare an appropriate reception for the good cause, avoiding all these, he rested content with

---

[3] Wolfe (1945). Gauss did write to Gerling about the appendix a month earlier, saying: "I find all my own ideas and results developed with greater elegance. . . . I regard this young geometer Bolyai as a genius of the first order." That makes it all the more puzzling why Gauss did not help further János' mathematical career.

pious wishes and complaints about the lack of adequate civilization. Verily, it is not this attitude we call life, work and merit.[4]

GAUSS

There is evidence that Gauss had anticipated some of J. Bolyai's discoveries, in fact, that Gauss had been working on non-Euclidean geometry since the age of 15, i.e., since 1792 (see Bonola, 1955, Chapter 3). In 1817, Gauss wrote to W. Olbers: "I am becoming more and more convinced that the necessity of our [Euclidean] geometry cannot be proved, at least not by human reason nor for human reason. Perhaps in another life we will be able to obtain insight into the nature of space, which is now inattainable." In 1824, Gauss answered F. A. Taurinus, who had attempted to investigate the theory of parallels:

> In regard to your attempt, I have nothing (or not much) to say except that it is incomplete. It is true that your demonstration of the proof that the sum of the three angles of a plane triangle cannot be greater than 180° is somewhat lacking in geometrical rigor. But this in itself can easily be remedied, and there is no doubt that the impossibility can be proved most rigorously. But the situation is quite different in the second part, that the sum of the angles cannot be less than 180°; this is the critical point, the reef on which all the wrecks occur. I imagine that this problem has not engaged you very long. I have pondered it for over thirty years, and I do not believe that anyone can have given more thought to this second part than I, though I have never published anything on it.
>
> The assumption that the sum of the three angles is less than 180° leads to a curious geometry, quite different from ours [the Euclidean], but thoroughly consistent, which I have developed to my entire satisfaction, so that I can solve every problem in it with the exception of the determination of a constant, which cannot be designated *a priori*. The greater one takes this constant, the nearer one comes to Euclidean geometry, and when it is chosen infinitely large the two coincide. The theorems of this geometry appear to be paradoxical and, to the uninitiated, absurd; but calm, steady reflection reveals that they contain

---

[4] Quoted in L. Fejes Tóth, *Regular Figures* (Macmillan, N.Y. 1964), pp. 98–99.

Carl Friedrich Gauss

nothing at all impossible. For example, the three angles of a triangle become as small as one wishes, if only the sides are taken large enough; yet the area of the triangle can never exceed a definite limit, regardless of how great the sides are taken, nor indeed can it never reach it.

All my efforts to discover a contradiction, an inconsistency, in this non-Euclidean geometry have been without success, and the one thing in it which is opposed to our conceptions is that, if it were true, there must exist in space a linear magnitude, determined for itself (but unknown to us). But it seems to me that we know, despite the say-nothing word-wisdom of the metaphysicians, too little, or too nearly nothing at all, about the true nature of space, to consider as *absolutely impossible* that which appears to us unnatural. If this non-Euclidean geometry were true, and it were possible to compare that constant with such magnitudes as we encounter in our measurements on the earth and in the heavens, it could then be determined *a posteriori*. Consequently, in jest I have sometimes expressed the wish that the Euclidean geometry were not true, since then we would have *a priori* an absolute standard of measure.

I do not fear that any man who has shown that he possesses a thoughtful mathematical mind will misunderstand what has been said above, but in any case consider it a private communication of which no public use or use leading in any way to publicity is to be made. Perhaps I shall myself, if I have at some future time more leisure than in my present circumstances, make public my investigations.[5]

It is amazing that, despite his great reputation, Gauss was actually afraid to make public his discoveries in non-Euclidean geometry. He wrote to F. W. Bessel in 1829 that he feared "the howl from the Boeotians" if he were to publish his revolutionary discoveries.[6] He told H. C. Schumacher that he had "a great antipathy against being drawn into any sort of polemic."

The "metaphysicians" referred to by Gauss in his letter to Taurinus were followers of Immanuel Kant, the supreme European philosopher in the late eighteenth century and much of the nineteenth century. Gauss' discovery of non-Euclidean geometry refuted Kant's position that Euclidean space is *inherent in the structure of our mind.* In his *Critique of Pure Reason* (1781) Kant declared that "the concept of [Euclidean] space is by no means of empirical origin, but is an inevitable necessity of thought."

Another reason that Gauss withheld his discoveries was that he was a perfectionist, one who published only completed works of art. His devotion to perfected work was expressed by the motto on his seal, *pauca sed matura* ("few but ripe"). There is a story that the distinguished mathematician K. G. J. Jacobi often came to Gauss to relate new discoveries, only to have Gauss pull out some papers from his desk drawer that contained the very same discoveries. Perhaps it is because Gauss was so preoccupied with original work in many branches of mathematics, as well as in astronomy, geodesy, and phys-

---

[5] Wolfe (1945), pp. 46–47.

[6] An allusion to dull, obtuse individuals. "Actually, the 'Boeotian' critics of non-Euclidean geometry — conceited people who claimed to have proved that Gauss, Riemann, and Helmholz were blockheads — did not show up before the middle of the 1870s. If you witnessed the struggle against Einstein in the Twenties, you may have some idea of [the] amusing kind of literature [produced by these critics]. . . . Frege, rebuking Hilbert like a schoolboy, also joined the Boeotians. . . . 'Your system of axioms,' he said to Hilbert, 'is like a system of equations you cannot solve.' " (Freudenthal, 1962)

ics (he coinvented an improved telegraph with W. Weber), that he did not have the opportunity to put his results on non-Euclidean geometry into polished form. The few results he wrote down were found among his private papers after his death.

Gauss has been called "the prince of mathematicians" because of the range and depth of his work. (See the biographies by Bell, 1934; Dunnington, 1955; and Hall, 1970.)

LOBACHEVSKY

Another actor in this historical drama came along to steal the limelight from both J. Bolyai and Gauss: the Russian mathematician Nikolai Ivanovich Lobachevsky (1792–1856). He was the first to actually

Nikolai Ivanovich Lobachevsky

publish an account of non-Euclidean geometry, in 1829. Lobachevsky initially called his geometry "imaginary," then later "pangeometry." His work attracted little attention on the continent when it appeared because it was written in Russian. The reviewer at the St. Petersburg Academy rejected it, and a Russian literary journal attacked Lobachevsky for "the insolence and shamelessness of false new inventions" (Boeotians howling, as Gauss predicted). Nevertheless, Lobachevsky courageously continued to publish further articles in Russian and then a treatise in 1840 in German,[8] which he sent to Gauss. In an 1846 letter to Schumacher, Gauss reiterated his own priority in developing non-Euclidean geometry but conceded that "Lobachevsky carried out the task in a masterly fashion and in a truly geometric spirit." At Gauss' recommendation, Lobachevsky was elected to the Göttingen Scientific Society. (Why didn't Gauss recommend János Bolyai?)

Lobachevsky openly challenged the Kantian doctrine of space as a subjective intuition. In 1835 he wrote: "The fruitlessness of the attempts made since Euclid's time . . . aroused in me the suspicion that the truth . . . was not contained in the data themselves; that to establish it the aid of experiment would be needed, for example, of astronomical observations, as in the case of other laws of nature." (Gauss privately agreed with this view, having written to Olbers that "we must not put geometry on a par with arithmetic that exists purely a priori but rather with mechanics." The great French mathematicians J. L. Lagrange (1736–1813) and J. B. Fourier (1768–1830) tried to derive the parallel postulate from the law of the lever in statics.)

Lobachevsky has been called "the great emancipator" by Eric Temple Bell; his name, said Bell, should be as familiar to every schoolboy as that of Michelangelo or Napoleon.[9] Unfortunately, Lobachevsky was not so appreciated in his lifetime; in fact, in 1846 he was fired from the University of Kazan, despite 20 years of outstanding service as a teacher and administrator. He had to dictate his last book in the year before his death, for by then he was blind.

It is amazing how similar are the approaches of J. Bolyai and Lobachevsky and how different they are from earlier work. Both developed the subject much further than Gauss. Both attacked plane geometry

---

[8] For a translation of this paper see Bonola (1955).
[9] Bell (1954, Chapter 14).

via the "horosphere" in three-space (it is the limit of an expanding sphere when its radius tends to infinity). Both showed that geometry on a horosphere, where "lines" are interpreted as "horocycles" (limits of circles), is Euclidean. Both showed that Euclidean spherical trigonometry is valid in neutral geometry and both constructed a mapping from the sphere to the non-Euclidean plane to derive the formulas of non-Euclidean trigonometry (including the formula Taurinus discovered — see Theorem 10.4, Chapter 10, for a simpler derivation using a plane model). Both had a constant in their formulas that they could not explain; the later work of Riemann showed it to be the *curvature* of the non-Euclidean plane.

SUBSEQUENT DEVELOPMENTS

It was not until after Gauss' death in 1855, when his correspondence was published, that the mathematical world began to take non-Euclidean ideas seriously. (Yet, as late as 1888 Lewis Carroll was poking fun at non-Euclidean geometry.) Some of the best mathematicians (Beltrami, Klein, Poincaré, and Riemann) took up the subject, extending it, clarifying it, and applying it to other branches of mathematics, notably complex function theory. In 1868 the Italian mathematician Beltrami settled once and for all the question of a proof for the parallel postulate. He proved that no proof was possible. He did this by exhibiting a Euclidean model of non-Euclidean geometry. (We will discuss his model in the next chapter.)

Bernhard Riemann, who was a student of Gauss, had the most profound insight into the geometry, not just the logic. In 1854, he built upon Gauss' discovery of the *intrinsic geometry* on a surface in Euclidean three-space. Riemann invented the concept of an abstract geometrical surface that need not be embeddable in Euclidean three-space yet on which the "lines" can be interpreted as geodesics and the intrinsic curvature of the surface can be precisely defined. Elliptic (and, of course, spherical) geometry "exist" on such surfaces that have constant positive curvature, while the hyperbolic geometry of Bolyai and Lobachevsky "exists" on such a *surface of constant negative curvature*. That is the view of geometers today about the "reality" of

Georg Friedrich Bernhard Riemann

those non-Euclidean planes. We will describe Gauss and Riemann's idea only in Appendix A, since it is too advanced for the level of this text. A further generalization of that idea provided the geometry for Einstein's general theory of relativity.

Interestingly, a direct relationship between the special theory of relativity and hyperbolic geometry was discovered by the physicist Arnold Sommerfeld in 1909 and elucidated by the geometer Vladimir Varičak in 1912. A model of hyperbolic plane geometry is a sphere of imaginary radius with antipodal points identified in the three-dimensional space-time of special relativity, vindicating Lambert's idea (see Rosenfeld, 1988, pp. 230 and 270; or Yaglom, 1979, p. 222 ff.). Moreover, Taurinus' technique of substituting ir for r to go from spherical trigonometry to hyperbolic trigonometry received a structural explanation in 1926–1927 when Élie Cartan developed his theory of Riemannian symmetric spaces: The Euclidean sphere of curvature $1/r^2$ is "dual" to the hyperbolic plane of curvature $-1/r^2$ (see Helgason, 1962, p. 206).

HYPERBOLIC GEOMETRY

Let us return to our elementary investigation of the particular non-Euclidean geometry discovered by Gauss, J. Bolyai, and Lobachevsky, nowadays called *hyperbolic geometry* (see Appendix A for a discussion of elliptic geometry and other geometries discovered by Riemann). *Hyperbolic geometry is, by definition, the geometry you get by assuming all the axioms for neutral geometry and replacing Hilbert's parallel postulate by its negation, which we shall call the "hyperbolic axiom."*

HYPERBOLIC AXIOM. In hyperbolic geometry there exist a line l and a point P not on l such that at least two distinct lines parallel to l pass through P (see Figure 6.1).

We can immediately see the flaw in Legendre's attempted proof of the parallel postulate (Chapter 1), namely that the entire line l lies in the interior of $\sphericalangle APB$ without meeting either side, a phenomenon Legendre assumed to be impossible.

The following lemma (preliminary result) is the first important consequence of the hyperbolic axiom.

LEMMA 6.1. Rectangles do not exist.

In fact, we saw in Chapter 5 that the existence of rectangles (Clairaut's axiom) implies Hilbert's parallel postulate, the negation of the hyperbolic axiom (the idea of the proof is due to Proclus).

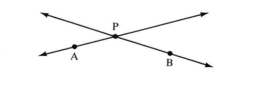

FIGURE 6.1

Using this lemma, we can establish a universal version of the hyperbolic axiom. The parallel postulate in Euclidean geometry states that for every line and for every point off the line, uniqueness of parallels holds. Its negation, the hyperbolic axiom, states that for *some* line *l* and *some* point P not on *l*, uniqueness of parallels *fails* to hold. Could it be possible that in hyperbolic geometry uniqueness of parallels fails for some *l* and P but *holds for other l* and P? We will show that this is impossible.

UNIVERSAL HYPERBOLIC THEOREM.   In hyperbolic geometry, for every line *l* and every point P not on *l* there pass through P at least two distinct parallels to *l*.

Proof:
Drop perpendicular $\overleftrightarrow{PQ}$ to *l* and erect line *m* through P perpendicular to $\overleftrightarrow{PQ}$. Let R be another point on *l*, erect perpendicular *t* to *l* through R, and drop perpendicular $\overleftrightarrow{PS}$ to *t*. (See Figure 6.2.) Now $\overleftrightarrow{PS}$ is parallel to *l*, since they are both perpendicular to *t* (Corollary 1 to Theorem 4.1). We claim that *m* and $\overleftrightarrow{PS}$ are distinct lines. Assume on the contrary that S lies on *m*. Then □PQRS is a rectangle. This contradicts Lemma 6.1. ∎

COROLLARY.   In hyperbolic geometry, for every line *l* and every point P not on *l*, there are infinitely many parallels to *l* through P.

Proof:
Just vary the point R in the above proof. ∎

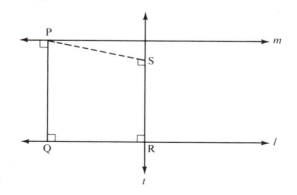

FIGURE 6.2

ANGLE SUMS (AGAIN)

Combining Lemma 6.1 with Theorem 4.7 yields the next theorem.

THEOREM 6.1. In hyperbolic geometry, all triangles have angle sum
< 180°. *not all same though*

If △ABC is any triangle, then 180° minus the angle sum of △ABC is
a *positive* number. This number we called the *defect* of the triangle, and
it plays a very important role in hyperbolic geometry (see Exercise 5
and Chapter 10).

COROLLARY. In hyperbolic geometry all convex quadrilaterals have
angle sum less than 360°.

Proof:
Given any quadrilateral □ABCD (Figure 6.3). Take diagonal AC
and consider triangles △ABC and △ACD; by the theorem, these
triangles have angle sum < 180°. The assumption that □ABCD is
convex implies that $\overrightarrow{AC}$ is between $\overrightarrow{AB}$ and $\overrightarrow{AD}$ and that $\overrightarrow{CA}$ is
between $\overrightarrow{CB}$ and $\overrightarrow{CD}$, so that $(\sphericalangle BAC)° + (\sphericalangle CAD)° = (\sphericalangle BAD)°$
and $(\sphericalangle ACB)° + (\sphericalangle ACD)° = (\sphericalangle BCD)°$ (by Theorem 4.3(3)). By
adding all six angles, we see that the angle sum of □ABCD is
< 360°. ■

SIMILAR TRIANGLES

Next we shall consider Wallis' postulate, which cannot hold in hyperbolic geometry because, as we saw in Chapter 5, it implies the Euclid-

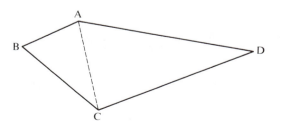

FIGURE 6.3

ean parallel postulate. Thus, under certain circumstances similar triangles do not exist (negation of Wallis' postulate). But we can prove even more: under *no* circumstances do similar noncongruent triangles exist!

THEOREM 6.2. In hyperbolic geometry if two triangles are similar, they are congruent. (In other words, AAA is a valid criterion for congruence of triangles.)

Proof:
Assume on the contrary that there exist triangles △ABC and △A'B'C' which are similar but not congruent. Then no corresponding sides are congruent; otherwise the triangles would be congruent (ASA). Consider the triples (AB, AC, BC) and (A'B', A'C', B'C') of sides of these triangles. One of these triples must contain at least two segments that are larger than the two corresponding segments of the other triple, e.g., AB > A'B' and AC > A'C'. Then (by definition of >) there exist points B″ on AB and C″ on AC such that AB″ ≅ A'B' and AC″ ≅ A'C' (see Figure 6.4). By SAS, △A'B'C' ≅ △AB″C″. Hence, corresponding angles are congruent: ⊰AB″C″ ≅ ⊰B', ⊰AC″B″ ≅ ⊰C'. By the hypothesis that △ABC and △A'B'C' are similar, we also have ⊰AB″C″ ≅ ⊰B, ⊰AC″B″ ≅ ⊰C (Axiom C-5). This implies that B⃡C ∥ B⃡″C″ (Theorem 4.1 and Exercise 32, Chapter 4), so that quadrilateral ☐BB″C″C is convex. Also, (⊰B)° + (⊰BB″C″)° = 180° = (⊰C)° + (⊰CC″B″)° (Theorem 4.3(2) and 4.3(5)). It follows that quadrilateral ☐BB″C″C has angle sum 360°. This contradicts the corollary to Theorem 6.1. ■

To sum up, in hyperbolic geometry it is impossible to magnify or shrink a triangle without distortion. In a hyperbolic world photography would be inherently surrealistic!

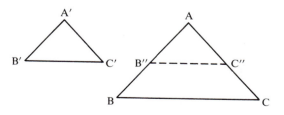

FIGURE 6.4

A startling consequence of Theorem 6.2 is that in hyperbolic geometry a segment can be determined with the aid of an angle; for example, an angle of an equilateral triangle determines the length of a side uniquely. This is sometimes stated more dramatically by saying that hyperbolic geometry has an *absolute unit of length* (see Gauss' letter to Taurinus, quoted earlier in this chapter). If the geometry of the physical universe were hyperbolic, it would no longer be necessary to keep a unit of length carefully guarded in the Bureau of Standards (the same is true for elliptic geometry).

PARALLELS THAT ADMIT A COMMON PERPENDICULAR

In Chapter 5, commenting on the flaw in Proclus' attempted proof of the parallel postulate, I remarked that it was presumptuous to assume that parallel lines looked like railroad tracks, that is, that they were everywhere equidistant from each other. Let us now make this remark more precise. Given lines l and l' and points A, B, C, . . . on l. Drop perpendiculars AA', BB', CC', . . . from these points to l'. We will say that points A, B, C, . . . are *equidistant* from l' if all these perpendicular segments are congruent to one another (Figure 6.5).

THEOREM 6.3.   In hyperbolic geometry if l and l' are any distinct parallel lines, then any set of points on l equidistant from l' has at most two points in it.

Proof:
Assume on the contrary there is a set of three points A, B, and C on l equidistant from l'. Then quadrilaterals □A'B'BA, □A'C'CA,

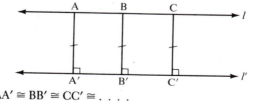

FIGURE 6.5   AA' ≅ BB' ≅ CC' ≅ . . . .

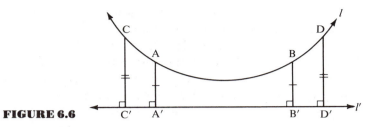

FIGURE 6.6

and □B'C'CB are Saccheri quadrilaterals (the base angles are right angles and the sides are congruent). In Exercise 1, Chapter 5, you showed that the summit angles of a Saccheri quadrilateral are congruent. Thus, ⊀A'AB ≅ ⊀B'BA, ⊀A'AC ≅ ⊀C'CA, and ⊀B'BC ≅ ⊀C'CB. By transitivity (Axiom C-5), it follows that the supplementary angles ⊀B'BA and ⊀B'BC are congruent to each other; hence, by definition, they are right angles. Therefore, these Saccheri quadrilaterals are all rectangles. But rectangles do not exist in hyperbolic geometry (Lemma 6.1). This contradiction shows that A, B, and C cannot be equidistant from *l'*. ∎

The theorem states that *at most* two points at a time on *l* can be equidistant from *l'*. It allows the possibility that there are pairs of points (A, B), (C, D), . . . on *l* such that each pair is equidistant from *l'* — thus, dropping perpendiculars, AA' ≅ BB' and CC' ≅ DD', but AA' is not congruent to CC'. A diagram for this might be Figure 6.6, which suggests that the point "in the middle" of *l* is closest to *l'*, with *l* moving away from *l'* symmetrically on either side of this middle point. We will prove that this is indeed the case (Theorems 6.4 and 6.5 and Exercises 4 and 10).

Note, however, that Theorem 6.3 allows another possibility, that there is no pair of points on *l* equidistant from *l'*! An diagram for this might be Figure 6.7, where the points on *l* are at varying distances from *l'*; *l* moves away from *l'* in one direction and approaches *l'* in the other direction without ever meeting it. Thus, different pairs of parallel lines need not resemble each other — some may look like the first diagram, some like the second.

THEOREM 6.4. In hyperbolic geometry if *l* and *l'* are parallel lines for which there exists a pair of points A and B on *l* equidistant from *l'*, then

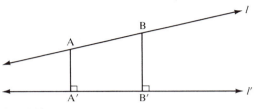

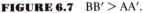

FIGURE 6.7 BB' > AA'.

l and *l'* have a common perpendicular segment that is also the shortest segment between *l* and *l'*.

Proof:
Suppose A and B on *l* are equidistant from *l'*. Then □A'B'BA is a Saccheri quadrilateral, where A' and B' are the feet on *l'* of the perpendiculars from A and B. Let M be the midpoint of AB and M' the midpoint of A'B' (Proposition 4.3; see Figure 6.8). The theorem will follow from the next lemma. ■

LEMMA 6.2. The segment joining the midpoints of the base and summit of a Saccheri quadrilateral is perpendicular to both the base and the summit, and this segment is shorter than the sides (see Figure 6.8).

Proof:
We know that ∢A ≅ ∢B (Exercise 1, Chapter 5). Hence, △A'AM ≅ △B'BM (SAS). Therefore, the corresponding sides A'M and B'M are congruent. This implies △A'M'M ≅ △B'M'M (SSS). Therefore, the corresponding angles ∢A'M'M and ∢B'M'M are congruent. Since these are supplementary angles, they must be right angles, proving MM' perpendicular to the base

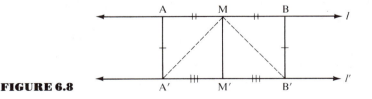

FIGURE 6.8

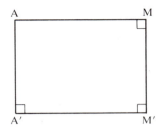

FIGURE 6.9

A′B′. From the two pairs of congruent triangles, we also have ∢A′MM′ ≅ ∢B′MM′ and ∢A′MA ≅ ∢B′MB. Adding the degrees of these angles, we have (∢AMM′)° = (∢BMM′)° (Theorem 4.3(3)), i.e., the supplementary angles ∢AMM′ and ∢BMM′ have the same number of degrees. Hence, they are right angles and MM′ is also perpendicular to the summit AB.

Consider next quadrilateral □A′M′MA (Figure 6.9). It has three right angles, so it is what we call a Lambert quadrilateral (Exercise 4, Chapter 5). In hyperbolic geometry the fourth angle must be acute, since rectangles do not exist (Lemma 6.1). You showed in Exercise 4(c), Chapter 5, that AA′ > MM′, i.e., that MM′ is shorter than AA′. The remainder of the proof that MM′ is shorter than any other segment between *l* and *l*′ is left for Exercise 3. ■

THEOREM 6.5. In hyperbolic geometry if lines *l* and *l*′ have a common perpendicular segment MM′, then they are parallel and MM′ is unique. Moreover, if A and B are any points on *l* such that M is the midpoint of segment AB, then A and B are equidistant from *l*′.

Proof:
The fact that *l* and *l*′ are parallel follows from the first corollary to the alternate interior angle theorem (Theorem 4.1). If *l* and *l*′ had another common perpendicular segment NN′, then □M′N′NM would be a rectangle, which cannot exist (Lemma 6.1). Suppose now that M is the midpoint of AB. Drop perpendiculars AA′ and BB′ to *l*. We must prove that AA′ ≅ BB′. (See Figure 6.10.) First, △AM′M ≅ △BM′M (SAS), AM′ ≅ BM′, and ∢AM′M ≅ ∢BM′M. Therefore, (∢A′M′A)° = 90° − (∢AM′M)° = 90° −

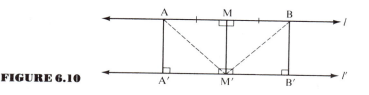

FIGURE 6.10

$(\sphericalangle BM'M)° = (\sphericalangle B'M'B)°$ (by Theorem 4.3) so that $\sphericalangle A'M'A \cong \sphericalangle B'M'B$. Hence, $\triangle AA'M' \cong \triangle BB'M'$ (AAS), so that the corresponding sides AA' and BB' are congruent. ■

LIMITING PARALLEL RAYS

Theorems 6.4 and 6.5 and Exercises 4 and 10 give us a good under-standing of the first type of parallel lines. We know that such lines actually exist from the usual construction: start with any line *l* and any point P not on it (Figure 6.11). Drop perpendicular $\overrightarrow{PQ}$ to *l* and let *m* be the perpendicular through P to $\overleftrightarrow{PQ}$. Then *m* and *l* have the common perpendicular segment PQ. Pairs of points on *m* situated symmetrically about $\overleftrightarrow{PQ}$ are equidistant from *l*. By the universal hyper-bolic theorem, there exist other lines *n* through P parallel to *l*. We can-not yet say that any such *n* is the second type of parallel, for *n* and *l* might have a common perpendicular going through a point other than P.

How then do we know that parallels of the second type exist? Here the axioms of continuity come in. The following is the intuitive idea (see Figure 6.12). Consider one ray $\overrightarrow{PS}$ of *m*, and consider various rays between $\overrightarrow{PS}$ and $\overrightarrow{PQ}$. Some of these rays, such as $\overrightarrow{PR}$, will intersect *l*; others, such as $\overrightarrow{PY}$, will not. A continuity argument shows that as R recedes endlessly on *l* from Q, $\overrightarrow{PR}$ will approach a certain limiting ray

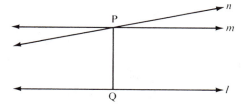

FIGURE 6.11

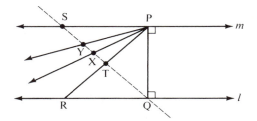

FIGURE 6.12

$\overrightarrow{PX}$ that does *not* meet *l*. The ray $\overrightarrow{PX}$ is "limiting" in the following precise sense: any ray between $\overrightarrow{PX}$ and $\overrightarrow{PQ}$ intersects *l*, whereas any other ray $\overrightarrow{PY}$ such that $\overrightarrow{PX}$ is between $\overrightarrow{PY}$ and $\overrightarrow{PQ}$ does not intersect *l*. The ray $\overrightarrow{PX}$ may be called the *left limiting parallel ray* to *l* through P. Similarly, there is a *right limiting parallel ray* on the opposite side of $\overrightarrow{PQ}$.

THEOREM 6.6. For every line *l* and every point P not on *l*, let Q be the foot of the perpendicular from P to *l*. Then there are two unique nonopposite rays $\overrightarrow{PX}$ and $\overrightarrow{PX'}$ on opposite sides of $\overleftrightarrow{PQ}$ that do not meet *l* and have the property that a ray emanating from P meets *l* if and only if it is between $\overrightarrow{PX}$ and $\overrightarrow{PX'}$. Moreover, these limiting rays are situated symmetrically about $\overleftrightarrow{PQ}$ in the sense that $\sphericalangle XPQ \cong \sphericalangle X'PQ$.

Proof:
To prove rigorously that $\overrightarrow{PX}$ exists, consider the line $\overleftrightarrow{SQ}$ (Figure 6.12). Let Σ_1 be the set of all points T on segment SQ such that $\overrightarrow{PT}$ meets *l*, together with all points on the ray opposite to $\overrightarrow{QS}$; let Σ_2 be the complement of Σ_1 (so $Q \in \Sigma_1$ and $S \in \Sigma_2$). By the crossbar theorem (Chapter 3), if point T on segment SQ belongs to Σ_1, then the entire segment TQ (in fact, $\overrightarrow{TQ}$) is contained in Σ_1. Hence, (Σ_1, Σ_2) is a Dedekind cut. By Dedekind's axiom (Chapter 3), there is a unique point X on $\overleftrightarrow{SQ}$ such that for P_1 and P_2 on $\overleftrightarrow{SQ}$, $P_1 * X * P_2$ if and only if $X \neq P_1$ or P_2, $P_1 \in \Sigma_1$, and $P_2 \in \Sigma_2$.

By definition of Σ_1 and Σ_2, rays below $\overrightarrow{PX}$ all meet *l* and rays above $\overrightarrow{PX}$ do not. We claim that $\overrightarrow{PX}$ does not meet *l* either. Assume on the contrary that $\overrightarrow{PX}$ meets *l* in a point U (Figure 6.13). Choose any point V on *l* to the left of U, i.e., $V * U * Q$ (Axiom B-2). Since V and U are on the same side of $\overleftrightarrow{SQ}$ (Exercise 9, Chapter 3), V and P are on opposite sides, so VP meets SQ in a point Y. We have

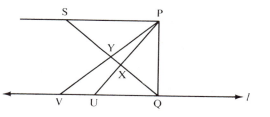

FIGURE 6.13

$Y * X * Q$ (Proposition 3.7), so $Y \in \Sigma_2$, contradicting the fact that $\overrightarrow{PY}$ meets l. It follows that $\overrightarrow{PX}$ is the left limiting parallel ray (we obtain the right limiting parallel ray in a similar manner).

To prove symmetry, assume on the contrary that angles ⊀XPQ and ⊀X'PQ are not congruent, e.g., $(\sphericalangle XPQ)° < (\sphericalangle X'PQ)°$. By Axiom C-4, there is a ray between $\overrightarrow{PX'}$ and $\overrightarrow{PQ}$ that intersects l (by definition of limiting ray) in a point R' such that ⊀R'PQ ≅ ⊀XPQ (Figure 6.14). Let R be the point on the opposite side of $\overrightarrow{PQ}$ from R' such that $R * Q * R'$ and $RQ \cong R'Q$ (Axiom C-1). Then $\triangle RPQ \cong \triangle R'PQ$ (SAS). Hence, ⊀RPQ ≅ ⊀R'PQ, and by transitivity (Axiom C-5), ⊀RPQ ≅ ⊀XPQ. But this is impossible, because $\overrightarrow{PR}$ is between $\overrightarrow{PX}$ and $\overrightarrow{PQ}$ (Axiom C-4). ■

Either of the congruent angles ⊀XPQ and ⊀X'PQ is called (by an abuse of language) the *angle of parallelism* at point P with respect to l. Its degree measure is usually denoted $\Pi(PQ)°$. Note that $\Pi(PQ)° < 90°$, for $\Pi(PQ)° = 90°$ would contradict the universal hyperbolic theorem (see Exercise 7(a)). It can be shown that as P varies, $\Pi(PQ)°$ takes on all possible values between 0° and 90° (see Major Exercise 9). One of the greatest discoveries by J. Bolyai and Lobachevsky is their formula for this number of degrees (see

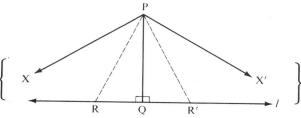

FIGURE 6.14

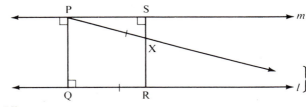

FIGURE 6.15

Theorem 10.2 in Chapter 10). A *natural unit segment* OI in hyperbolic geometry is any segment OI such that $\Pi(OI)° = 45°$. Major Exercise 5 shows that all such segments are congruent to each other.

We have proved the existence of limiting parallel rays by a continuity argument conforming to today's standards of rigor. Gauss, J. Bolyai, and Lobachevsky took this existence for granted, but J. Bolyai discovered a simple *straightedge-and-compass construction* for the limiting rays. Let Q be the foot of the perpendicular from P to *l*, *m* the line through P perpendicular to $\overleftrightarrow{PQ}$, R any point on *l* different from Q, and S the foot of the perpendicular from R to *m* (Figure 6.15). Then PR > QR (Exercise 3) and PS < QR (Exercise 4(c), Chapter 5, on Lambert quadrilaterals). By the elementary continuity principle, a compass with center P and radius congruent to QR will intersect segment SR in a unique point X between S and R. It can be proved that $\overrightarrow{PX}$ is the right limiting parallel ray to *l* through P! (The proof is complicated. See p. 269 in Chapter 7, Project 4 in this chapter, or Theorem 10.9, Chapter 10.)

CLASSIFICATION OF PARALLELS

We have discussed two types of parallels to a given *l*. The first type consists of parallels *m* such that *l* and *m* have a common perpendicular; *m* diverges from *l* on both sides of the common perpendicular. The second type consists of parallels that approach *l* asymptotically in one direction (i.e., they contain a limiting parallel ray in that direction) and which diverge from *l* in the other direction. If *m* is the second type of parallel, Exercises 7 and 8 show that *l* and *m* do not have a common perpendicular. We have implied that these two are the only types of parallels, and this is the content of the next theorem.

THEOREM 6.7. Given m parallel to l such that m does not contain a limiting parallel ray to l in either direction. Then there exists a common perpendicular to m and l (which is unique by Theorem 6.5).

This theorem is proved by Borsuk and Szmielew (1960, p. 291) by a continuity argument, but their proof gives you no idea of how to actually find the common perpendicular. There is an easy way to find it in the Klein and Poincaré models, discussed in the next chapter. Hilbert gave a direct construction, which we will sketch. (Project 1 gives another.)

Proof:
Hilbert's idea is to find two points H and K on l that are equidistant from m, for once these are found, the perpendicular bisector of segment HK is also perpendicular to m (see Lemma 6.2). Choose any two points A and B on l and suppose that the perpendicular segment AA′ from A to m is longer than the perpendicular segment BB′ from B to m. (See Figure 6.16.) Let E be the point between A′ and A such that A′E ≅ B′B. On the same side of $\overleftrightarrow{AA'}$ as B, let $\overrightarrow{EF}$ be the unique ray such that ∢A′EF ≅ ∢B′BG, where A * B * G. The key point that will be proved in Major Exercises 2–6 is that $\overrightarrow{EF}$ intersects $\overrightarrow{AG}$ in a point H. Let K be the unique point on $\overrightarrow{BG}$ such that EH ≅ BK. Drop perpendiculars $\overleftrightarrow{HH'}$ and $\overleftrightarrow{KK'}$ to m. The upshot of these constructions is that □EHH′A′ is congruent to □BKK′B′ (just divide them into triangles). Hence, the corresponding sides HH′ and KK′ are congruent, so that the points H and K on l are equidistant from m, as required. ∎

To sum up, given a point P not on l, there exist exactly two limiting parallel rays to l through P, one in each direction. There are infinitely many lines through P that do not enter the region between the limiting

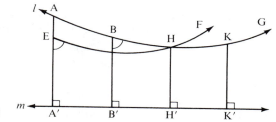

FIGURE 6.16

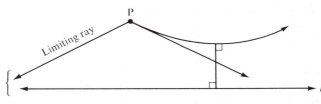

FIGURE 6.17

rays and *l*. Each such line is divergently parallel to *l* and admits a unique common perpendicular with *l* (for one of these lines the common perpendicular will go through P, but for all the rest the common perpendicular will pass through other points).

A Note on Terminology. In most books on hyperbolic geometry the word "parallel" is used only for lines that contain limiting parallel rays in our sense. The other lines, which admit a common perpendicular, have various names in the literature: "nonintersecting," "ultraparallel," "hyperparallel," and "superparallel." We will continue to use the word "parallel" to mean "nonintersecting." A parallel to *l* that contains a limiting parallel ray to *l* will be called an *asymptotic parallel*, and a parallel to *l* that admits a common perpendicular to *l* will be called a *divergently parallel line*. Rays that are limiting parallel will be denoted by a brace in diagrams (see Figure 6.17).

STRANGE NEW UNIVERSE?

In this chapter we have only begun to investigate the "strange new universe" of hyperbolic geometry. You can develop more of this geometry by doing the exercises, reading Chapter 10, and examining works in the bibliography at the end of the book. You will encounter new entities such as asymptotic triangles, ideal and ultra-ideal points, equidistant curves, horocycles, and pseudospheres.

If you consider this geometry too "far out" to pursue, you are in for a surprise. We will see in the next chapter that if the undefined terms of hyperbolic geometry are suitably interpreted, hyperbolic geometry can be considered a part of Euclidean geometry!

Meanwhile, notice how we have deepened our understanding of the role of the parallel postulate P in Euclidean geometry. Any statement S in the language of our geometry that is a theorem in Euclidean geometry $(P \Rightarrow S)$ and whose negation is a theorem in hyperbolic geometry $(\sim P \Rightarrow \sim S)$ is equivalent (in neutral geometry) to the parallel postulate (by Logic Rule 9(c)). For example, by Exercise 14, Chapter 5, "The angle sum of every triangle is 180°" is such a statement. By Exercise 12 in this chapter, "Every triangle has a circumscribed circle" is another such statement. By Theorem 5.1, "Every point interior to an acute angle lies on a line intersecting both sides of the angle in two distinct points" is a third such statement. I leave the enjoyment of providing a long list of such statements to you (Exercise 15). I urge your instructor to give a prize to the student(s) with the longest list.

REVIEW EXERCISE

Which of the following statements are correct?

(1) The negation of Hilbert's parallel postulate states that for every line l and every point P not on l there exist at least two lines through P parallel to l.

(2) It is a theorem in neutral geometry that if lines l and m meet on a given side of a transversal t, then the sum of the degrees of the interior angles on that given side of t is less than 180°.

(3) Gauss began working on non-Euclidean geometry when he was 15 years old.

(4) The philosopher Kant taught that our minds could not conceive of any geometry other than Euclidean geometry.

(5) The first mathematician to publish an account of hyperbolic geometry was the Russian Lobachevsky.

(6) The crossbar theorem asserts that a ray emanating from a vertex A of $\triangle$ABC and interior to $\sphericalangle$A must intersect the opposite side BC of the triangle.

(7) It is a theorem in hyperbolic geometry that for any segment AB there exists a square having AB as one of its sides.

(8) Every Saccheri quadrilateral is a convex quadrilateral.

(9) In hyperbolic geometry if $\triangle$ABC and $\triangle$DEF are equilateral triangles and $\sphericalangle$A $\cong$ $\sphericalangle$D, then the triangles are congruent.

(10) In hyperbolic geometry, given a line l and a fixed segment AB, the set of all points on a given side of l whose perpendicular segment to l is congruent to AB equals the set of points on a line parallel to l.

(11) In hyperbolic geometry any two parallel lines have a common perpendicular.

(12) In hyperbolic geometry the fourth angle of a Lambert quadrilateral is obtuse.

(13) In hyperbolic geometry some triangles have angle sum less than $180°$ and some triangles have angle sum equal to $180°$.

(14) In hyperbolic geometry if point P is not on line l and Q is the foot of the perpendicular from P to l, then the angle of parallelism for P with respect to l is the angle that a limiting parallel ray to l emanating from P makes with $\overrightarrow{PQ}$.

(15) J. Bolyai showed how to construct limiting parallel rays using the elementary continuity principle instead of Dedekind's axiom.

(16) In hyperbolic geometry if $l \parallel m$, then there exist three points on m that are equidistant from l.

(17) In hyperbolic geometry if m is any line parallel to l, then there exist two points on m which are equidistant from l.

(18) In hyperbolic geometry if P is a point not lying on line l, then there are exactly two lines through P parallel to l.

(19) In hyperbolic geometry if P is a point not lying on line l, then there are exactly two lines through P perpendicular to l.

(20) In hyperbolic geometry if $l \parallel m$ and $m \parallel n$, then $l \parallel n$ (transitivity of parallelism).

(21) In hyperbolic geometry if m contains a limiting parallel ray to l, then l and m have a common perpendicular.

(22) In hyperbolic geometry if l and m have a common perpendicular, then there is one point on m that is closer to l than any other point on m.

(23) In hyperbolic geometry if m does not contain a limiting parallel ray to l and if m and l have no common perpendicular, then m intersects l.

(24) In hyperbolic geometry the summit angles of a Saccheri quadrilateral are right angles.

(25) Every valid theorem of neutral geometry is also valid in hyperbolic geometry.

(26) In hyperbolic geometry opposite angles of any parallelogram are congruent to each other.

(27) In hyperbolic geometry opposite sides of any parallelogram are congruent to each other.

(28) In hyperbolic geometry, let $\sphericalangle$P be any acute angle, let X be any point on one side of this angle, and let Y be the foot of the perpendicular from X to the other side. If X recedes endlessly from P along its side, then Y will recede endlessly from P along its side.

(29) In hyperbolic geometry if three points are not collinear, there is always a circle that passes through them.
(30) In hyperbolic geometry there exists an angle and there exists a line that lies entirely within the interior of this angle.

EXERCISES

The following are exercises in hyperbolic geometry. You are to assume the hyberbolic axiom and you can use the theorems presented in this chapter as well as any theorems of neutral geometry. However, do not use the Euclidean theorems stated in either Exercises 18–26 or the Major Exercises of Chapter 5. (We can now assert that the theorems of neutral geometry are exactly those statements that are valid in *both* hyperbolic geometry and Euclidean geometry.)

1. Prove that if □A′B′BA is a Saccheri quadrilateral (∢A′ and ∢B′ are right angles and AA′ ≅ BB′), then the summit AB is greater than the base A′B′. (Hint: Join the midpoints M and M′ and apply Exercise 4, Chapter 5, to the Lambert quadrilaterals □A′M′MA and □M′B′BM.)
2. Suppose that lines *l* and *l′* have a common perpendicular MM′. Let A and B be points on *l* such that M is *not* the midpoint of segment AB. Prove that A and B are not equidistant from *l′*.
3. Assume that the parallel lines *l* and *l′* have a common perpendicular segment MM′. Prove that MM′ is the shortest segment between any point of *l* and any point of *l′*. (Hint: In showing MM′ < AA′; first dispose of the case in which AA′ is perpendicular to *l′* by means of Exercise 4, Chapter 5, and take care of the other case by Exercise 27, Chapter 4.)
4. Again, assume that MM′ is the common perpendicular segment between *l* and *l′*. Let A and B be any points of *l* such that M * A * B, and drop perpendiculars AA′ and BB′ to *l′*. Prove that AA′ < BB′. (Hint: Use Exercise 3, Chapter 5; see Figure 6.18.)

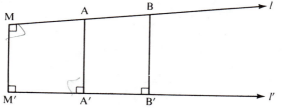

FIGURE 6.18

5. We have seen that in hyperbolic geometry the defect of any triangle is positive (Theorem 6.1). In Euclidean geometry all triangles have the same defect, namely, zero. In hyperbolic geometry could all triangles have the same defect? Assume that they do and use the additivity of the defect (Theorem 4.6) to derive a contradiction. Is there an upper bound for the defect of a triangle?

6. Given parallel lines l and m. Given points A and B that lie on the opposite side of m from l; i.e., for any point P on l, A and P are on opposite sides of m and B and P are on opposite sides of m. Prove that A and B lie on the same side of l.

7. (a) Prove that the angle of parallelism is acute by showing precisely how "$\Pi(PQ)° = 90°$" implies that there is a unique parallel to l through P, contradicting the universal hyperbolic theorem.

   (b) Let $\overrightarrow{PY}$ be a limiting parallel ray to l through P, and let X be a point on this ray between P and Y (Figure 6.19). It may seem intuitively obvious that $\overrightarrow{XY}$ is a limiting parallel ray to l *through* X, but this requires proof. Justify the steps that have not been justified.

Proof:
(1) We must prove that any ray $\overrightarrow{XS}$ between $\overrightarrow{XY}$ and $\overrightarrow{XR}$ meets l, where R is the foot of the perpendicular from X to l. (2) S and Y are on the same side of $\overleftrightarrow{XR}$. (3) P and Y are on opposite sides of $\overleftrightarrow{XR}$. (4) By Exercise 6, S and Y are on the same side of $\overleftrightarrow{PQ}$. (5) S and R are on the same side of $\overleftrightarrow{XY} = \overleftrightarrow{PY}$. (6) Q and R are on the same side of $\overleftrightarrow{PY}$. (7) Q and S are on the same side of $\overleftrightarrow{PY}$. (8) Thus, $\overrightarrow{PS}$ lies between $\overrightarrow{PY}$ and $\overrightarrow{PQ}$, so it intersects l in a point T. (9) Point X is exterior to $\triangle PQT$. (10) $\overrightarrow{XS}$ does not intersect PQ. (11) Hence $\overrightarrow{XS}$ intersects QT (Proposition 3.9(a)), so $\overrightarrow{XS}$ meets l. ■

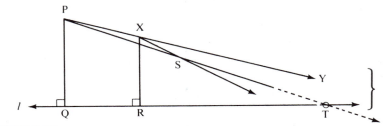

FIGURE 6.19

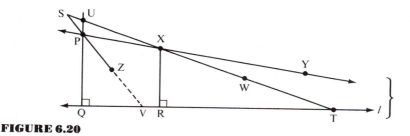

FIGURE 6.20

(c) Let us assume instead that $\overrightarrow{XY}$ is limiting parallel to l, with P * X * Y. Prove that $\overrightarrow{PY}$ is limiting parallel to l. (Hint: See Figure 6.20. You must show that $\overrightarrow{PZ}$ meets l in a point V. Choose any S such that S * P * Z. Show that SX meets $\overleftrightarrow{PQ}$ in a point U such that U * P * Q. Choose any W such that U * X * W, and show that $\overrightarrow{XW}$ is between $\overrightarrow{XY}$ and $\overrightarrow{XR}$ so that $\overrightarrow{XW}$ meets l in a point T. Apply Pasch's theorem to get V.)

8. Let $\overrightarrow{PX}$ be the right limiting parallel ray to l through P, and let Q and X′ be the feet of the perpendiculars from P and X, respectively, to l (Figure 6.21). Prove that PQ > XX′. (Hint: Use Exercise 7 to show that ⦡X′XY is acute and that ⦡X′XP is obtuse, so that Exercise 3, Chapter 5, can be applied to □PQX′X.) This exercise shows that the distance from X to l decreases as X recedes from P along a limiting parallel ray. In fact, one can prove that the distance from X to l approaches zero (see Major Exercise 11).

9. Let △ABC be any triangle, and let L, M, and N be the midpoints of BC, AB, and AC, respectively. Prove that △AMN is *not* similar to △ABC. (See Figure 6.22.) (Hint: Otherwise defect □MBCN = 0.) Prove that MN is *not* congruent to BL by assuming the contrary and deducing that △ABC has angle sum 180°. (Hint: Choose D such that M * N * D and ND ≅ MN. Show that △ANM ≅ △CND, then that △MDC ≅ △CBM. Substitute appropriately in the equation 180° = (⦡BMC)° + (⦡CMD)° + (⦡AMN)° to get the result.)

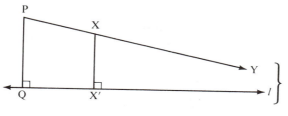

FIGURE 6.21

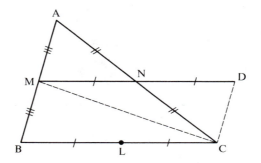

FIGURE 6.22

10. Assume that the parallel lines l and l' have a common perpendicular $\overleftrightarrow{PQ}$. For any point X on l, let X' be the foot of the perpendicular from X to l'. Prove that as X recedes endlessly from P on l, the segment XX' increases indefinitely; see Figure 6.23. (Hint: We saw that it increases in Exercise 4. Drop a perpendicular XY to the limiting parallel ray between $\overrightarrow{PX}$ and $\overrightarrow{PX'}$. Use the crossbar theorem to show that $\overrightarrow{PY}$ intersects XX' in a point Z. Use Proposition 4.5 to show that $XZ \geq XY$. Conclude by applying Exercise 6, Chapter 5, to show that XY increases indefinitely as X recedes from P.)

11. This problem has five parts. In the first part we will construct a Saccheri quadrilateral associated to any triangle; we will then apply this construction.

    (a) Given $\triangle ABC$, let I, J, and K be the midpoints of BC, CA, and AB, respectively. Drop perpendiculars AD, BE, and CF from the vertices to $\overleftrightarrow{IJ}$. Prove that $AD \cong CF \cong BE$, and, hence, that $\square EDAB$ is a Saccheri quadrilateral (with the same area as $\triangle ABC$). (See Figure 6.24.)

    (b) Prove that the perpendicular bisector of AB (i.e., the perpendicular through K) is also perpendicular to $\overleftrightarrow{IJ}$, and, hence, that $\overleftrightarrow{IJ}$ is divergently parallel to $\overleftrightarrow{AB}$.

    (c) Recall that we denote the length of a segment by a bar; e.g., the

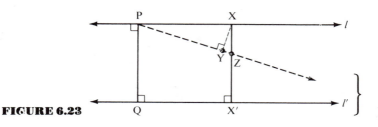

FIGURE 6.23

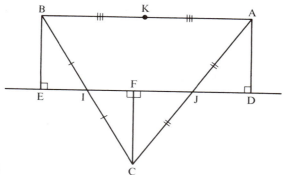

FIGURE 6.24

length of MN is $\overline{MN}$ (Theorem 4.3B). Prove that $\overline{IJ} = \frac{1}{2}\overline{ED}$ (a separate argument is needed in case ∢A or ∢B is obtuse and the diagram is different). Deduce that in hyperbolic geometry $\overline{IJ} < \frac{1}{2}\overline{AB}$.

(d) Suppose now that ∢C is a right angle. Prove that the Pythagorean theorem does not hold in hyperbolic geometry. (Hint: If the theorem were valid for right triangles △BCA and △ICJ, then $\overline{IJ} = \frac{1}{2}\overline{AB}$ could be proved, contradicting part (c) of this exercise.)

(e) Suppose instead that AC ≅ BC. Prove that K, F, and C are collinear but F is not the midpoint of CK (use Lemma 6.2 and part (a) of this exercise). For application of this result to mechanics, see Adler (1966), pp. 192 and 253–257.

12. In Exercise 9, Chapter 5, we saw the elder Bolyai's false proof of the parallel postulate. The flaw in his argument was the assumption that *every* triangle has a circumscribed circle, i.e., that there is a circle passing through the three vertices of the triangle. The idea of the Euclidean proof of this assumption is to show that the perpendicular bisectors of the sides of the triangle meet in a point, and that this point is the center of the circumscribed circle. Figure out how Euclid's fifth postulate is used to prove that two of the perpendicular bisectors *l* and *m* have a common point (use Proposition 4.10) and then argue by congruent triangles to prove that the third perpendicular bisector passes through that point and that the point is equidistant from the three vertices. (Hint: Join the common point D to the midpoint N of the third side, and prove that $\overleftrightarrow{DN}$ is perpendicular to the third side; see Figure 6.25.)

13. Part of the argument in Exercise 12 works for hyperbolic geometry; that is, *if* two of the perpendicular bisectors have a common point, then the third perpendicular bisector also passes through that point. In hyperbolic geometry there will be triangles for which two of the perpendicular

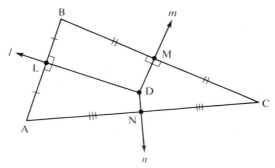

FIGURE 6.25

bisectors are parallel (otherwise the elder Bolyai's proof would be correct). Moreover, these perpendicular bisectors can be parallel in two different ways. Suppose that they are divergently parallel; that is, suppose that the perpendicular bisectors *l* and *m* have a common perpendicular *t* (see Figure 6.26). Prove that the third perpendicular bisector *n* is also perpendicular to *t*. (Hint: Let A′, B′, and C′ be the feet on *t* of the perpendiculars dropped from A, B, and C, respectively. Let *l* bisect AB at L and be perpendicular to *t* at L′, and let *m* bisect BC at M and be perpendicular to *t* at M′. Let N be the midpoint of AC. Show by Theorem 6.5 that AA′ ≅ BB′ and CC′ ≅ BB′. Hence, □C′A′AC is a Saccheri quadrilateral with N the midpoint of its summit AC. If N′ is the midpoint of the base A′C′, use Lemma 6.2 to show that $n = \overleftrightarrow{NN'}$ is

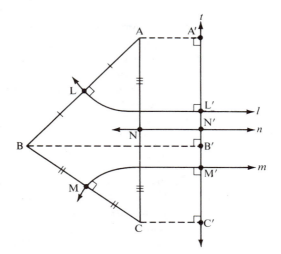

FIGURE 6.26

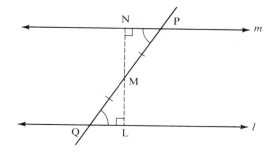

FIGURE 6.27

perpendicular to t and $\overleftrightarrow{AC}$; see Major Exercise 7 for the asymptotically parallel case.)

14. In Theorem 4.1 it was proved in neutral geometry that if alternate interior angles are congruent, then the lines are parallel. Strengthen this result in hyperbolic geometry by proving that the lines are divergently parallel, i.e., that they have a common perpendicular. (Hint: Let M be the midpoint of transversal segment PQ and drop perpendiculars MN and ML to lines m and l; see Figure 6.27. Prove that L, M, and N are collinear by the method of congruent triangles.)

15. *Make a long list of statements equivalent in neutral geometry to Hilbert's parallel postulate.* This list is a reward for all the work you have done.

16. Although the circumscribed circle may not exist for some triangles in hyperbolic geometry, prove that the inscribed circle always exists. (Hint: Verify that the usual Euclidean proof — that the angle bisectors meet in a point equidistant from the sides — still works. Use the crossbar theorem.)

17. Comment on the following injunction by Saint Augustine: "The good Christian should beware of mathematicians and all those who make empty prophesies. The danger already exists that the mathematicians have made a covenant with the devil to darken the spirit and to confine man in the bonds of Hell."

MAJOR EXERCISES

1. Let A, D be points on the same side of line $\overleftrightarrow{BC}$ such that $\overleftrightarrow{BA} \parallel \overleftrightarrow{CD}$. Then the figure consisting of segment BC (called the *base*) and rays $\overrightarrow{BA}$ and $\overrightarrow{CD}$ (called the *sides*) is called the *biangle* [ABCD with *vertices* B and C

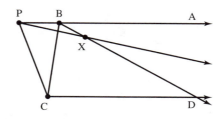

FIGURE 6.28

(see Figure 6.28). The *interior* of [ABCD is the intersection of the interiors of its *angles* ∢ABC and ∢DCB; if P lies in the interior and X is either vertex, ray $\overrightarrow{XP}$ is called an *interior ray*. We write $\overrightarrow{BA}\,|\,\overrightarrow{CD}$ when these rays are sides of a biangle and when every interior ray emanating from B intersects $\overrightarrow{CD}$; in that case, we say that $\overrightarrow{BA}$ is *limiting parallel* to $\overrightarrow{CD}$, generalizing the previous definition which required ∢DCB to be a right angle, and we say that the biangle [ABCD is *closed* at B. Given $\overrightarrow{BA}\,|\,\overrightarrow{CD}$, prove the following generalization of Exercise 7: If P * B * A or if B * P * A, then $\overrightarrow{PA}\,|\,\overrightarrow{CD}$.

2. *Symmetry of limiting parallelism.* If $\overrightarrow{BA}\,|\,\overrightarrow{CD}$, then $\overrightarrow{CD}\,|\,\overrightarrow{BA}$. (In that case we say simply that biangle [ABCD is *closed.*) Justify the unjustified steps in the proof (see Figure 6.29).

Proof:
(1) Assume that [ABCD is not closed at C. (2) Then some interior ray $\overrightarrow{CE}$ does not intersect $\overrightarrow{BA}$. (3) Point E, which so far is just a label, can be chosen so that ∢BEC < ∢ECD, by the important corollary to Aristotle's axiom, Chapter 3. (4) Segment BE does not intersect $\overrightarrow{CD}$. (5) Interior ray $\overrightarrow{BE}$ intersects $\overrightarrow{CD}$ in a point F, and B * E * F. (6) Since ∢BEC is an exterior angle for △EFC, ∢BEC > ∢ECF. (7) Contradiction. (I am indebted to George E. Martin for this simple proof.) ∎

3. *Transitivity of limiting parallelism.* If $\overrightarrow{AB}$ and $\overrightarrow{CD}$ are both limiting parallel to $\overrightarrow{EF}$, then they are limiting parallel to each other. Justify the steps in the proof. (See Figure 6.30.)

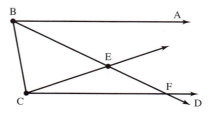

FIGURE 6.29

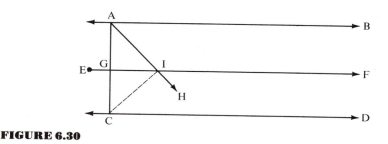

FIGURE 6.30

Proof:

(1) $\overleftrightarrow{AB}$ and $\overleftrightarrow{CD}$ have no point in common. (2) Hence, there are two cases, depending on whether $\overleftrightarrow{EF}$ is between $\overleftrightarrow{AB}$ and $\overleftrightarrow{CD}$ or $\overleftrightarrow{AB}$ and $\overleftrightarrow{CD}$ are both on the same side of $\overleftrightarrow{EF}$. (3) In case $\overleftrightarrow{EF}$ is between $\overleftrightarrow{AB}$ and $\overleftrightarrow{CD}$, let G be the intersection of AC with $\overleftrightarrow{EF}$. We may assume G lies on ray $\overrightarrow{EF}$; otherwise we can consider $\overrightarrow{GF}$. (4) Any ray $\overrightarrow{AH}$ interior to $\sphericalangle GAB$ must intersect $\overleftrightarrow{EF}$ in a point I. (5) $\overrightarrow{IH}$, lying interior to $\sphericalangle CIF$, must intersect $\overleftrightarrow{CD}$. (6) Hence, any ray $\overrightarrow{AH}$ interior to $\sphericalangle CAB$ must intersect $\overleftrightarrow{CD}$, so $\overrightarrow{AB}$ is limiting parallel to $\overleftrightarrow{CD}$. ∎

Step (7) is the following sublemma. That this requires such a long proof was overlooked even by Gauss. The proof (for which I am indebted to Edwin E. Moise) uses our hypotheses of limiting parallelism. If we had made the weaker hypothesis of just parallel lines, the sublemma would not follow, as you will show in Exercise K-2(c) of Chapter 7.

SUBLEMMA. If $\overleftrightarrow{AB}$ and $\overleftrightarrow{CD}$ are both on the same side of $\overleftrightarrow{EF}$, we may assume that $\overleftrightarrow{CD}$, for example, is between $\overleftrightarrow{AB}$ and $\overleftrightarrow{EF}$ (see Figure 6.31).

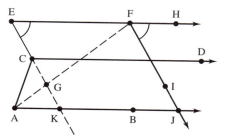

FIGURE 6.31

Proof of Sublemma:

(1) It suffices to prove there is a line transversal to the three rays $\overrightarrow{AB}$, $\overrightarrow{CD}$, $\overrightarrow{EF}$. (2) In case A and F are on the same side of $\overleftrightarrow{EC}$, then ray $\overrightarrow{EA}$ is interior to ⊀E. (3) Then $\overrightarrow{EA}$ intersects $\overleftrightarrow{CD}$, by symmetry. (4) So $\overrightarrow{EA}$ is our transversal. (5) In case A and F are on opposite sides of $\overleftrightarrow{EC}$, let G be the point at which AF meets $\overleftrightarrow{EC}$. (6) Choosing H such that E * F * H, we have $\overrightarrow{FH} \,|\, \overrightarrow{AB}$. (7) ⊀HFG > ⊀E. (8) Therefore there is a ray $\overrightarrow{FI}$ interior to ⊀HFA = ⊀HFG such that ⊀HFA ≅ ⊀E. (9) $\overrightarrow{FI}$ meets $\overrightarrow{AB}$ at a point J. (10) $\overleftrightarrow{FJ} \,\|\, \overleftrightarrow{EC}$. (11) $\overleftrightarrow{EC}$ intersects side AF and does not intersect side FJ of △AFJ. (12) Hence $\overleftrightarrow{EC}$ intersects AJ and is our transversal. ■

Conclusion of Proof (see Figure 6.32):

(8) Then AE intersects $\overleftrightarrow{CD}$ in a point G, which we may assume lies on ray $\overrightarrow{CD}$. (9) Any ray $\overrightarrow{AH}$ interior to ⊀GAB intersects $\overrightarrow{EF}$ in a point I. (10) Since $\overleftrightarrow{CD}$ enters △AEI at G and does not intersect side EI, it must intersect AI. (11) Therefore, $\overleftrightarrow{CD}$ is limiting parallel to $\overrightarrow{AB}$. ■

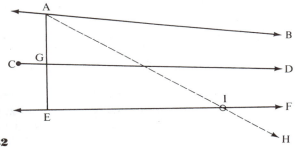

FIGURE 6.32

Note 1. The last four steps did not use the hypothesis that $\overrightarrow{CD} \,|\, \overrightarrow{EF}$; they therefore prove that *any line between two asymptotically parallel lines is asymptotically parallel to both and in the same direction.*

Note 2. Given rays *r* and *s*, define *r* ~ *s* to mean that either *r* ⊂ *s* or *s* ⊂ *r* or *r* | *s*. Major Exercises 1–3 show that this is an equivalence relation among rays. An equivalence class of rays is called an *ideal point,* or an *end,* and we adopt the convention that it lies on all (and only those) lines containing the rays making up the class. Since a point on a line breaks the line into two opposite rays and opposite rays are not equivalent, we see that *every line has two ends lying on it.* The set of all ideal points was named by Cayley *the absolute.* (This is the beginning of constructing a hyperbolic analogue of the projective completion of an affine plane described in Chapter 2; we continue the construction in Major Exercise 13. The absolute is analogous to the line at infinity of the affine plane, but the absolute could not be a new line, because it intersects each old line in two points; it will turn out to be a *conic* in the projective completion.)

If R, S are the vertices of *r*, *s*, where *r* | *s*, and Ω is the ideal point determined by these rays, we write *r* = PΩ and *s* = SΩ and refer to the closed biangle with sides *r*, *s* as the *singly asymptotic triangle* △RSΩ. The next two exercises show that these triangles have some properties in common with ordinary triangles. (You can similarly define as an exercise *doubly* (two ideal points) and *triply* (three ideal points) *asymptotic triangles.*)

4. *Exterior angle theorem.* If △PQΩ is a singly asymptotic triangle, the exterior angles at P and Q are greater than their respective opposite interior angles. Justify the steps in the proof.

Proof (see Figure 6.33):
(1) Given R * Q * P. We must show that ∡RQΩ is greater than ∡QPΩ.
(2) Let $\overrightarrow{QD}$ be the unique ray on the same side of $\overleftrightarrow{PQ}$ as ray QΩ such that ∡RQD ≅ ∡QPΩ. (3) If U * Q * D, then ∡UQP ≅ QPΩ. (4) By Exercise 14, $\overleftrightarrow{QD}$ is divergently parallel to $\overleftrightarrow{PΩ}$. (5) Hence, $\overrightarrow{QD}$ is between $\overrightarrow{QR}$ and $\overrightarrow{QΩ}$. (6) ∡RQΩ > ∡QPΩ. ∎

5. *Congruence theorem.* If in asymptotic triangles △ABΩ and △A′B′Ω′ we have ∡BAΩ ≅ ∡B′A′Ω′, then ∡ABΩ ≅ ∡A′B′Ω′ if and only if

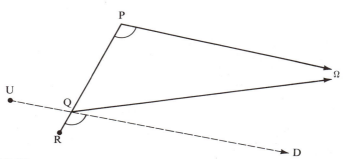

FIGURE 6.33

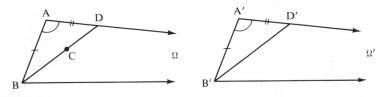

FIGURE 6.34

$AB \cong A'B'$. Justify the steps in the proof and deduce as a corollary that $PQ \cong P'Q'$ if and only if $\Pi(PQ)° = \Pi(P'Q')°$.

Proof (see Figure 6.34):
(1) Assume $AB \cong A'B'$ and on the contrary $\sphericalangle AB\Omega > \sphericalangle A'B'\Omega'$. (2) There is a unique ray $\overrightarrow{BC}$ between $B\Omega$ and $\overrightarrow{BA}$ such that $\sphericalangle ABC \cong \sphericalangle A'B'\Omega'$. (3) $\overrightarrow{BC}$ intersects $A\Omega$ in a point D. (4) Let D′ be the unique point on $A'\Omega'$ such that $AD \cong A'D'$. (5) Then $\triangle BAD \cong \triangle B'A'D'$. (6) Hence, $\sphericalangle A'B'D' \cong \sphericalangle A'B'\Omega'$, which is absurd. (7) Assume conversely that $\sphericalangle AB\Omega \cong \sphericalangle A'B'\Omega'$ and on the contrary $A'B' < AB$. (8) Let C be the point on AB such that $BC \cong B'A'$, and let $C\Omega$ be the ray from C limiting parallel to $A\Omega$ (see Figure 6.35). (9) Then $C\Omega$ is also limiting parallel to $B\Omega$. (10) By the first part of the proof, $\sphericalangle BC\Omega \cong \sphericalangle B'A'\Omega'$; hence, $\sphericalangle BC\Omega \cong \sphericalangle BA\Omega$. (11) But $\sphericalangle BC\Omega > \sphericalangle BA\Omega$, which is a contradiction. ∎

6. *Conclusion of the proof of theorem 6.7.* We wish to show that $\overrightarrow{EF}$ intersects $\overrightarrow{AG}$ (see Figure 6.36). Justify the steps in the proof.

Proof:
(1) Let $\overrightarrow{A'M}$ be limiting parallel to $\overrightarrow{EF}$, $\overrightarrow{A'N}$ limiting parallel to $\overrightarrow{AG}$, and $\overrightarrow{B'P}$ limiting parallel to $\overrightarrow{BG}$. (2) Since $EA' \cong BB'$ and $\sphericalangle A'EF \cong \sphericalangle B'BG$,

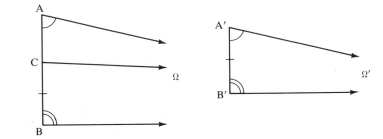

FIGURE 6.35

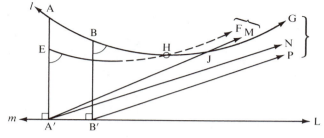

FIGURE 6.36

we have $\sphericalangle EA'M \cong \sphericalangle BB'P$. (3) $\overrightarrow{B'L}$ differs from $\overrightarrow{B'P}$ and $\overrightarrow{A'L}$ differs from $\overrightarrow{A'N}$. (4) $\sphericalangle MA'L \cong \sphericalangle PB'L$. (5) $\overrightarrow{B'P}$ is limiting parallel to $\overrightarrow{A'N}$. (6) Hence, $\sphericalangle NA'L$ is smaller than $\sphericalangle PB'L$. (7) It follows that $\overrightarrow{A'M}$ lies between $\overrightarrow{A'N}$ and $\overrightarrow{A'A}$, so it must intersect $\overrightarrow{AG}$ in a point J. (8) J is on the same side of $\overleftrightarrow{EF}$ as A'; hence, it is on the side opposite from A. (9) Thus, AJ intersects $\overleftrightarrow{EF}$ in a point H, which must be on $\overrightarrow{EF}$ because H is on the same side of $\overleftrightarrow{AA'}$ as J. ∎

Where was the hypothesis of this theorem used?

7. In Exercises 12 and 13 we considered the perpendicular bisectors of the sides of $\triangle ABC$ and we showed that (1) if two of them have a common point, the third passes through that point; (2) if two of them have a common perpendicular, the third has that same perpendicular. It follows that if two of them are asymptotically parallel, then any two of them are asymptotically parallel. This result can be strengthened as follows: if perpendicular bisectors l and m are asymptotically parallel in the direction of ideal point Ω, then the third perpendicular bisector n is asymptotically parallel to l and m in the *same direction* Ω. Give the proof and justify each step. The proof is based on the following two lemmas:

LEMMA 6.3. Given $\triangle ABC$. Let l, m, and n be the perpendicular bisectors of sides $\underline{AB}$, $\underline{BC}$, and $\underline{AC}$ at their midpoints L, M, and N, respectively. Let $\overline{AC} \geqq \overline{AB}$ and $\overline{AC} \geqq \overline{BC}$ (AC is the longest side). Then l, m, and n all intersect AC.

Proof:
(1) $(\sphericalangle B)° \geqq (\sphericalangle A)°$ and $(\sphericalangle B)° \geqq (\sphericalangle C)°$. (2) Hence, there is a point L' on AC such that $\sphericalangle A \cong \sphericalangle L'BA$, and a point M' on AC such that $\sphericalangle C \cong \sphericalangle M'BC$. (See Figure 6.37.) (3) Then $AL' \cong BL'$ and $CM' \cong BM'$. (4)

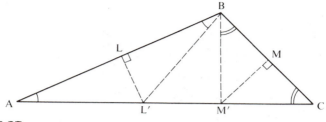

FIGURE 6.37

Thus, l is the line joining L to L′ and $m = \overleftrightarrow{MM'}$. (5) It follows that all three perpendicular bisectors cut AC. ∎

LEMMA 6.4.   No line intersects all three sides of a trebly asymptotic triangle.

Proof:
(1) Suppose that a line t cuts l at Q and m at P. (2) Then ray $\overleftrightarrow{PQ}$ of t lies between the rays $P\Omega_2$ and $P\Omega_1$, which are limiting parallel to l. (See Figure 6.38.) (3) $P\Omega_3$, the other ray through P that is limiting parallel to n, is opposite to $P\Omega_2$. (4) Hence, $P\Omega_1$ lies between $\overrightarrow{PQ}$ and $P\Omega_3$. (5) Thus, $\overrightarrow{PQ}$ does not intersect n. (6) Similarly, $\overrightarrow{QP}$ does not intersect n. ∎

8. Given any angle $\sphericalangle$A′OA. It is a theorem in hyperbolic geometry that there is a unique line l called the *line of enclosure* of this angle such that l is limiting parallel to both sides $\overrightarrow{OA'}$ and $\overrightarrow{OA}$. Only the idea of the proof is given here; see if you can fill in the details (Wolfe, 1945, p. 97):
Assume that A and A′ are chosen so that $OA \cong OA'$ (see Figure 6.39). Let A′Ω be the limiting parallel ray to $\overrightarrow{OA}$ through A′, and AΣ the

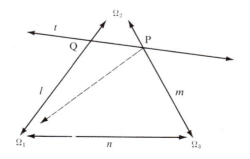

FIGURE 6.38

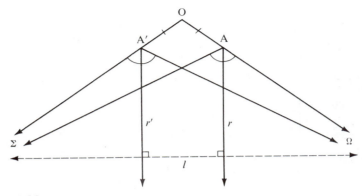

FIGURE 6.39

limiting parallel ray to $\overrightarrow{OA'}$ through A. Let the rays r and r' be the bisectors of ⊀ΣAΩ and ⊀ΩA'Σ, respectively. The idea of the proof is to show that the lines m and m' containing these rays are neither intersecting nor asymptotically parallel, so that, by Theorem 6.7, they have a unique common perpendicular l that turns out to be the line of enclosure of ⊀A'OA. (See also Exercise K-11, Chapter 7; the advantage of this complicated proof is that it yields a straightedge-and-compass construction.)

9. Use the result of the previous exercise to prove that every acute angle is an angle of parallelism, i.e., given an acute angle ⊀BOA, there is a unique line l perpendicular to $\overleftrightarrow{BO}$ and limiting parallel to $\overrightarrow{OA}$. (Hint: Reflect across $\overleftrightarrow{OB}$.)

Alternatively, fill in the details of the following continuity proof of Lobachevsky. First show that there exist perpendiculars to $\overrightarrow{OB}$ that fail to intersect $\overrightarrow{OA}$ by the following argument. In Figure 6.40, B is the

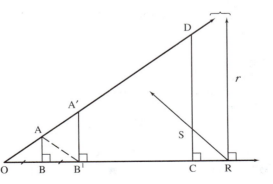

FIGURE 6.40

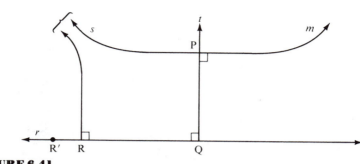

FIGURE 6.41

foot of the perpendicular from A and OB $\cong$ BB'. If the perpendicular at B' intersects $\overrightarrow{OA}$ at A', then

$$\delta OA'B' = \delta OAB' + \delta AA'B' = 2\delta OAB + \delta AA'B' > 2\delta OAB.$$

If we iterate this doubling along $\overrightarrow{OB}$ and the perpendicular always hits $\overrightarrow{OA}$, the defects of the resulting triangles will increase indefinitely. So we must eventually arrive at a point where the perpendicular fails to intersect $\overrightarrow{OA}$.

Second, apply Dedekind's axiom to obtain "the first" such perpendicular ray r emanating from R.

Finally, show that $r \mid \overrightarrow{OA}$. For any interior ray $\overrightarrow{RS}$, let C be the foot of the perpendicular from S; show that $\overrightarrow{CS}$ hits $\overrightarrow{OA}$ at some point D and apply Pasch's theorem to $\triangle OCD$.

10. Let l and m be divergently parallel lines and let t be their common perpendicular cutting l at Q and m at P (Figure 6.41). Let r be a ray of l emanating from Q and s the ray of m emanating from P on the same side of t as r. Prove that there is a unique point R on r such that the perpendicular to l through R is limiting parallel to s. Prove also that for every point R' on r such that R' * R * Q, the perpendicular to l through R' is divergently parallel to m. (Hint: Use Major Exercises 3 and 9.)

11. Let ray r emanating from point P be limiting parallel to line l and let Q be the foot of the perpendicular from P to l (Figure 6.42). Justify the terminology "asymptotically parallel" by proving that for any point R between P and Q there exists a point R' on ray r such that R'Q' $\cong$ RQ, where Q' is the foot of the perpendicular from R' to l. (Hint: Use Major Exercise 3 and Theorem 6.6 to prove that the line through R that is asymptotically parallel to l in the opposite direction from r intersects r at a point S. Show that if T is the foot of the perpendicular from S to l, the point R' obtained by reflecting R across line $\overleftrightarrow{ST}$ is the desired point.)

Similarly, show that the lines diverge in the other direction.

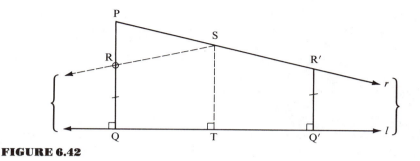

FIGURE 6.42

12. Let l and n be divergently parallel lines and PQ their common perpendicular segment. The midpoint S of PQ is called *the symmetry point of l*
and n. Let m be the perpendicular to PQ through S. Let Ω and Ω' be the
ideal points of l, and let Σ and Σ' be the ideal points of n (labeled as in
Figure 6.43). By Major Exercise 8, there are unique lines "joining"
these ideal points. Prove that (a) $\Omega\Sigma'$ and $\Sigma\Omega'$ meet at S; (b) m is
perpendicular to both $\Omega\Sigma$ and $\Omega'\Sigma'$. (Hint: Use Major Exercise 5 and
the symmetry part of Theorem 6.6.)

13. *Projective completion of the hyperbolic plane.* The ideal points were defined
in Note 2 after Major Exercise 3. By adding them as ends to our lines, we
ensure that asymptotically parallel lines meet at an ideal point; Major
Exercise 11 shows that the lines do converge in the direction of that
common end. We need to add more "points at infinity" to ensure that
divergently parallel lines will meet. Two divergently parallel lines have
a unique common perpendicular t. A third line perpendicular to t can be
considered to have "the same direction" as the first two, so all three
should meet at the same point, just as in the projective completion of the
Euclidean plane. We therefore define the *pole* $P(t)$ to be the set of all
lines perpendicular to t and specify that $P(t)$ lies on all those lines and no
others; poles of lines are called *ultra-ideal points*. Note that $t \neq
u \Rightarrow P(t) \neq P(u)$ (uniqueness of the common perpendicular), unlike

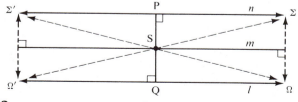

FIGURE 6.43

the Euclidean case. A "point" of the projective completion $\mathscr{P}$ is defined to be either a point of the hyperbolic plane (called "ordinary") or an ideal point or an ultra-ideal point.

We also add new "lines at infinity" as follows. The *polar* $p(A)$ of an ordinary point A is the set of all poles of lines through A, and the only points incident with $p(A)$ are those poles; polars of ordinary points are called *ultra-ideal lines*. The polar $p(\Omega)$ of an ideal point Ω consists of Ω and all poles of lines having Ω as an end; again, the incidence relation is $\in$, and $p(\Omega)$ is called an *ideal line*. The polar of an ultra-ideal point $P(t)$ is just t. A "line" of $\mathscr{P}$ is defined to be a polar of a point of $\mathscr{P}$. We have defined incidence already. The pole of $p(A)$ is A and of $p(\Omega)$ is Ω.

THEOREM. $\mathscr{P}$ is a projective plane and p is a polarity (an isomorphism of $\mathscr{P}$ onto its dual plane).

Since the ideal points are the only points of $\mathscr{P}$ that lie on their polars, the absolute γ is by definition the conic determined by polarity p and $p(\Omega)$ is the tangent line to γ at Ω (see Project 2, Chapter 2). If Ω and Σ are the two ends of ordinary line t, then, by definition, the point of intersection of the two tangent lines $p(\Omega)$ and $p(\Sigma)$ is $P(t)$, which gives geometric meaning to the rather abstract $P(t)$. Moreover, the interior of γ is the set of ordinary points, since every line through an ordinary point is ordinary and intersects γ twice.

Your exercise is to prove this theorem. To get you started, we show that Axiom I-1 holds for $\mathscr{P}$:

  (i) Two ordinary points A, B lie on ordinary line $\overleftrightarrow{AB}$ and do not lie on any "extraordinary" lines by definition of the latter.
 (ii) Given ordinary A and ideal Ω, they are joined by the ordinary line containing ray $A\Omega$ (which exists and is unique by Theorem 6.6).
(iii) Given ideal points Ω and Σ, let A be any ordinary point and consider the rays $A\Sigma$ and $A\Omega$. If these are opposite, then the line containing them joins Ω and Σ; otherwise, the line of enclosure (Major Exercise 8) of the angle determined by these coterminal rays joins Ω and Σ. Uniqueness of line $\Omega\Sigma$ follows from the fact that the angle of parallelism is acute.
 (iv) Given ordinary A and ultra-ideal $P(t)$, the line joining them is the perpendicular to t through A.
  (v) Given ideal Ω and ultra-ideal $P(t)$. If Ω lies on t, these points lie on $p(\Omega)$; by definition of incidence, they do not lie on any other extraordinary line, and they could not lie on an ordinary line u because u would then be both asymptotically parallel to and perpendicular to t. If Ω does not lie on t, let A be a point on t. If ray $A\Omega$

is at right angles to t, the line containing $A\Omega$ joins Ω to $P(t)$; otherwise, Major Exercise 9 ensures that there is a unique line $u \perp t$ such that $A\Omega$ is limiting parallel to u and u joins Ω to $P(t)$.

(vi) Given ultra-ideal points $P(t)$ and $P(u)$, t meets u either at ordinary point A, in which case $p(A)$ is the join, or at ideal point Ω, in which case $p(\Omega)$ is the join, or, by Theorem 6.7, at ultra-ideal point $P(m)$, in which case m (the common perpendicular to t and u) joins $P(t)$ and $P(u)$.

PROJECTS

1. Here is another construction for the common perpendicular between divergently parallel lines l and n. It suffices to locate their symmetry point S, for a perpendicular can then be dropped from S to both lines. Take any segment AB on l. Construct point C on l such that B is the midpoint of AC and lay off any segment A′B′ on n congruent to AB. Let M, M′, N, and N′ be the midpoints of AA′, BB′, BA′, and CB′, respectively. Then the lines $\overleftrightarrow{MM'}$ and $\overleftrightarrow{NN'}$ are distinct and intersect at S. (The proof follows from the theory of *glide reflections*; see Exercises 21 and 22 in Chapter 9; also see Coxeter, 1968, p. 269, where it is deduced from Hjelmslev's midline theorem. Beware that Coxeter's description of midlines is partially wrong; e.g., no midline through S cuts l and n.)

2. M. Pieri has shown that the foundations of geometry can be built on the single undefined term "point" and the single undefined relation "point A is *equidistant* from points B and C." It is obviously possible to define "A, B, C are collinear" in terms of "betweenness," namely, "A * B * C or B * A * C or A * C * B." What is not obvious is that in hyperbolic geometry it is possible to define "betweenness" in terms of "collinearity," as was done by F. P. Jenks, and that "collinearity" can in fact be taken as the single undefined relation for a hyperbolic geometry based on the elementary continuity principle (see Exercise K-21, Chapter 7). Report on all of these results, using as a reference H. Royden's paper "Remarks on Primitive Notions for Elementary Euclidean and Non-Euclidean Plane Geometry" in Henkin, Suppes, and Tarski (1959), with corrections in W. Schwabhäuser's paper "Metamathematical Methods in Foundations of Geometry," in *Logic, Methodology and Philosophy of Science,* Y. Bar-Hillel, ed., Amsterdam: North Holland, 1965; and Blumenthal and Menger (1970), p. 220.

3. Hilbert showed that all of plane hyperbolic geometry can be deduced from the incidence, betweenness, and congruence axioms, and a continuity axiom asserting the existence of two nonopposite limiting parallel rays emanating from a given point not on a given line. Report on the proof of Saccheri's acute angle hypothesis from these axioms (see Wolfe, 1945, p. 78; Archimedes' axiom is not needed in this proof). Report also on the introduction of coordinates on the basis of these axioms (see W. Szmielew, "A new analytic approach to hyperbolic geometry," *Fundamenta Mathematicae,* **50** (1961): 129–158), and the use of such coordinates to prove the circular continuity principle (see J. Strommer, "Ein elementar Beweis des Kreisaxiome der hyperbolischen geometrie," *Acta Scientiarum Mathematicarum Szeged,* **22,** (1961): 190–195).

4. If Dedekind's axiom is dropped from our axioms for hyperbolic geometry, then it is impossible to prove the existence of limiting parallel rays, for W. Pejas has constructed a "semielliptic" Archimedean geometry in which the hyperbolic axiom holds but any pair of parallel lines have a unique common perpendicular (see *Mathematische Annalen,* **143** (1961): 233). If Dedekind's axiom is replaced with the elementary continuity principle, then a proof of the existence of limiting parallel rays has been given by embedding in a metric projective plane (see Hessenberg and Diller, 1967, p. 239). Report on these results. If you could apply János Bolyai's construction (p. 198) for a more direct proof, you would probably be awarded a Ph.D. (See Appendix B and M. J. Greenberg, "On J. Bolyai's Parallel Construction," *Journal of Geometry,* **12/1** (1979): 45–64.)

5. In Euclidean geometry, it is impossible to trisect every angle using straightedge and compass alone; in hyperbolic geometry, not only is it impossible to trisect every angle but it is also impossible to trisect every segment using straightedge and compass alone! In Euclidean geometry, it is impossible to construct with straightedge and compass alone a regular 4-gon having the same area as a given circle; in hyperbolic geometry, however, this construction is possible. Report on these results. (Use Martin, 1982, Chapter 34.)

CHAPTER 7

INDEPENDENCE OF THE PARALLEL POSTULATE

All my efforts to discover a contradiction, an inconsistency, in this non-Euclidean geometry have been without success. . . .

C. F. GAUSS

CONSISTENCY OF HYPERBOLIC GEOMETRY

In the previous chapter you were introduced to hyperbolic geometry and presented with some theorems that must seem very strange to someone accustomed to Euclidean geometry. Even though you may admit that the proofs of these theorems are correct, given our assumptions, you may feel that the basic assumption of hyperbolic geometry — the hyperbolic axiom — is a false assumption. Let's examine what might be meant by saying it's false.

Suppose I assume that when I drop some object, say, a stone, it will "fall" upward. I can go out and drop rocks and, unless I have rocks in my head, I will discover that my assumption was false.

Now what sort of experiment could I perform to show that the hyperbolic assumption is false, or, equivalently, to show that its negation, the Euclidean parallel postulate, is true? First of all, I would have to understand what this statement means. In the above example I understood very well the meaning of "stone" and what it means to "drop" one, so I could act upon this understanding. But what does it mean that *l* is a "line," that P is a "point" not "on" *l*, or that there is a "unique parallel" to *l* through P? I might represent "points" and

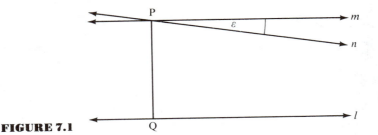

FIGURE 7.1

"lines" with paper, pencil, and straightedge. Suppose I draw $\overleftrightarrow{PQ}$ perpendicular to l and m through P perpendicular to $\overleftrightarrow{PQ}$, and then draw a line n through P, making a very small angle of $\epsilon°$ with m. Using Euclidean trigonometry, I can calculate exactly how far out on n I would have to go to get to the point where n is supposed to intersect l, but if ϵ is small enough, that point might be millions of miles away. Thus, I could not physically perform the experiment to prove that the hyperbolic axiom is false.

But is geometry about lines that we can draw? Applied geometry (engineering) is; but pure geometry is about ideal lines, which are concepts, not objects. The only experiments we can perform on these ideal lines are thought-experiments. So the question should be: Can we *conceive* of a non-Euclidean geometry? Kant said no, that any geometry other than Euclidean is inconceivable. At the time, of course, no one had yet conceived of a different geometry. It is in this sense that Gauss, J. Bolyai, and Lobachevsky created a "new universe."

Other questions can be raised. Mathematicians reject many of their own ideas because they either lead to contradictions or do not lead anywhere, i.e., do not prove fruitful, useful, or interesting. Does the hyperbolic axiom lead to a contradiction? Saccheri thought it would, and tried to prove the parallel postulate that way. Is hyperbolic geometry fruitful, useful, or interesting?

Let us postpone this question and take up the former: Is hyperbolic geometry *consistent?* Nowadays we refer to this as a question in *metamathematics*, i.e., a question outside of a mathematical system about the system itself. The question is not about lines or points or other geometric entities; it is a question about the *whole system* of hyperbolic geometry.

If hyperbolic geometry were inconsistent, an ordinary mathematical argument could derive a contradiction. Saccheri tried to do this and failed. Could it be that he wasn't clever enough, that someday some genius will find a contradiction?

On the other hand, can it be proved that hyperbolic geometry is *consistent* — can it be proved that there is no possible way to derive a contradiction?

We might ask the same question about Euclidean geometry — how do we *know* it is consistent? Of course, this was never a burning question before the discovery of non-Euclidean geometry simply because everyone *believed* Euclidean geometry to be consistent. Remarkably enough, if we make this belief an explicit assumption (albeit a metamathematical assumption), it is possible to give a proof that hyperbolic geometry is consistent. Let us state this possibility as a theorem:

METAMATHEMATICAL THEOREM 1.   If Euclidean geometry is consistent, then so is hyperbolic geometry.   iff

Granting this result for the moment, we get the following important corollary.

COROLLARY.   If Euclidean geometry is consistent, then no proof or disproof of the parallel postulate from the rest of Hilbert's postulates will ever be found, i.e., the parallel postulate is independent of the other postulates.

To prove the corollary, assume on the contrary that a proof of the parallel postulate exists. Then hyperbolic geometry would be inconsistent, since the hyperbolic axiom contradicts a proved result. But Metamathematical Theorem 1 asserts that hyperbolic geometry is consistent relative to Euclidean geometry. This contradiction proves that no proof of the parallel postulate exists (RAA). The hypothesis that Euclidean geometry is consistent ensures that no *dis*proof exists either. ■

Thus, 2000 years of efforts to prove Euclid V were in vain. There is no more hope of proving it than there is of finding a method for trisecting every angle using straightedge and compass alone.

Of course, when we say this, we are *assuming* the consistency of the venerable Euclidean geometry. Had Saccheri, Legendre, F. Bolyai, or any of the dozens of other scholars succeeded in proving Euclid V from the other axioms, with the noble intention of making Euclidean geometry more secure and elegant, they would have instead completely destroyed Euclidean geometry as a consistent body of thought! (I urge you, dear reader, to go over the preceding statements very carefully to make sure you have understood them. If you have not understood, you have missed the main point of this book.)

In the form given here, Metamathematical Theorem 1 is due to Eugenio Beltrami (1835–1900); a different proof was later given by Felix Klein (1849–1925).[1] Beltrami proved the relative consistency of hyperbolic geometry in 1868 using differential geometry (see The Pseudosphere, Chapter 10). Klein recognized that projective geometry could be used to give another proof. In 1871 he applied the method Arthur Cayley used (in 1859) to express distance and angle measure by projective coordinates.

To prove Metamathematical Theorem 1, we have to again ask ourselves, What is a "line" in hyperbolic geometry — in fact, what is the hyperbolic plane? The honest answer is that we don't know; it is just an abstraction. A hyperbolic "line" is an undefined term describing an abstract concept that resembles the concept of a Euclidean line except for its parallelism properties. Then how shall we visualize hyperbolic geometry? In mathematics, as in any other field of research, posing the right question is just as important as finding answers.

The question of "visualizing" means finding Euclidean objects that represent hyperbolic objects. This means finding a Euclidean *model* for hyperbolic geometry. In Chapter 2 we discussed the idea of models for an axiom system; there we showed that the Euclidean parallel postulate is independent of the axioms for incidence geometry by exhibiting three-point and five-point models of incidence geometry that are not Euclidean. Here we want to know whether the parallel postulate is independent of a much *larger* system of axioms, namely,

---

[1] Beltrami made important contributions to differential geometry. Klein was a master of many branches of mathematics and an influential teacher. His book on the history of nineteenth-century mathematics shows how familiar he was with all aspects of the subject. Klein's famous inaugural address in 1872, his *Erlanger Programme*, made the study of groups of transformations and their invariants the key to geometry (see Chapter 9).

Eugenio Beltrami

neutral geometry. We can show that it is, and by the same method —
by exhibiting models for hyperbolic geometry.[2]

THE BELTRAMI-KLEIN MODEL

For brevity, we will refer to this first model (the Beltrami-Klein
model) as the "Klein model." We fix once and for all a circle γ in the
Euclidean plane (which Cayley referred to as "the absolute"). If O is

---

[2] Unlike the situation for incidence geometry, we cannot construct a model for neutral
geometry in which the elliptic parallel property holds, because it is a theorem in neutral
geometry that parallel lines exist (see Corollary 2 to Theorem 4.1). If you worked through Major
Exercise 13, Chapter 6, you will easily understand the motivation for the Beltrami-Klein model.
It is the projective completion of the hyperbolic plane!

Felix Klein

the center of γ and OR is a radius, the *interior* of γ by definition consists of all points X such that OX $<$ OR (see Figure 7.2). In Klein's model the points in the interior of γ represent the points of the hyperbolic plane.

Recall that a chord of γ is a segment AB joining two points A and B on γ. We wish to consider the segment without its endpoints, which we will call an *open chord* and denote by A) (B. In Klein's model the open chords of γ represent the lines of the hyperbolic plane. The relation "lies on" is represented in the usual sense: P lies on A) (B means that P lies on the Euclidean line $\overleftrightarrow{AB}$ and P is between A and B. The hyperbolic relation "between" is represented by the usual Euclidean relation "between." This much is easy. The representation of "congruence" is much more complicated, and we will discuss it later in this chapter (The Projective Nature of the Beltrami-Klein Model).

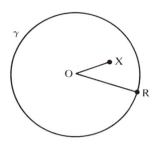

FIGURE 7.2

It is immediately clear from Figure 7.3 that the hyperbolic axiom holds in this representation.

Here the two open chords *m* and *n* through P are both parallel to the open chord *l*—for what does "parallel" mean in this representation? The definition of "parallel" states that two lines are parallel if they have no point in common. In Klein's representation this becomes: two open chords are parallel if they have no point in common (in the definition of "parallel," replace the word "line" by "open chord"). The fact that the three chords, when extended, may meet outside the circle γ is irrelevant — points outside of γ do not represent points of the hyperbolic plane. So let us summarize the Beltrami-Klein proof of the relative consistency of hyperbolic geometry as follows:

First, a glossary is set up to "translate" the five undefined terms ("point," "line," "lies on," "between," and "congruent") into their interpretations in the Euclidean model (we have done this for the first four terms). All the defined terms are then interpreted by "translating" all occurrences of undefined terms. For instance, the defined term "parallel" was interpreted by replacing every occurrence of the

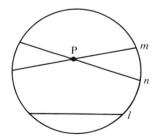

FIGURE 7.3

word "line" in the definition by "open chord." Once all the defined terms have been interpreted, we have to interpret the axioms of the system. Incidence Axiom 1, for example, has the following interpretation in the Klein model:

INCIDENCE AXIOM 1 (Klein). Given any two distinct points A and B in the interior of circle γ. There exists a unique open chord l of γ such that A and B both lie on l.

We must prove that this is a theorem in Euclidean geometry (and similarly, prove the interpretations of all the other axioms). Once all the interpreted axioms have been proved to be theorems in Euclidean geometry, any proof of a contradiction within hyperbolic geometry could be translated by our glossary into a proof of a contradiction in Euclidean geometry. From our assumption that Euclidean geometry is consistent, it follows that no such proof exists. Thus, if Euclidean geometry is consistent, so is hyperbolic geometry.

We must now backtrack and prove that the interpretations of the axioms of hyperbolic geometry in the Klein model are theorems in Euclidean geometry. Let us prove Axiom I-1 (Klein) stated above:

Proof
Given A and B interior to γ. Let $\overleftrightarrow{AB}$ be the Euclidean line through them (see Figure 7.4). This line intersects γ in two distinct points C and D. Then A and B lie on the open chord C) (D, and, by Axiom I-1 for Euclidean geometry, this is the only open chord on which they both lie. ■

In the second step of the proof we used a theorem from Euclidean geometry that states that a line passing through the interior of a circle

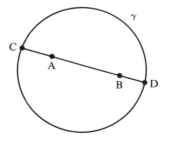

FIGURE 7.4

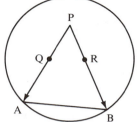

FIGURE 7.5 Limiting parallel rays.

intersects the circle in two distinct points. This can be proved from the circular continuity principle (see Major Exercise 1, Chapter 4). Verifications of the interpretations of the other incidence axioms, the betweenness axioms, and Dedekind's axiom are left as exercises; the congruence axioms are verified later in this chapter.

One nice aspect of the Klein model is that it is easy to visualize the limiting parallel rays (see Figure 7.5). Let P be a point interior to γ and not on the open chord A)(B. A and B are points on the circle and therefore do not represent points in the hyperbolic plane; they are said to represent *ideal points* and are called the *ends* of the hyperbolic line represented by A)(B (see Note 2 following Major Exercise 3, Chapter 6). Then the limiting parallel rays to A)(B from P are represented by the segments PA and PB with the endpoints A and B omitted. It is clear that any ray between these limiting parallel rays intersects the open chord A)(B, whereas all other rays emanating from P do not. The symmetry and transitivity of limiting parallelism (which, as you saw in Major Exercises 2 and 3, Chapter 6, were tricky to prove) are utterly obvious in the Klein model, as is the fact that every angle has a *line of enclosure* (given ⊀QPR, if A is the end of $\overrightarrow{PQ}$ and B is the end of $\overrightarrow{PR}$, then A)(B is the line of enclosure of ⊀QPR — see Major Exercise 8, Chapter 6).

Let us conclude this section by considering the interpretation in the Klein model of "congruence," the subtlest part of the model. One method of interpretation is to use a system of numerical measurement of angle degrees and segment lengths. Two angles would then be interpreted as congruent if they had the same number of degrees, and two segments would be interpreted as congruent if they had the same length (compare Theorem 4.3). The catch is that Euclidean methods of measuring degrees and lengths cannot be used. If we use Euclidean length, for example, then every line (i.e., open chord) would have a finite length less than or equal to the length of a diameter of γ. This would invalidate the interpretations of Axioms B-2 and C-1, which ensure that lines are infinitely long.

We will further discuss the matter in this chapter (in the sections Perpendicularity in the Beltrami-Klein Model and The Projective Nature of the Beltrami-Klein Model), but first let's consider the Poincaré models, in which congruence of angles is easier to describe.

THE POINCARÉ MODELS

A disk model due to Henri Poincaré (1854–1912)[3] also represents points of the hyperbolic plane by the points *interior* to a Euclidean circle γ, but lines are represented differently. First, all open chords that pass through the center O of γ (i.e., all *open diameters l* of γ) represent lines. The other lines are represented by *open arcs of circles orthogonal to γ*. More precisely, let δ be a circle orthogonal to γ (at each point of intersection of γ and δ the radii of γ and δ through that point are perpendicular). Then intersecting δ with the interior of γ gives an open arc m, which by definition represents a hyperbolic line in the Poincaré model. So we will call *Poincaré line*, or "P-line," either an open diameter l of γ or an open circular arc m orthogonal to γ (see Figure 7.6).

A point interior to γ "lies on" a Poincaré line if it lies on it in the Euclidean sense. Similarly, "between" has its usual Euclidean interpretation (for A, B, and C on an open arc coming from an orthogonal circle δ with center P, B is *between* A and C if $\overrightarrow{PB}$ is between $\overrightarrow{PA}$ and $\overrightarrow{PC}$).

The interpretation of *congruence for segments* in the Poincaré model is complicated, being based on a way of measuring length that is different from the usual Euclidean way, just as in the Klein model (see

---

[3] Poincaré was the cousin of the president of France. Like Gauss, Poincaré made profound discoveries in many branches of mathematics and physics; he even started a new branch of mathematics, algebraic topology. He used his models of hyperbolic geometry to discover new theorems about automorphic functions of a complex variable. Poincaré is also important as a philosopher of science (see Chapter 8).

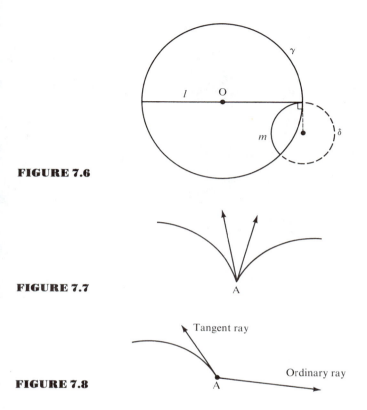

FIGURE 7.6

FIGURE 7.7

FIGURE 7.8

p. 248). *Congruence for angles* has the usual Euclidean meaning, how-ever, and this is the main advantage of the Poincaré model over the Klein model.[4] Specifically, if two directed circular arcs intersect at a point A, the number of degrees in the *angle* they make is by definition the number of degrees in the angle between their tangent rays at A (see Figure 7.7). Or, if one directed circular arc intersects an ordinary ray at A, the number of degrees in the *angle* they make is by definition the number of degrees in the angle between the tangent ray and the ordinary ray at A (see Figure 7.8).

Having interpreted all the undefined terms of hyperbolic geometry in the Poincaré model, we get (by substitution) interpretations of all

---

[4] Technically, we say that the Poincaré model is *conformal*—it represents angles accurately —while the Klein model is not. Another example of a conformal model is Mercator's map of the surface of the earth.

Henri Poincaré

the defined terms. For example, two Poincaré lines are *parallel* if and only if they have no point in common. Then all the axioms of hyperbolic geometry get translated into statements in Euclidean geometry, and it will be shown in the section Inversion in Circles later in this chapter that these interpretations are theorems in Euclidean geometry. Hence, the Poincaré model furnishes another proof that if Euclidean geometry is consistent, so is hyperbolic geometry.

The limiting parallel rays in the Poincaré model are illustrated in Figure 7.9. Here we have chosen l to be an open diameter A) (B; the rays are circular arcs that meet $\overleftrightarrow{AB}$ at A and B and are tangent to this line at those points. You can see how these rays approach l asymptotically as you move out toward the ideal points represented by A and B.

Figure 7.10 illustrates two parallel Poincaré lines with a common perpendicular. The diagram shows how m diverges from l on either side of the common perpendicular PO.

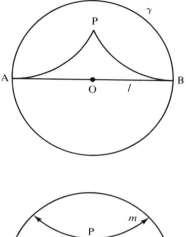

FIGURE 7.9

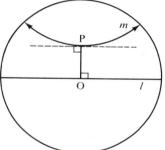

FIGURE 7.10

Figure 7.11 illustrates a Lambert quadrilateral. You can see that the fourth angle is acute. By adding the mirror image of this Lambert quadrilateral we get a diagram illustrating a Saccheri quadrilateral (Figure 7.12).

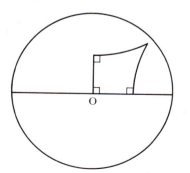

FIGURE 7.11

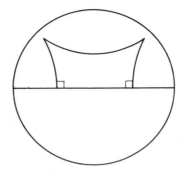

FIGURE 7.12

You may be surprised that we have two different models of hyperbolic geometry, one due to Klein and the other to Poincaré. (There is a third model, also due to Poincaré, soon to be described.) Yet you may have the feeling that these models are not "essentially different." In fact, these models are *isomorphic* in the technical sense that one-to-one correspondences can be set up between the "points" and "lines" in one model and the "points" and "lines" in the other so as to preserve the relations of incidence, betweenness, and congruence. Such isomorphism is illustrated in Figure 7.13. We start with the Klein model and consider, in Euclidean three-space, a sphere of the same radius sitting on the plane of the Klein model and tangent to it at the origin. We project upward orthogonally the entire Klein model onto the lower hemisphere of this sphere; by this projection, the chords in the Klein model become arcs of circles orthogonal to the equator. We then project stereographically from the north pole of the sphere onto the

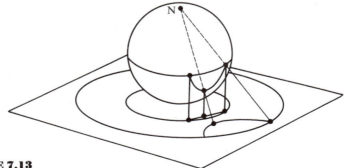

FIGURE 7.13

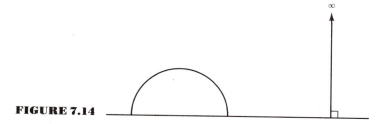

FIGURE 7.14

original plane. The equator of the sphere will project onto a circle larger than the one used in the Klein model, and the lower hemisphere will project stereographically onto the inside of this circle. Under these successive transformations, the chords of the Klein model will be mapped one-to-one onto the diameters and orthogonal arcs of the Poincaré model. In this way the isomorphism of the models may be established.

   One can actually prove that *all possible models of hyperbolic geometry are isomorphic to one another*, i.e., that the axioms for hyperbolic geometry are *categorical*. The same is true for Euclidean geometry. The categorical nature of Euclidean geometry is established by introducing Cartesian coordinates into the Euclidean plane. Analogously, the categorical nature of hyperbolic geometry is established by introducing Beltrami coordinates into the hyperbolic plane (for which hyperbolic trigonometry must first be developed).[5]

   In the other Poincaré model mentioned here, the points of the hyperbolic plane are represented by the points of one of the Euclidean half-planes determined by a fixed Euclidean line. If we use the Cartesian model for the Euclidean plane, it is customary to make the x axis the fixed line and then to use for our model the upper half-plane consisting of all points (x, y) with $y > 0$. Hyperbolic lines are represented in two ways:

1. As rays emanating from points on the x axis and perpendicular to the x axis;
2. As semicircles in the upper half-plane whose center lies on the x axis (see Figure 7.14).

Incidence and betweenness have the usual Euclidean interpretation.

[5] See Chapter 10 as well as Borsuk and Szmielew (1960), Chapter 6.

This model is conformal also (degrees of angles are measured in the Euclidean way). Measurement of lengths will be discussed later.

To establish isomorphism with the previous models, choose a point E on the equator of the sphere in Figure 7.13, and let Π be the plane tangent to the sphere at the point diametrically opposite to E. Stereographic projection from E to Π maps the equator onto a line in Π and the lower hemisphere onto the lower half-plane determined by this line. Notice that the points on this line represent ideal points. However, one ideal point is missing: the point E got lost in the stereographic projection. It is customary to imagine an ideal "point at infinity" ∞ that corresponds to E; it is the common end of all the vertical rays.

PERPENDICULARITY IN THE BELTRAMI-KLEIN MODEL

The Klein model is not conformal. Congruence of angles is interpreted differently from the usual Euclidean way, and will be explained later in this chapter (p. 260). Here we will describe only those angles that are congruent to their supplements, namely, right angles.

Let l and m be open chords of y. To describe when $l \perp m$ in the Klein model, there are two cases to consider:

Case 1. One of l and m is a diameter. Then $l \perp m$ in the Klein sense if and only if $l \perp m$ in the Euclidean sense. (See Figure 7.15.)

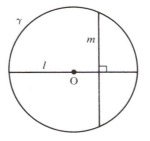

FIGURE 7.15

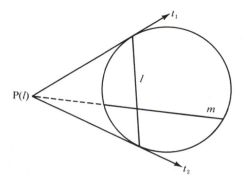

FIGURE 7.16

Case 2. Neither *l* nor *m* is a diameter. In this case we associate to *l* a certain point P(*l*) outside of *γ* called the *pole* of *l*, defined as follows. Let t_1 and t_2 be the tangents to *γ* at the endpoints of *l*. Then by definition P(*l*) is the unique point common to t_1 and t_2 (t_1 and t_2 are not parallel, because *l* is not a diameter); see Figure 7.16.

It turns out that *l is perpendicular to m in the sense of the Klein model if and only if the Euclidean line extending m passes through the pole of l.*

This description of perpendicularity will be justified later (pp. 260–261). We can use it to see more easily why divergently parallel lines have a common perpendicular—Theorem 6.7. In the hypothesis of Theorem 6.7 we are given two parallel lines that do not contain limiting parallel rays. In the Klein model this means that we are given open chords *l* and *m* that do not have a common end. The conclusion of Theorem 6.7 is that *l* and *m* have a common perpendicular *k*. How do we find *k*? Let's discuss case 2, leaving case 1 as an exercise. By the above description of perpendicularity if *k* were perpendicular to both *l* and *m*, the extension of *k* would have to pass through the pole of *l* and the pole of *m*. Hence, to construct *k*, we need only join these poles by a Euclidean line and take *k* to be the open chord of *γ* cut out by this line (Figure 7.17).[6]

There is a nice language that describes the behavior of pairs of lines in the Klein model. Let us call the points inside circle *γ* (which

---

[6] If *l* and *m* did have a common end Ω, the Euclidean line joining P(*l*) to P(*m*) would be tangent to *γ* at Ω. That is why Saccheri claimed that asymptotically parallel lines have "a common perpendicular at infinity," and this he found repugnant.

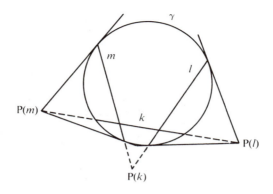

FIGURE 7.17

represent all the points in the hyperbolic plane) *ordinary points*. We already called the points on the circle γ *ideal points*. Let us call the points outside γ *ultra-ideal points*. Finally, for every diameter of γ, let us imagine another point "at infinity" such that all the Euclidean lines parallel in the Euclidean sense to this diameter meet in this point at infinity, just as railroad tracks appear to meet at the horizon. These points at infinity will also be called *ultra-ideal*. We can then say that two Klein lines "meet" at an ordinary point, an ideal point, or an ultra-ideal point, depending on whether they are intersecting, asymptotically parallel, or divergently parallel, respectively. The ultra-ideal point at which divergently parallel Klein lines *l* and *m* "meet" is the pole P(*k*) of their common perpendicular *k* (see Figure 7.17).

This language is suggestive of further theorems in hyperbolic geometry. For example, we know that two ordinary points determine a unique line, and we have seen that two ideal points also determine a unique line, the line of enclosure of Major Exercise 8, Chapter 6. We can ask the same question about two points that are ultra-ideal or about two points of different species. For example, an ordinary point and an ideal or ultra-ideal point always determine a unique line, but two ultra-ideal points may or may not (see Figure 7.18). Let us translate back from this language, say, in the case of an ordinary point O and an ultra-ideal point P(*l*) that is the pole of a Klein line *l*. What is the Klein line "joining" O to P(*l*)? It is the unique Klein line *m* through O that is perpendicular in the sense of the Klein model to the line *l* (see Figure 7.16). We leave the other cases for exercises.

If you did most of the exercises in hyperbolic geometry in Chapter 6, deriving results without having reliable diagrams to guide you, the

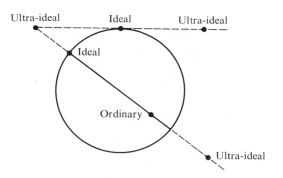

FIGURE 7.18

Klein and Poincaré models must come as a great relief. It is a useful exercise to take an absurd diagram like Figure 6.26 and draw those divergently parallel perpendicular bisectors of the triangle more accurately in one of the models. It is amazing that J. Bolyai and Lobachevsky were able to visualize hyperbolic geometry without such models, especially since they worked in three dimensions. They must have had non-Euclidean eyesight.

A MODEL OF THE HYPERBOLIC PLANE FROM PHYSICS

This model comes from the theory of special relativity. In Cartesian three-space $\mathbb{R}^3$, with coordinates denoted x, y and t (for *time*), distance will be measured by the Minkowski metric

$$ds^2 = dx^2 + dy^2 - dt^2.$$

Then with respect to the Minkowski metric, the surface of equation

$$x^2 + y^2 - t^2 = -1$$

is a "sphere" centered at the origin $O = (0, 0, 0)$ of imaginary radius $i = \sqrt{-1}$. (As was mentioned in Chapter 5, Lambert was the first to wonder if such a model existed.) In Euclidean terms, it is a two-sheeted *hyperboloid* (surface of revolution obtained by rotating the hyperbola $t^2 - x^2 = 1$ around the x axis). We choose the sheet Σ: $t \geq 1$ as our model. It looks like an infinite bowl (see Figure 7.19).

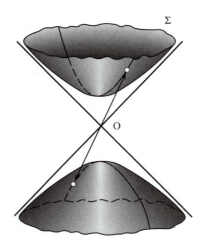

FIGURE 7.19

Analogously with our interpretation of "lines" on a sphere in Chapter 2, Exercise 10(c), "lines" are interpreted to be the sections of Σ cut out by planes through O; thus a "line" is one branch of a *hyperbola* on Σ.

Here is an isomorphism of Σ with the Beltrami-Klein model Δ. The plane $t = 1$ is tangent to Σ at the point $C = (0, 0, 1)$. Let Δ be the unit disk centered at C in this plane. Projection from O gives a one-to-one correspondence between the points of Δ and the points of Σ (i.e., point P of Δ corresponds to the point P′ at which ray $\overrightarrow{OP}$ pierces Σ). Similarly, each chord m of Δ lies on a unique plane Π through O, and m corresponds to the section $m′$ of Σ cut out by Π. This isomorphism of incidence models can be used to interpret betweenness and congruence on Σ. Alternatively, they can be defined in terms of the measurement of arc length induced on Σ by the Minkowski metric; then further argument is needed to verify that our correspondence is indeed an isomorphism of models of hyperbolic geometry. Another justification of Σ as a model of the hyperbolic plane will be given analytically in Chapter 10 (see the discussion of Weierstrass coordinates in the section Coordinates in the Hyperbolic Plane).

Note. From the point of view of Einstein's special relativity theory, Σ can be identified with the set of plane uniform motions, and the hyperbolic distance can be identified with the relative velocity of one motion with respect to the other. A glossary can be set up to translate every theorem of hyperbolic geometry into a theorem of relativistic kinematics, and conversely. See Yaglom (1979), p. 225 ff.

INVERSION IN CIRCLES

In order to define congruence in the Poincaré models and verify the axioms of congruence, we must study the operation of inversion in a Euclidean circle; this operation will turn out to be the interpretation of reflection across a line in the hyperbolic plane. *This theory is part of Euclidean geometry,* so we may use the theorems you proved in Exercises 18–26, Chapter 5.

DEFINITION.   Let γ be a circle of radius r, center O. For any point $P \neq O$ the *inverse* P' of P with respect to γ is the unique point P' on ray $\overrightarrow{OP}$ such that $(\overline{OP})\ (\overline{OP'}) = r^2$ (where $\overline{OP}$ denotes the length of segment OP with respect to a fixed unit of measurement); see Figure 7.20.

The following properties of inversion are immediate from the definition:

PROPOSITION 7.1.   (a) $P = P'$ if and only if P lies on the circle of inversion γ. (b) If P is inside γ then P' is outside γ, and if P is outside γ then P' is inside γ. (c) $(P')' = P$.

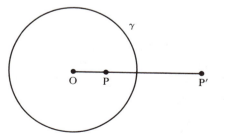

FIGURE 7.20

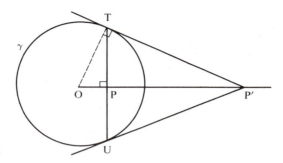

FIGURE 7.21

The next two propositions tell how to construct the inverse point with a straightedge and compass.

PROPOSITION 7.2. Suppose P is inside γ. Let TU be the chord of γ through P which is perpendicular to $\overleftrightarrow{OP}$. Then the inverse P′ of P is the pole of chord TU, i.e., the point of intersection of the tangents to γ at T and U. (See Figure 7.21).

Proof:

Suppose the tangent to γ at T cuts $\overrightarrow{OP}$ at point P′. Right triangle △OPT is similar to right triangle △OTP′ (since they have ⦟TOP in common and the angle sum is 180°). Hence, corresponding sides are proportional (Exercise 18, Chapter 5). As $\overline{OT} = r$, we get $(\overline{OP})/r = r/(\overline{OP'})$, which shows that P′ is inverse to P. Reflecting across $\overleftrightarrow{OP}$ (Major Exercise 2, Chapter 3), we see that the tangent to γ at U also passes through P′, so P′ is indeed the pole of TU. ■

PROPOSITION 7.3. If P is outside γ, let Q be the midpoint of segment OP. Let σ be the circle with center Q and radius $\overline{OQ} = \overline{QP}$. Then σ cuts γ in two points T and U, $\overleftrightarrow{PT}$ and $\overleftrightarrow{PU}$ are tangent to γ, and the inverse P′ of P is the intersection of TU and OP. (See Figure 7.22.)

Proof:

By the circular continuity principle (Chapter 3), σ and γ do meet in two points T and U. Since ⦟OTP and ⦟OUP are inscribed in semicircles of σ, they are right angles (Exercise 24, Chapter 5); hence, $\overleftrightarrow{PT}$ and $\overleftrightarrow{PU}$ are tangent to γ. If TU meets OP in a point P′, then P is the inverse of P′ (Proposition 7.2); hence, P′ is the inverse of P in γ. ■

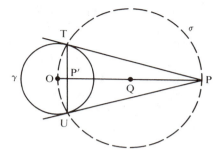

FIGURE 7.22

The next proposition shows how to construct the Poincaré line joining two ideal points — the line of enclosure.

PROPOSITION 7.4. Let T and U be points on γ that are not diametrically opposite and let P be the pole of TU. Then $PT \cong PU$, $\sphericalangle PTU \cong \sphericalangle PUT$, $\overleftrightarrow{OP} \perp \overleftrightarrow{TU}$, and the circle δ with center P and radius $\overline{PT} = \overline{PU}$ cuts γ orthogonally at T and U. (See Figure 7.23.)

Proof:
By definition of pole, $\sphericalangle OTP$ and $\sphericalangle OUP$ are right angles, so by the hypotenuse-leg criterion, $\triangle OTP \cong \triangle OUP$. Thus, $PT \cong PU$, $\sphericalangle OPT \cong \sphericalangle OPU$. The base angles $\sphericalangle PTU$ and $\sphericalangle PUT$ of the isosceles triangle $\triangle TPU$ are then congruent, and the angle bisector $\overrightarrow{PO}$ is perpendicular to the base TU. The circle δ is then well defined because $\overline{PT} = \overline{PU}$ and δ cuts γ orthogonally by our hypothesis that $\overleftrightarrow{PT}$ and $\overleftrightarrow{PU}$ are tangent to γ. ∎

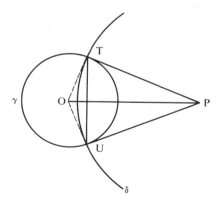

FIGURE 7.23

LEMMA 7.1.   Given that point O does not lie on circle δ. (a) If two lines through O intersect δ in pairs of points (P_1, P_2) and (Q_1, Q_2), respectively, then $(\overline{OP_1})(\overline{OP_2}) = (\overline{OQ_1})(\overline{OQ_2})$. This common product is called the *power* of O with respect to δ when O is outside δ, and minus this number is called the power of O when O is inside δ. (b) If O is outside δ and a tangent to δ from O touches δ at point T, then $(\overline{OT})^2$ equals the power of O with respect to δ.

Proof:

(a) Since angles that are inscribed in a circle and subtend the same arc are congruent (Exercise 25, Chapter 5), we have

$$\sphericalangle P_2P_1Q_2 \cong \sphericalangle P_2Q_1Q_2$$
$$\sphericalangle P_1Q_2Q_1 \cong \sphericalangle P_1P_2Q_1$$

(see Figure 7.24). It follows that $\triangle OP_1Q_2$ and $\triangle OQ_1P_2$ are similar, so that $(\overline{OP_1})/(\overline{OQ_1}) = (\overline{OQ_2})/(\overline{OP_2})$, as asserted.

(b) Let C be the center of δ and let line OC cut δ at P_1 and P_2, with $O * P_1 * C * P_2$. By the Pythagorean theorem (Exercise 21, Chapter 5),

$$\begin{aligned}
(\overline{OT})^2 &= (\overline{OC})^2 - (\overline{CT})^2 \\
&= (\overline{OC} - \overline{CT})(\overline{OC} + \overline{CT}) \\
&= (\overline{OC} - \overline{CP_1})(\overline{OC} + \overline{CP_2}) \\
&= (\overline{OP_1})(\overline{OP_2})
\end{aligned}$$

(see Figure 7.25). ■

PROPOSITION 7.5.   Let P be any point which does not lie on circle γ and which does not coincide with the center O of γ, and let δ be a circle through P. Then δ cuts γ orthogonally if and only if δ passes through the inverse point P' of P with respect to γ.

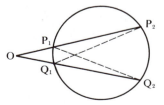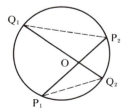

FIGURE 7.24

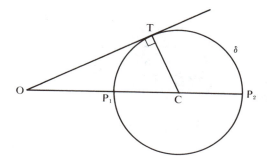

FIGURE 7.25

Proof:

Suppose first that δ passes through P′. Then the center C of δ lies on the perpendicular bisector of PP′ (Exercise 17, Chapter 4); hence, $\overline{CO} > \overline{CP}$ (Exercise 27, Chapter 4) and O lies outside δ. Therefore, there is a point T on δ such that the tangent to δ at T passes through O (Proposition 7.3). Lemma 7.1(b) then gives $(\overline{OT})^2 = (\overline{OP})(\overline{OP'}) = r^2$ so that T also lies on γ and δ cuts γ orthogonally.

Conversely, let δ cut γ orthogonally at points T and U. Then the tangents to δ at T and U meet at O, so that O lies outside δ. It follows that $\overrightarrow{OP}$ cuts δ again at a point Q. By Lemma 7.1(b), $r^2 = (\overline{OT})^2 = (\overline{OP})(\overline{OQ})$ so that Q = P′, the inverse of P in γ. ∎

Proposition 7.5 can be used to construct the P-line joining two points P and Q inside γ that do not lie on a diameter of γ. First, construct the inverse point P′, using Proposition 7.2. Then construct the circle δ determined by the three noncollinear points P, Q, and P′ (use Exercise 12, Chapter 6). By Proposition 7.5, δ will be orthogonal to γ; intersecting δ with the interior of γ gives the desired P-line. This verifies the interpretation of **Axiom I-1** for the Poincaré disk model. The verification is even simpler for the Poincaré upper half-plane model: Given two points P and Q that do not lie on a vertical ray, let the perpendicular bisector of Euclidean segment PQ cut the x axis at C. Then the semicircle centered at C and passing through P and Q is the desired P-line.

We could also have verified the interpretations of the incidence axioms, the betweenness axioms, and Dedekind's axiom by using isomorphism with the Klein model (where the verifications are trivial).

We turn now to the congruence axioms. Since angles are measured in the Euclidean sense in the Poincaré models, the interpretation of **Axiom C-5** is trivially verified. Consider **Axiom C-4,** the laying off of a congruent copy of a given angle at some vertex A (for the disk model). If A is the center of γ, the angle is formed by diameters and the laying off is accomplished in the Euclidean way. If A is not the center O of γ, then the verification is a matter of finding a unique circle δ through A that is orthogonal to γ and tangent to a given Euclidean line l that passes through A and not through O (since the tangents determine the angle measure). By Proposition 7.5, δ must pass through the inverse A' of A with respect to γ. The center C of δ must lie on the perpendicular bisector of chord AA' (Exercise 17, Chapter 4); call this bisector m. If δ is to be tangent to l at A, then C must also lie on the perpendicular n to l at A. So δ must be the circle whose center is the intersection C of m and n and whose radius is CA (see Figure 7.26).

To define congruence of segments in the disk model, we introduce the following definition of length:

DEFINITION. Let A and B be points inside γ, and let P and Q be the ends of the P-line through A and B. We define the *cross-ratio* (AB, PQ) by

$$(AB,PQ) = \frac{(\overline{AP})\,(\overline{BQ})}{(\overline{BP})\,(\overline{AQ})}$$

(where, for example, $\overline{AP}$ is the Euclidean length of the Euclidean segment AP). We then define the *Poincaré length* d(AB) by

$$d(AB) = |\log(AB,PQ)|.$$

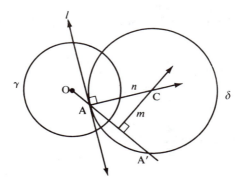

FIGURE 7.26

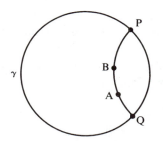

FIGURE 7.27

Notice first of all that this length does not depend on the order in which we write P and Q. For if (AB, PQ) $= x$, then (AB, QP) $= 1/x$, and $|\log(1/x)| = |-\log x| = |\log x|$. Furthermore, since (AB, PQ) $=$ (BA, QP), we see that d(AB) also does not depend on the order in which we write A and B.

We may therefore interpret the Poincaré segments AB and CD to be *Poincaré-congruent* if d(AB) $= d$(CD). With this interpretation, **Axiom C-2** is immediately verified.

Suppose we fix the point A on the P-line from P to Q and let point B move continuously from A to P, where Q * A * B * P, as in Figure 7.27. The cross-ratio (AB, PQ) will increase continuously from 1 to ∞, since $(\overline{AP})/(\overline{AQ})$ is constant, $\overline{BP}$ approaches zero, and $\overline{BQ}$ approaches $\overline{PQ}$. If we fix B and let A move continuously from B to Q, we get the same result. It follows immediately that for any Poincaré ray $\overrightarrow{CD}$, there is a unique point E on $\overrightarrow{CD}$ such that d(CE) $= d$(AB), where A and B are given in advance. This verifies **Axiom C-1.**

We next verify **Axiom C-3.** This will follow immediately from the additivity of the Poincaré length, which asserts that if A * C * B in the sense of the disk model, then d(AC) $+ d$(CB) $= d$(AB). To prove this additivity, label the ends so that Q * A * B * P. Then the cross-ratios (AB, PQ), (AC, PQ), and (CB, PQ) are all greater than 1 (because $\overline{AP} > \overline{BP}, \overline{BQ} > \overline{AQ}$, etc.); their logs are thus positive and we can drop the absolute value signs. We have

$$d(AC) + d(CB) = \log (AC, PQ) + \log (CB, PQ)$$
$$= \log [(AC, PQ)(CB, PQ)],$$

but (AC, PQ)(CB, PQ) $=$ (AB, PQ), as can be seen by canceling terms.

Finally, to verify **Axiom C-6 (SAS),** we must study the effect of inversions on the objects and relations in the disk model.

DEFINITION. Let O be a point and k a positive number. The *dilation* with *center* O and *ratio* k is the transformation of the Euclidean plane that fixes O and maps a point P ≠ O onto the unique point P* on $\overrightarrow{OP}$ such that $\overline{OP^*} = k(\overline{OP})$ (so that points are moved radially from O a distance k times their original distance).

LEMMA 7.2. Let δ be a circle with center C ≠ O and radius s. Under the dilation with center O and ratio k, δ is mapped onto the circle δ^* with center C* and radius ks. If Q is a point on δ, the tangent to δ^* at Q* is parallel to the tangent to δ at Q.

Proof:
Choose rectangular coordinates so that O is the origin. Then the dilation is given by $(x, y) \rightarrow (kx, ky)$. The image of the line having equation $ax + by = c$ is the line having equation $ax + by = kc$; hence, the image is parallel to the original line. In particular, $\overleftrightarrow{CQ}$ is parallel to $\overleftrightarrow{C^*Q^*}$, and their perpendiculars at Q and Q*, respectively, are also parallel. If δ has equation $(x - c_1)^2 + (y - c_2)^2 = s^2$, then δ^* has equation $(x - kc_1)^2 + (y - kc_2)^2 = (ks)^2$, from which the lemma follows. ∎

PROPOSITION 7.6. Let γ be a circle of radius r and center O, δ a circle of radius s and center C. Assume that O lies outside δ; let p be the power of O with respect to δ (see Lemma 7.1). Let $k = r^2/p$. Then the image δ' of δ under inversion in γ is the circle of radius ks whose center is the image C* of C under the dilation from O of ratio k. If P is any point on δ and P' is its inverse in γ, then the tangent t' to δ' at P' is the reflection across the perpendicular bisector of PP' of the tangent to δ at P (see Figure 7.28).

Proof:
Since O is outside δ, $\overrightarrow{OP}$ either cuts δ in another point Q or is tangent to δ at P (in which case let Q = P). Then

$$\frac{\overline{OP}}{\overline{OQ}} = \frac{\overline{OP'}}{\overline{OQ}} \cdot \frac{\overline{OP}}{\overline{OP}} = \frac{r^2}{p},$$

which shows that P' is the image of Q under the dilation from O of ratio $k = r^2/p$. Hence, $\delta^* = \delta'$. By Lemma 7.2, the tangent t' to δ' at P' is parallel to the tangent u to δ at Q. Let t be tangent to δ at P. By

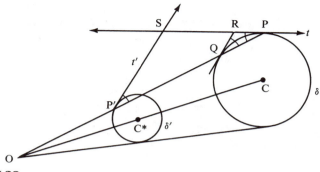

FIGURE 7.28

Proposition 7.4, t and u meet at a point R such that $\sphericalangle RQP \cong \sphericalangle RPQ$. By Exercises 4, 5, and 32, Chapter 4, t and t' meet at a point S such that $\sphericalangle SP'P \cong \sphericalangle SPP'$. Since $\triangle PSP'$ is an isosceles triangle (base angles are congruent), S lies on the perpendicular bisector of PP'. Hence, t' is the reflection of t across this perpendicular bisector. ■

COROLLARY. Circle δ is orthogonal to circle γ if and only if δ is mapped onto itself by inversion in γ.

Proof:
If δ is orthogonal to γ and P lies on δ then $p = (\overline{OP})(\overline{OP'}) = r^2$ (Proposition 7.5 and Lemma 7.1), so $k = 1$ and $\delta = \delta'$. Conversely, if $\delta = \delta'$, then $p = r^2$ and δ passes through the inverse P' of P in γ, so that by Proposition 7.5, δ is orthogonal to γ. ■

LEMMA 7.3. Let O be the center of circle γ, let P and Q be two points that are not collinear with O, and let P' and Q' be their inverses in γ. Then $\triangle POQ$ is similar to $\triangle Q'OP'$ (Figure 7.29).

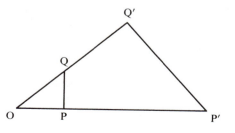

FIGURE 7.29

Proof:
The triangles have ∡POQ in common and $\overline{(OP)}\,\overline{(OP')} = r^2 = \overline{(OQ)}\,\overline{(OQ')}$. Thus, the SAS similarity criterion is satisfied (Exercise 20, Chapter 5). ∎

PROPOSITION 7.7. Let *l* be a line not passing through the center O of circle γ. The image of *l* under inversion in γ is a punctured circle with missing point O. The diameter through O of the completed circle δ is (when extended) perpendicular to *l*. (See Figure 7.30).

Proof:
Let A be the foot of the perpendicular from O to *l*, P be any other point on *l*, and A′ and P′ their inverses in γ. By Lemma 7.3, △OP′A′ is similar to △OAP. Hence, ∡OP′A′ is a right angle, so that P′ must lie on the circle δ having OA′ as diameter (Exercise 26, Chapter 5). Conversely, if we start with any point P′ on δ other than O and let $\overrightarrow{OP'}$ cut *l* in P, then reversing the above argument shows that P′ is the inverse of P in γ. ∎

PROPOSITION 7.8. Let δ be a circle passing through the center O of γ. The image of δ minus O under inversion in γ is a line *l* not through O; *l* is parallel to the tangent to δ at O.

Proof:
Let A′ be the point on δ diametrically opposite to O, let A be its inverse in γ, and *l* the line perpendicular to $\overleftrightarrow{OA}$ at A (see Figure 7.30). By the proof of Proposition 7.7, inversion in γ maps *l* onto δ minus O; hence, it must map δ minus O onto *l* (Proposition 7.1(c)). ∎

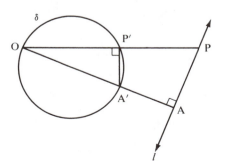

FIGURE 7.30

It is obvious that reflection in a Euclidean line preserves the magnitude but reverses the sense of directed angles. The next proposition generalizes this to inversions.

PROPOSITION 7.9.   A directed angle of intersection of two circles is preserved in magnitude but reversed in sense by an inversion. The same applies to the angle of intersection of a circle and a line or of two lines.

Proof:
Suppose that circles δ and σ intersect at point P with tangents l and m there. Let P' be the inverse of P in γ, let δ' and σ' be the images of δ and σ under inversion in γ, and let l' and m' be their respective tangents at P'. The first assertion then follows from the fact that l' and m' are the reflections of l and m across the perpendicular bisector of PP' (Proposition 7.6). The other cases follow from Propositions 7.7 and 7.8. ∎

The next proposition shows that inversion preserves the cross-ratio used to define Poincaré length.

PROPOSITION 7.10.   If A, B, P, Q are four points distinct from the center O of γ and A', B', P', Q' are their inverses in γ, then $(AB, PQ) = (A'B', P'Q')$.

Proof:
By Lemma 7.3, $(\overline{AP})/(\overline{OA}) = (\overline{A'P'})/(\overline{OP'})$ and $(\overline{AQ})/(\overline{OA}) = (\overline{A'Q'})/(\overline{OQ'})$, whence:

(1)
$$\frac{\overline{AP}}{\overline{AQ}} = \frac{\overline{AP}}{\overline{OA}} \cdot \frac{\overline{OA}}{\overline{AQ}} = \frac{\overline{OQ'}}{\overline{OP'}} \cdot \frac{\overline{A'P'}}{\overline{A'Q'}}.$$

Similarly,

(2)
$$\frac{\overline{BQ}}{\overline{BP}} = \frac{\overline{OP'}}{\overline{OQ'}} \frac{\overline{B'Q'}}{\overline{B'P'}}.$$

Multiplying equations (1) and (2) gives the result. ∎

PROPOSITION 7.11.   Let circle δ be orthogonal to circle γ. Then inversion in δ maps γ onto γ and maps the interior of γ onto itself. Inversion in δ preserves incidence, betweenness, and congruence in the sense of the Poincaré disk model inside γ.[7]

Proof:
The corollary to Proposition 7.6 tells us that γ is mapped onto itself. Suppose that P is inside γ and P' is its inverse in δ. Let C be the center and s the radius of δ. Let $\overrightarrow{CP}$ cut γ at Q and Q', so that by Proposition 7.5 $(\overline{CQ})(\overline{CQ'}) = s^2 = (\overline{CP})(\overline{CP'})$. Since P lies between Q and Q', we have the inequalities CQ < CP < CQ'. Taking the reciprocal reverses inequalities, and we get $s^2/\overline{CQ} > s^2/\overline{CP} > s^2/\overline{CQ'}$, which is the same as CQ' > CP' > CQ. Thus, P' lies between Q and Q' and therefore is inside γ.

By Propositions 7.6, 7.8, and 7.9, inversion in δ maps any circle σ orthogonal to γ either onto another circle σ' orthogonal to γ or onto a line σ' orthogonal to γ, i.e., a line through the center O of γ. Obviously, the line σ joining O to C is mapped onto itself and any other line σ through O is mapped onto a circle σ' punctured at C, which is orthogonal to γ (by Propositions 7.7 and 7.9). In all these cases the above argument shows that the part of σ inside γ maps onto the part of σ' inside γ. Hence, P-lines are mapped onto P-lines.

If A and B are inside γ and P and Q are the ends of the P-line through A and B, then inversion in δ maps P and Q onto the ends of the P-line through A' and B'. By Proposition 7.10, $d(AB) = d(A'B')$, so congruence of segments is preserved. Proposition 7.9 shows that congruence of angles is also preserved. Furthermore, Poincaré betweenness is also preserved because B is between A and D if and only if A, B, and D are Poincaré-collinear and $d(AD) = d(AB) + d(BD)$. ∎

We come finally to the verification of the **SAS axiom.** We are given two Poincaré triangles $\triangle ABC$ and $\triangle XYZ$ inside γ such that $\angle A \cong \angle X$, $d(AC) = d(XZ)$, and $d(AB) = d(XY)$ (Figure 7.31). We must prove that the triangles are Poincaré-congruent. We first reduce to the case where $A = X = O$ (the center of γ): let δ be the circle orthogonal to γ through A and B, and σ the circle orthogonal to γ through A and C.

---

[7] It is easy to see that if, in the statement of Proposition 7.11, δ is taken to be a line through the center O of γ and "inversion" is replaced by "reflection," then the conclusion of Proposition 7.11 still holds.

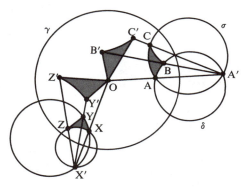

FIGURE 7.31

Then δ again meets σ at point A' outside γ, which is inverse to A in γ (Proposition 7.5). Let ε be the circle centered at A' of radius s, where $s^2 = (\overline{AA'})(\overline{A'O})$. Since $\overline{AA'} = \overline{A'O} - \overline{AO}$, $s^2 = (\overline{A'O})^2 - (\overline{AO})(\overline{A'O}) = (\overline{A'O})^2 - r^2$, where r is the radius of γ. This equation shows that ε is orthogonal to γ (converse of the Pythagorean theorem). By definition of ε, O is the inverse of A in ε, and by Proposition 7.11, inversion in ε maps the Poincaré triangle $\triangle ABC$ onto a Poincaré-congruent Poincaré triangle $\triangle OB'C'$. In the same way, Poincaré triangle $\triangle XYZ$ can be mapped by inversion onto a Poincaré-congruent Poincaré triangle $\triangle OY'Z'$ (see Figure 7.31).

LEMMA 7.4. If $d(OB) = d$, then $\overline{OB} = r(e^d - 1)/(e^d + 1)$, where e is the base of the natural logarithm and r is the radius of γ.

Proof:
If P and Q are the ends of the diameter of γ through B, labeled so that Q * O * B * P, then $d = \log(OB, PQ)$. Exponentiating both sides of this equation gives

$$e^d = (OB, PQ) = \frac{\overline{OP}}{\overline{OQ}} \cdot \frac{\overline{BQ}}{\overline{BP}} = \frac{\overline{BQ}}{\overline{BP}} = \frac{r + \overline{OB}}{r - \overline{OB}}$$

and solving this equation for $\overline{OB}$ gives the result. ■

Returning to the proof of SAS, we have shown that we may assume that $A = X = O$. By Lemma 7.4 and the SAS hypothesis, we get $\overline{OB} = \overline{OY}, \overline{OC} = \overline{OZ}$, and $\angle BOC \cong \angle YOZ$. Hence, a suitable Euclidean rotation about O—combined, if necessary, with reflection in a diameter — will map Euclidean triangle △OBC onto Euclidean triangle △OYZ.[8] This transformation maps γ onto itself and the orthogonal circle through B and C onto the orthogonal circle through Y and Z, preserving Poincaré length and angle measure. Hence, the Poincaré triangles △OBC and △OYZ are Poincaré congruent. ■

This verification of SAS actually proves the following geometric description of Poincaré congruence:

THEOREM 7.1. Two triangles in the Poincaré disk model are Poincaré-congruent if and only if they can be mapped onto each other by a succession of inversions in circles orthogonal to γ and/or reflections in diameters of γ.

We will now apply the Poincaré model to determine the formula of J. Bolyai and Lobachevsky for the angle of parallelism. Let $\Pi(d)$ denote the number of *radians* in the angle of parallelism corresponding to the Poincaré distance d (the number of radians is $\pi/180$ times the number of degrees).

THEOREM 7.2. In the Poincaré disk model the formula for the angle of parallelism is $e^{-d} = \tan[\Pi(d)/2]$.

In this formula e is the base for the natural logarithm. The trigonometric tangent function is defined analytically as sin/cos, where the sine and cosine functions are defined by their Taylor series expansions (the tangent is *not* to be interpreted as the ratio of opposite to adjacent for a right triangle in the hyperbolic plane!).

Proof:
By definition of the angle of parallelism, d is the Poincaré distance $d(PQ)$ from some point P to some Poincaré line l, and $\Pi(d)$ is the number of radians in the angle that a limiting parallel ray to l through P makes with $\overrightarrow{PQ}$. We may choose l to be a diameter of γ and Q to be the center of γ, so that P lies on the perpendicular diameter. A limiting parallel ray through P is then an arc of a circle δ orthogonal to γ such that δ is tangent to l at one end Σ. The tangent line to δ at P therefore meets l at some interior point R that is the

---

[8] This is "intuitively obvious," but will be justified by Proposition 9.5 of Chapter 9.

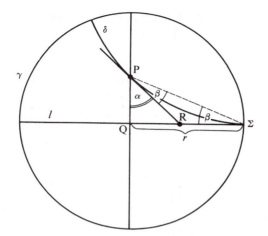

FIGURE 7.32

pole of chord PΣ of δ, and, by Proposition 7.4, ⦠RPΣ and ⦠RΣP both have the same number of radians β (see Figure 7.32). Let α = Π(d), which is the number of radians in ⦠RPQ. Since 2β is the number of radians in ⦠PRQ (exterior to △PRΣ), we get α + 2β = π/2, or β = π/4 − α/2. The Euclidean distance $\overline{PQ}$ is r tan β, so that, by the proof of Lemma 7.4,

$$e^d = \frac{1 + \tan\beta}{1 - \tan\beta}$$

Using the formula for β and the trigonometric identity

$$\tan(\pi/4 - \alpha/2) = \frac{1 - \tan(\alpha/2)}{1 + \tan(\alpha/2)},$$

we get the desired formula after some algebra. ■

We have developed only enough of the geometry of inversion in circles to verify the axioms in the Poincaré disk model. You will find further developments in the exercises and in Chapters 9 and 10. Inversion has many other applications in geometry, notably in Feuerbach's famous theorem on the nine-point circle of a triangle, the problem of Apollonius, and construction of linkages that change linear motion into curvilinear motion (see Eves, 1972; Kay, 1969; Kutuzov, 1960; and Pedoe, 1970).

THE PROJECTIVE NATURE OF THE BELTRAMI-KLEIN MODEL

Having verified that the Poincaré disk interpretation is indeed a model of hyperbolic geometry, it follows from the isomorphism previously discussed that the Klein interpretation is also a model.

To be more explicit, consider the unit sphere Σ in Cartesian three-dimensional space given by the equation $x_1^2 + x_2^2 + x_3^2 = 1$. Let γ be the unit circle in the equatorial plane of Σ, determined by the equation $x_3 = 0$ and the equation for Σ. We will represent both the Poincaré disk and the Klein disk by the set Δ of points inside γ, and we will take as our isomorphism F the composite of two mappings: If N is the north pole $(0, 0, 1)$ of Σ, first project Δ onto the southern hemisphere of Σ stereographically from N. Then project orthogonally back upward to the disk Δ (see Figure 7.33).

The isomorphism F will be considered to go from the Poincaré model to the Klein model. By an easy exercise in similar triangles, you can show that F is given in coordinates by

$$F(x_1, x_2, 0) = \left(\frac{2x_1}{1 + x_1^2 + x_2^2}, \frac{2x_2}{1 + x_1^2 + x_2^2}, 0 \right)$$

Or, if we ignore the third (zero) coordinate and use the single complex coordinate $z = x_1 + ix_2$, then F is given by

$$F(z) = \frac{2z}{1 + |z|^2}.$$

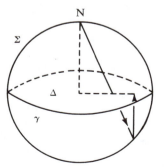

FIGURE 7.33

It is clear that F maps the diameter of γ with ends P and Q onto the same diameter (but moving the points on the diameter out toward the circle). Let δ be a circle orthogonal to γ and cutting γ at points P and Q. We claim that F maps the Poincaré line with ends P and Q onto the open chord P)(Q. In fact, *if* A *is on the arc of δ from P to Q inside γ, then* $F(A)$ *is the point at which* $\overrightarrow{OA}$ *hits chord* PQ (see Figure 7.34).

Proof:
We can prove this as follows. Suppose the center C of δ has coordinates (c_1, c_2). By Proposition 7.3, the points P and Q are the intersections with γ of the circle having CO as diameter. After simplifying, the equation of this circle turns out to be

(1)
$$x_1^2 - c_1 x_1 + x_2^2 - c_2 x_2 = 0.$$

Combining this equation with the equation $x_1^2 + x_2^2 = 1$ for γ gives the equation

(2)
$$c_1 x_1 + c_2 x_2 = 1$$

for the line joining P to Q (called the *polar* of C with respect to γ). Since δ is orthogonal to γ, $\sphericalangle OQC$ is a right angle, and the Pythagorean theorem gives

(3)
$$\overline{CQ}^2 = \overline{CO}^2 - \overline{OQ}^2 = c_1^2 + c_2^2 - 1$$

for the square of the radius of δ. Hence, δ is the circle

$$(x_1 - c_1)^2 + (x_2 - c_2)^2 = c_1^2 + c_2^2 - 1,$$

which simplifies to

(4)
$$x_1^2 + x_2^2 = 2c_1 x_1 + 2c_2 x_2 - 1.$$

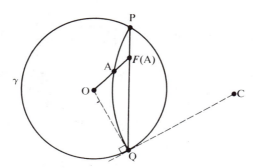

FIGURE 7.34

If now $A = (a_1, a_2)$ lies on δ and $F(A) = (b_1, b_2)$ is its image under F, we have for $j = 1, 2$

(5) $$b_j = 2a_j/(1 + a_1^2 + a_2^2),$$
(6) $$b_j = a_j/(c_1a_1 + c_2a_2).$$

It follows that

(7) $$c_1b_1 + c_2b_2 = 1,$$

and hence, $F(A)$ lies on the polar of C, as asserted. ∎

We now use the isomorphism F to define congruence in the Klein model. Two segments (respectively, two angles) are interpreted to be *Klein-congruent* if their inverse images under F in the Poincaré model are Poincaré-congruent (as was defined before). With this interpretation, the verification of the congruence axioms is immediate. (It follows from this interpretation that the Klein model is conformal only at O.)

Next, let us *justify the previous description of perpendicularity in the Klein model* (p. 239). According to the above definition, two Klein lines l and m are Klein-perpendicular if and only if their inverse images $F^{-1}(l)$ and $F^{-1}(m)$ are perpendicular Poincaré lines. There are three cases to consider.

Case I. Both l and m are diameters. In this case it is clear that perpendicularity has its usual Euclidean meaning.

Case II. Only l is a diameter. Then $F^{-1}(l) = l$. The only way $F^{-1}(m)$, an arc of an orthogonal circle δ, can be perpendicular to l is if the Euclidean line extending l passes through the center C of δ (see Figure 7.35). In that case the extension of l is the perpendicular bisector of chord m (Exercise 17, Chapter 4). Conversely, if l is perpendicular to m in the Euclidean sense, l bisects m, and hence, the extension of l goes through C and l is then perpendicular to arc $F^{-1}(m)$.

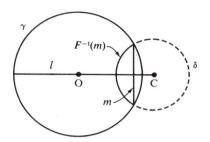

FIGURE 7.35

Case III. Neither l nor m is a diameter. Then $F^{-1}(l)$ and $F^{-1}(m)$ are arcs of circles δ and σ orthogonal to γ. Suppose δ is orthogonal to σ. By Proposition 7.4, the centers of these circles are the poles $P(l)$ and $P(m)$ of l and m, since these circles meet γ at the ends of l and m. Let P and Q be the ends of m. Inversion in δ interchanges P and Q, since this inversion maps both γ and σ onto themselves (corollary to Proposition 7.6). But if P and Q are inverse in δ, the Euclidean line joining them has to pass through the center $P(l)$ of δ (see Figure 7.36).

Conversely, if the extension of m passes through $P(l)$, then P and Q are inverse to each other in δ (since points on γ are mapped onto γ by inversion in δ). By Proposition 7.5, σ is orthogonal to δ. ∎

Next, let us describe the interpretation of reflections in the Klein model. In both Euclidean and hyperbolic geometries the *reflection* in a line m is the transformation R_m of the plane, which leaves each point of

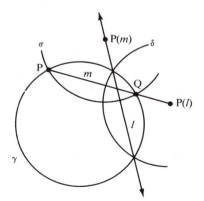

FIGURE 7.36

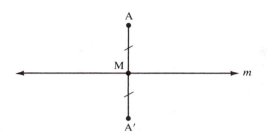

FIGURE 7.37

m fixed and transforms a point A not on *m* as follows. Let M be the foot of the perpendicular from A to *m*. Then, by definition R_m(A) is the unique point A′ such that A′ ∗ M ∗ A and A′M ≅ MA (Figure 7.37). In Major Exercise 2, Chapter 3, you showed that reflection preserves incidence, betweenness, and congruence.

Returning to the Klein model, assume first that *m* is not a diameter of γ and let P be its pole. To drop a Klein perpendicular from A to *m*, we draw the line joining A and P. Let it cut *m* at M and let *t* be the chord of γ cut out by this Euclidean line. Let Q be the pole of *t* and draw the line joining Q and A. Let this line cut γ at Σ and Σ′ and let *n* be the open chord Σ) (Σ′. Draw the line joining Σ′ and M and let it cut γ again at point Ω. If we now join Ω and Q, we obtain a line that cuts *t* at A′ and γ again at Ω′ (see Figure 7.38).

CONTENTION: The point A′ just constructed is the reflection in the Klein model of A across *m*. The Euclidean lines extending ΩΣ and Ω′Σ′ meet at P and ΩΣ′ meets Ω′Σ at point M.

One justification for this construction is given in Major Exercise 12, Chapter 6. Here is another. Start with divergently parallel Klein lines *l* = ΩΩ′ and *n* = ΣΣ′ and their common perpendicular *t*. Let *l* meet *t* in A′ and *n* meet *t* in A, and let M be the midpoint of AA′ in the sense of the model. Let *m* be the Klein line through M Klein-perpendicular to *t*; *m* is obtained by joining M to the pole Q of *t*. Ray MΣ′ is limiting parallel to *n*. If we reflect across *m*, then *n* is mapped onto the line through A′ Klein-perpendicular to *t*, namely, the line *l*. The end Σ′ is mapped onto the end of *l* on the same side of *t* as Σ′, namely, the point Ω′. Hence, ray MΣ′ is mapped onto ray MΩ′. Now reflect across *t*; Ω′ is sent to Ω, so MΩ′ is mapped to MΩ. But successive reflections in the Klein-perpendicular lines *m* and *t* combine to give the 180° rota-

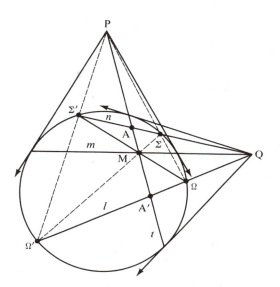

FIGURE 7.38

tion about M. Hence, MΩ is the ray opposite to MΣ′. Similarly, MΣ is the ray opposite to MΩ′. Since reflection in *m* sent Σ′ to Ω′ and Σ to Ω, Σ′Ω′ and ΣΩ must both be Klein-perpendicular to *m* and their Euclidean extensions meet at the pole P of *m*.

Second, let us describe the Klein reflection for the case in which *m* is a diameter of *γ*. In this case P is a point at infinity, *t* is perpendicular to *m* in the Euclidean sense, and M is the Euclidean midpoint of chord *t* (since a diameter perpendicular to a chord bisects it). Chord ΩΣ was shown to be perpendicular to diameter *m* in the argument above, so Ω is the Euclidean reflection of Σ across *m*. Hence, $\overleftrightarrow{Q\Omega}$ is the Euclidean reflection of $\overleftrightarrow{Q\Sigma}$ and we deduce that A′ *is the ordinary Euclidean reflection of* A *across diameter m* (see Figure 7.39).

In order to describe the Klein reflection more succinctly, let us return to the notion of cross-ratio (AB, CD) defined by the formula

$$(AB, CD) = \frac{\overline{AC}}{\overline{AD}} \cdot \frac{\overline{BD}}{\overline{BC}}$$

DEFINITION. If A, B, C, and D are four distinct *collinear* points in the Euclidean plane such that (AB, CD) = 1, we say that C and D are *harmonic conjugates* with respect to AB and that ABCD is a *harmonic*

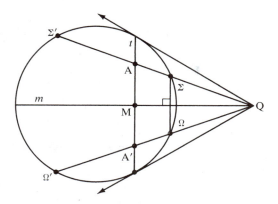

FIGURE 7.39

tetrad. By symmetry of the cross-ratio, A and B are then also harmonic conjugates with respect to CD.

Another way to write the condition for a harmonic tetrad is $\overline{AC}/\overline{AD} = \overline{BC}/\overline{BD}$. Since C and D are distinct, one must be inside segment AB and the other outside (so that "C and D divide AB internally and externally in the same ratio"). Moreover, given AB, then C and D determine each other uniquely. For example, suppose A * C * B, and let $k = \overline{AC}/\overline{CB}$. If $k < 1$, then D is the unique point such that D * A * B and $\overline{DB} = \overline{AB}/(1 - k)$, whereas if $k > 1$, then D is the unique point such that A * B * D and $\overline{DB} = \overline{AB}/(k - 1)$; see Figure 7.40. The case $k = 1$ is indeterminate, for there is no point D outside AB such that $\overline{AD} = \overline{BD}$. Thus, the midpoint M of AB has no harmonic conjugate. This exception can be removed by completing the Euclidean plane to the real projective plane by adding a "line at infinity" (see Chapter 2). Then the harmonic conjugate of M is defined to be the "point at infinity" on $\overleftrightarrow{AB}$.

There is a nice way of constructing the harmonic conjugate of C with respect to AB with straightedge alone: Take any two points I and J collinear with C but not lying on $\overleftrightarrow{AB}$. Let $\overleftrightarrow{AJ}$ meet $\overleftrightarrow{BI}$ at point K and

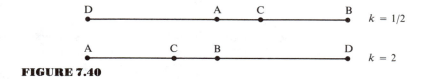

FIGURE 7.40

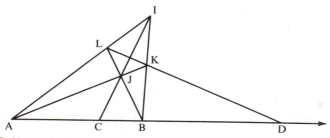

FIGURE 7.41

let $\overleftrightarrow{AI}$ meet $\overleftrightarrow{BJ}$ at point L. *Then $\overleftrightarrow{AB}$ meets $\overleftrightarrow{KL}$ at the harmonic conjugate* D *of* C (Figure 7.41).

We will justify this *harmonic construction* on p. 266. Meanwhile, as a device to help remember the construction, "project" line $\overleftrightarrow{ID}$ to infinity. Then our figure becomes Figure 7.42. Since □A′B′K′L′ is now a parallelogram, we see that C′ is the midpoint of A′B′ and its harmonic conjugate is the "point at infinity" D′ on $\overleftrightarrow{A'B'}$. (This mnemonic device can be turned into a proof based on projective geometry — see Eves, 1972, Chapter 6.)

If you will now refer back to Figure 7.38, where the Klein reflection A′ of A was constructed, you will see that A′ *is the harmonic conjugate of* A *with respect to* MP. Just relabel the points in Figure 7.38 by the correspondences I-Σ′, J-Σ, K-Ω, L-Ω′, A-P, B-M, C-A, and D-A′ to obtain a figure for constructing the harmonic conjugate.

DEFINITION. Let *m* be a line and P a point not on *m*. A transformation of the Euclidean plane called *the harmonic homology with center* P *and axis m* is defined as follows. Leave P and every point on *m* fixed. For

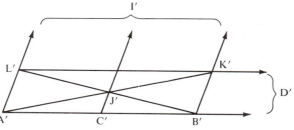

FIGURE 7.42

any other point A let the line *t* joining P to A meet *m* at M. Assign to A the unique point A′ on *t*, which is the harmonic conjugate of A with respect to MP.

With this definition we can restate our result.

THEOREM 7.3. Let *m* be a Klein line that is not a diameter of *γ* and let P be its pole. Then reflection across *m* is interpreted in the Klein model as restriction to the interior of *γ* of the harmonic homology with center P and with axis the Euclidean line extending *m*. If *m* is a diameter of *γ*, then reflection across *m* has its usual Euclidean meaning.

To justify the harmonic construction, we need the notion of a *perspectivity*. This is the mapping of a line *l* onto a line *n* obtained by projecting from a point P not on either line (Figure 7.43). It assigns to point A on *l* the point A′ of intersection of $\overleftrightarrow{PA}$ with *n*. (Should $\overleftrightarrow{PA}$ be parallel to *n*, the image of A is the point at infinity on *n*.) P is called the *center* of this perspectivity.

LEMMA 7.5. A perspectivity preserves the cross-ratio of four collinear points; i.e., if A, B, C, and D are four points on line *l* and A′, B′, C′, and D′ are their images on line *n* under the perspectivity with center P, then (AB, CD) = (A′B′, C′D′).

Proof:
By Exercise 23, Chapter 5, we have

$$\frac{\overline{AC}}{\overline{BC}} = \frac{\overline{AP}\,\sin\sphericalangle APC}{\overline{BP}\,\sin\sphericalangle BPC}$$

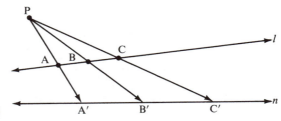

FIGURE 7.43

and

$$\frac{\overline{BD}}{\overline{AD}} = \frac{\overline{BP} \, \sin \, \sphericalangle BPD}{\overline{AP} \, \sin \, \sphericalangle APD},$$

which gives

$$(AB, CD) = \frac{(\sin \, \sphericalangle APC)(\sin \, \sphericalangle BPD)}{(\sin \, \sphericalangle BPC)(\sin \, \sphericalangle APD)}.$$

But $\sin \, \sphericalangle APC = \sin \, \sphericalangle A'PC'$, $\sin \, \sphericalangle PBD = \sin \, \sphericalangle B'PD'$, and so on, so we obtain the same formula for $(A'B', C'D')$. ■

Now refer back to Figure 7.41. Let $\overleftrightarrow{IJ}$ meet $\overleftrightarrow{KL}$ at point M. Using the perspectivity with center I, Lemma 7.5 gives us $(AB, CD) =$ (LK, MD), whereas using the perspectivity with center J we get $(AB, CD) = (KL, MD)$. But $(KL, MD) = 1/(LK, MD)$, by definition of cross-ratio. Hence, (AB, CD) is its own reciprocal, which means $(AB, CD) = 1$, i.e., ABCD is a harmonic tetrad, as asserted. This justifies the harmonic construction on p. 265. ■

Next, we will apply Theorem 7.3 to calculate the length of a segment in the Klein model. According to our general procedure, length in the Klein model is defined by pulling back to the Poincaré model via the inverse of the isomorphism F and using the definition of length already given there. Thus, the length $d'(AB)$ of a segment in the Klein model is given by $d'(AB) = d(ZW) = |\log(ZW, PQ)|$, where $A = F(Z), B = F(W)$, and P and Q are the ends of the Poincaré line through Z and W. By our earlier result illustrated in Figure 7.34, P and Q are also the ends of the Klein line through A and B.

The next theorem shows how to calculate $d'(AB)$ directly in terms of A, B, P, and Q. In its proof we will need the remark "the cross-ratio (AB, PQ) is preserved by any Klein reflection." This is clear if we are reflecting in a diameter of γ. Otherwise, by Theorem 7.3, the Klein reflection is a harmonic homology whose center R lies outside γ. A reflection in the hyperbolic plane preserves collinearity, so for any Klein line l the mapping of l onto its Klein reflection n is just the perspectivity with center R. Therefore, Lemma 7.5 ensures that the cross-ratio is preserved.

THEOREM 7.4. If A and B are two points inside γ and P and Q are the ends of the chord of γ through A and B, then the *Klein length* of segment AB is given by the formula

$$d'(AB) = \tfrac{1}{2}|\log(AB, PQ)|.$$

Proof:
We saw in the verification of the SAS axiom for the Poincaré disk model that any Poincaré line can be mapped onto a diameter by an inversion in a suitable orthogonal circle. Proposition 7.10 guarantees that cross-ratios are preserved by inversions. The transformation of the Klein model that corresponds to this inversion under our isomorphism F is a harmonic homology (Theorem 7.3), and this preserves cross-ratios of collinear points by the above remark. Hence, we may assume that A and B lie on a diameter.

Let $A = F(Z)$ and $B = F(W)$, so that, by definition, $d'(AB) = d(ZW)$. After a suitable rotation (which preserves cross-ratios), we may assume that the given diameter is the real axis. Its ends P and Q then have coordinates $-1, +1$. If Z and W have real coordinates z and w, then

$$(ZW, PQ) = \frac{1+z}{1-z} \cdot \frac{1-w}{1+w}$$

$$(AB, PQ) = \frac{1+F(z)}{1-F(z)} \cdot \frac{1-F(w)}{1+F(w)}.$$

But

$$1 - F(z) = 1 - \frac{2z}{1+|z|^2} = \frac{1 - 2z + |z|^2}{1 + |z|^2}$$

$$1 + F(z) = \frac{1 + 2z + |z|^2}{1 + |z|^2}$$

$$\frac{1 + F(z)}{1 - F(z)} = \frac{1 + 2z + |z|^2}{1 - 2z + |z|^2}.$$

Since z is real, $z = \pm|z|$ and we get

$$\frac{1 + F(z)}{1 - F(z)} = \left(\frac{1+z}{1-z}\right)^2.$$

From this and the formula obtained from it by substituting w for z, it follows that $(AB, PQ) = (ZW, PQ)^2$, and taking logarithms of both sides proves the theorem. ∎

Finally, let us apply our results to justify J. Bolyai's construction of the limiting parallel ray (p. 198). We are given a Klein line l and a point P not on it. Point Q on l is the foot of the Klein perpendicular t from P to l, and m is the Klein perpendicular to t through P. Let R be any other point on l and S the foot on m of the Klein perpendicular from R. Bolyai's construction is based on the contention that if the limiting parallel ray to l from P in the direction $\overrightarrow{QR}$ meets RS at X, then PX is Klein-congruent to QR.

Let T and M be the poles of t and m. Let Ω and Ω' be the ends of l. If we join these ends to M, the intersections Σ and Σ' with γ will be the ends of the Klein reflection n of l across m.

As Figure 7.44 shows, the collinear points Ω, X, P, and Σ' are in perspective with the collinear points Ω, R, Q, and Ω' (in that order), the center of the perspectivity being M. By Lemma 7.5, such a perspectivity preserves cross-ratios, so that $(XP, \Omega\Sigma') = (RQ, \Omega\Omega')$. Theorem 7.4 tells us that $d'(XP) = d'(RQ)$, justifying Bolyai's contention. (In case m is a diameter of γ, M is a point at infinity; then instead of Lemma 7.5 we use the parallel projection theorem (preceding Exercises 18–26, Chapter 5) to deduce the above equality of cross-ratios.) ∎

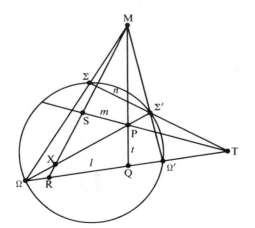

FIGURE 7.44

Note: The method used to prove Theorems 7.2 and 7.4 is very useful for solving other problems in the Klein and Poincaré models. The idea is that the figure being studied can be moved, by a succession of hyperbolic reflections, to a special position where one or more of the hyperbolic lines is represented by a diameter of the absolute circle γ and one point is the center O of γ. The movement to this special position does not alter the geometric properties of the figure, and in that special position, elementary arguments and calculations based on Euclidean geometry can be used to solve the problem.

For example, if P, P' $\neq$ O, then the statement OP $\cong$ OP' has the same truth value whether interpreted in the Euclidean, Poincaré, or Klein senses (according to Lemma 7.4 and Theorem 7.4), and $\sphericalangle$POP' has the same measure in all three senses. In particular, a hyperbolic circle with hyperbolic center O is represented in both models by a Euclidean circle with Euclidean center O.

You will see some nice applications of this method in Exercises K-15, K-17 through K-20, and P-5, and in Chapters 9 and 10. The general study of geometric motions is in Chapter 9.

REVIEW EXERCISE

Which of the following statements are correct?

(1) Although 2000 years of efforts to prove the parallel postulate as a theorem in neutral geometry have been unsuccessful, it is still possible that someday some genius will succeed in proving it.

(2) If we add to the axioms of neutral geometry the elliptic parallel postulate (that no parallel lines exist), we get another consistent geometry called elliptic geometry.

(3) All the ultra-ideal points in the Klein model are points in the Euclidean plane outside γ.

(4) Both the Klein and Poincaré models are "conformal" in the sense that congruence of angles has the usual Euclidean meaning.

(5) In the Poincaré model "lines" are represented by all open diameters of a fixed circle γ and by all open arcs inside γ of circles intersecting γ.

(6) For any chord A)(B whatever of circle γ, the tangents to γ at the endpoints A and B of the chord meet in a unique point called the *pole* of that chord.

(7) In the Poincaré model two Poincaré lines are interpreted as "perpendicular" if and only if they are perpendicular in the usual Euclidean sense.

(8) In the Klein model two open chords are interpreted to be "perpendicular" if and only if they are perpendicular in the usual Euclidean sense.

(9) Inversion in a given circle maps all circles onto circles.

(10) Ultra-ideal points have no representation in the Poincaré models.

(11) Four points in the Euclidean plane form a harmonic tetrad if they are collinear and their cross-ratio equals 1.

(12) If point O is outside circle δ and a tangent from O to δ touches δ at point T, then the power of O with respect to δ is equal to the square of the distance from O to T.

(13) Let point P lie on circle δ and let P′ and δ' be their inverses in another circle γ such that γ does not pass through P or the center of δ. Then the tangent to δ' at P′ is parallel to the tangent to δ at P.

(14) The inverse of the center of a circle δ is the center of the inverted circle δ'.

(15) In order for the midpoint M of segment AB to have a harmonic conjugate with respect to AB, for all A and B, the Euclidean plane must be extended to the real projective plane by adding a line of points at infinity.

(16) If a statement in plane hyperbolic geometry holds when interpreted in the Klein or Poincaré model, then that statement is a theorem in hyperbolic geometry.

The following exercises (all of which are major exercises) will be divided into three categories: (1) K-exercises, on the Klein model; (2) P-exercises, on the Poincaré models and on circles; (3) H-exercises, on harmonic tetrads and theorems of Menelaus, Ceva, Gergonne, and Desargues.

K-EXERCISES

K-1. Verify the interpretations of the incidence axioms, the betweenness axioms, and Dedekind's axiom for the Klein model (Archimedes' axiom follows from Dedekind's – see Chapter 3; see Exercise 31, Chapter 4, for the interpretation of B-4.)

K-2. (a) Let *l* be a diameter of γ and let *m* be an open chord of γ that does not meet *l* and whose endpoints differ from the endpoints of *l*. Draw a diagram showing the common perpendicular *k* to *l* and *m* in the Klein model. (Hint: Use the pole of *m* and the description of perpendicularity in case 1, p. 238.)

(b) Let *l* and *m* be intersecting open chords of γ. It is a valid theorem in hyperbolic geometry that for any two intersecting nonperpendicular lines there exists a third line perpendicular to one of them and asymptotically parallel to the other (see Major Exercise 9, Chapter 6). Draw the two lines in the Klein model that are perpendicular to *l* and asymptotically parallel to *m* (on the left and right, respectively). This shows that the angle of parallelism can be any acute angle whatever. Explain.

(c) In the Euclidean plane any three parallel lines have a common transversal. Draw three parallel lines in the Klein model that do *not* have a common transversal.

K-3. (a) In the Klein model an ideal point and an ordinary point always determine a unique Klein line. Translate this back into a theorem in hyperbolic geometry about limiting parallel rays.

(b) Suppose the ultra-ideal points P(*l*) and P(*m*) are poles of Klein lines *l* and *m*, respectively. You saw in Figure 7.18 that the Euclidean line joining P(*l*) and P(*m*) need not cut through the circle γ, and hence need not determine a Klein line. Show that the only case in which there is a Klein line joining P(*l*) and P(*m*) is when *l* and *m* are divergently parallel.

(c) Suppose the ultra-ideal point P(*l*) is the pole of a Klein line *l* and Ω is an ideal point; Ω is uniquely determined by a ray *r* in the direction of Ω. State the necessary and sufficient conditions on *r* and *l* in order that P(*l*) and Ω determine a Klein line. Translate this into a theorem in hyperbolic geometry.

K-4. Given chords *l* and *m* of γ that are not diameters. Suppose the line extending *m* passes through the pole of *l*. Prove that the line extending *l* passes through the pole of *m*. (Hint: Use either Equation (2), p. 259, or the theory of orthogonal circles.)

K-5. Use the Klein model to show that in the hyperbolic plane there exists a pentagon with five right angles and there exists a hexagon with six right angles. (Hint: Begin with two lines having a common perpendicular. Locate the poles of these two lines, then draw an appropriate line through each of the poles, etc.) Does there exist, for all $n \geq 5$, an *n*-sided polygon with *n* right angles?

K-6. Justify the following construction of the Klein reflection A′ of A across *m*, which is simpler than the one in Figure 7.38. Let Λ be an end of *m* and let P be the pole of *m*. Join Λ to A and let this line cut γ again at Φ. Join Φ to P and let this line cut γ at Φ′. Then A′ is the intersection of $\overleftrightarrow{AP}$ with ΛΦ′. (See Figure 7.45.)

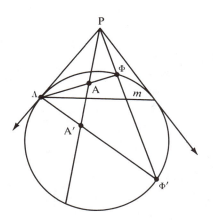

FIGURE 7.45

K-7. Given a segment AA′ in the Klein model. Show how to construct its hyperbolic midpoint with straightedge and compass (see Figures 7.38 and 7.39).

K-8. Construct triangles in the Klein model such that the perpendicular bisectors of the sides are (a) divergently parallel and (b) asymptotically parallel. (See Exercise 13 and Major Exercise 7, Chapter 6.)

K-9. Prove the formula

$$F(z) = \frac{2z}{1 + |z|^2}$$

for the isomorphism F of the Poincaré model onto the Klein model (see Figure 7.33). What is the formula for the inverse isomorphism? Angle measure in the Klein model is defined so that F preserves angle measure; draw the diagram which illustrates this.

K-10. Let $A = (0, 0)$, $B = (0, \frac{1}{2})$, and let l be the diameter of γ cut out by the x axis.
   (a) Find the Klein length $d'(AB)$.
   (b) Find the coordinates of the point M on segment AB that represents its midpoint in the Klein model.
   (c) Find the equation of the locus of points whose perpendicular Klein distance from l equals $d'(AB)$. (This locus is an "equidistant curve" — see Chapter 10, p. 393.)

K-11. Let Ω and Ω' be distinct ideal points and A an ordinary point. Let P be the pole of chord $\Omega\Omega'$, and let Euclidean ray $\overrightarrow{AP}$ cut γ at Σ. Prove that $A\Sigma$ represents the bisector of $\sphericalangle\Omega A\Omega'$ in the Klein model (see Figure 7.46). Apply this result to justify the construction of the line of enclosure given in Major Exercise 8, Chapter 6. (Hint: Use Theorem 6.6.)

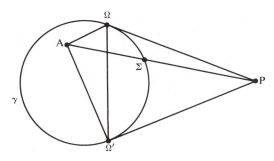

FIGURE 7.46

K-12. In Exercise 16, Chapter 6, you proved the theorem that the angle bisectors of a triangle in hyperbolic geometry (in fact, in neutral geometry) are concurrent. Using the construction of angle bisectors given in the previous exercise and the glossary of the Klein model, translate this theorem into a famous theorem in Euclidean geometry due to Brianchon (see Figure 7.47). This gives a hyperbolic proof of a Euclidean theorem (for a Euclidean proof, see Coxeter and Greitzer, 1967, p. 77).

K-13. It is a theorem in hyperbolic geometry that inside every trebly asymptotic triangle $\triangle \Sigma \Omega \Lambda$ there is a unique point G equidistant from all sides. Show that in the Klein model this theorem is a consequence of

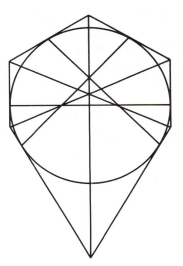

FIGURE 7.47 Brianchon's theorem.

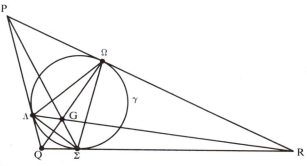

FIGURE 7.48

Gergonne's theorem in Euclidean geometry, which asserts that if the inscribed circle of $\triangle PQR$ touches the sides at points Λ, Σ, and Ω, then segments $P\Sigma$, $Q\Omega$, and $R\Lambda$ are concurrent (see Figure 7.48 and Exercise H-9). Show that $(\angle\Lambda G\Sigma)° = 120°$ in the sense of degree measure for the Klein model. (Hint: To take care of the special case where one side of $\triangle\Sigma\Omega\Lambda$ is a diameter, apply a harmonic homology to transform to the case where Gergonne's theorem applies.)

K-14. In order to express the Klein length $d'(AB) = \frac{1}{2}|\log(AB, PQ)|$ in terms of the coordinates (a_1, a_2) of A and (b_1, b_2) of B, prove that with a suitable ordering of the ends P and Q of the Klein line through A and B you have the formula

$$(AB, PQ)$$
$$= \frac{a_1b_1 + a_2b_2 - 1 - \sqrt{(a_1 - b_1)^2 + (a_2 - b_2)^2 - (a_1b_2 - a_2b_1)^2}}{a_1b_1 + a_2b_2 - 1 + \sqrt{(a_1 - b_1)^2 + (a_2 - b_2)^2 - (a_1b_2 - a_2b_1)^2}}$$

(Hint: If A and B have complex coordinates z and w, then P and Q have complex coordinates $tz + (1 - t)w$ and $uz + (1 - u)w$, where t and u are roots of a quadratic equation $Dx^2 + 2Ex + F = 0$ expressing the fact that P and Q lie on the unit circle. Find the coefficients $D, E,$ and F and show that

$$(AB, PQ) = \frac{t(1 - u)}{u(1 - t)} = \frac{E + F - \sqrt{E^2 - DF}}{E + F + \sqrt{E^2 - DF}}.$$

K-15. Use the formula for Klein length given in Theorem 7.4 to derive a proof of the Bolyai-Lobachevsky formula in Theorem 7.2 for the Klein model. (Hint: Take the vertex of the angle of parallelism α to be the center O of the absolute and show that the Klein distance d'

corresponding to α is given by

$$d' = \tfrac{1}{2} \log\frac{1 + \cos \alpha}{1 - \cos \alpha}.$$

Then use a half-angle formula from trigonometry.)

K-16. (a) Show that a Cartesian line l of equation $Ax + By + C = 0$ is a secant of the unit circle if and only if

$$A^2 + B^2 - C^2 > 0.$$

We will denote the expression on the left of this inequality by $|l|^2$.

(b) Prove that if $P' = (x', y')$ is the Klein reflection of $P = (x, y)$ across l, then

$$x' = \frac{|l|^2 x - 2A(Ax + By + C)}{|l|^2 + 2C(Ax + By + C)}$$

$$y' = \frac{|l|^2 y - 2B(Ax + By + C)}{|l|^2 + 2C(Ax + By + C)}.$$

(Hint: Use Theorem 7.3. In case $C = 0$, the Euclidean reflection is easy to calculate. If $C \neq 0$, the pole L of l has coordinates $(-A/C, -B/C)$, according to Equation (2), p. 259; you must calculate the coordinates of the point M where line $\overleftrightarrow{LP}$ meets l and then calculate the coordinates of the harmonic conjugate P' of P with respect to L and M.)

K-17. The line perpendicular to the bisector of $\sphericalangle A$ at A is called the *external bisector* of $\sphericalangle A$ (because its rays emanating from A bisect the two supplementary angles to $\sphericalangle A$). You proved (in Exercise 16, Chapter 6) that the (internal) bisectors of the angles of $\triangle ABC$ concur in the center I of the inscribed circle — this is a theorem in neutral geometry.

(a) Prove that in Euclidean geometry the internal bisector of $\sphericalangle A$ is concurrent with the external bisectors of $\sphericalangle B$ and $\sphericalangle C$.

(b) Deduce from the Klein model that in hyperbolic geometry, the internal bisector of $\sphericalangle A$ is "concurrent" with the external bisectors of $\sphericalangle B$ and $\sphericalangle C$ in a point which may be ordinary, ideal, or ultra-ideal. (See Figure 7.49.) (Hint for (a): Use the facts that the bisector of an angle is the locus of interior points equidistant from the sides, and that external bisectors are not parallel. Hint for (b): Take I to be the center O of the absolute γ and notice, using K-11, that the hyperbolic internal bisectors, being diameters of γ, coincide with the Euclidean internal bisectors. Hence, the hyperbolic external bisectors, being perpendicular to diameters of γ, coincide with the Euclidean external bisectors.)

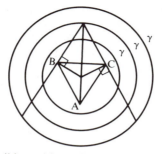

FIGURE 7.49   Three possible positions of the absolute.

K-18. It is a theorem in Euclidean geometry that the altitudes of an acute triangle are concurrent and the lines containing the altitudes of an obtuse triangle are concurrent (see Problem 8, Chapter 9). Applying this theorem to the Klein model, deduce that in hyperbolic geometry the altitudes of an acute triangle are concurrent and that the lines containing the altitudes of an obtuse triangle are "concurrent" in a point which may be ordinary, ideal, or ultra-ideal. (Hint: Place the triangle so that one vertex is O; show that the Klein lines containing the altitudes then coincide with the Euclidean perpendiculars from the vertices to the opposite sides. Use the crossbar and exterior angle theorems to verify that for acute triangles the point of concurrence is ordinary.)

K-19. It is a theorem in Euclidean geometry that the medians of a triangle are concurrent (see Exercise 69, Chapter 9). Show that this theorem also holds in hyperbolic geometry by a special position argument in the Klein model.[9] (Hint: If O is the hyperbolic midpoint of AB, it is also the Euclidean midpoint; if P, Q are the hyperbolic midpoints of AC, BC, use Exercise 11(b), Chapter 6, to show that $\overleftrightarrow{PQ}$ is Euclidean-parallel to $\overleftrightarrow{AB}$ — that is, $\overleftrightarrow{PQ}$ "meets" $\overleftrightarrow{AB}$ in the harmonic conjugate of O with respect to A and B. The result then follows from the harmonic construction in Figure 7.50.)

K-20. We have defined a *parallelogram* to be a quadrilateral in which the lines containing opposite sides are parallel.

   (a) Prove that in Euclidean geometry, a quadrilateral is a parallelo-

[9] Melissa Schmitz, an undergraduate student at the State University of New York at Geneseo, sent me a three-dimensional neutral geometry proof of this theorem due to F. Busulini. She also used a computer to discover that in hyperbolic geometry, the centroid does not lie two-thirds of the distance from each vertex to the midpoint of the opposite side (as it does in Euclidean geometry). Next question: Does the Euler line (Exercise 69, Chapter 9) exist in the projective completion of the hyperbolic plane?

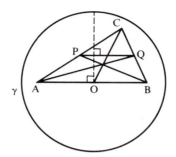

FIGURE 7.50

gram if and only if opposite sides are congruent. Show that in hyperbolic geometry, the opposite sides of a parallelogram need not be congruent.

For the remainder of this exercise, the geometry will be *hyperbolic*.

(b) Given □ABCD with opposite sides congruent. Prove that opposite angles are congruent and that the lines containing opposite sides are divergently parallel (use Exercise 14, Chapter 6). Such a quadrilateral will be called a *symmetric parallelogram*.

(c) Prove that the diagonals of a symmetric parallelogram □ABCD have the same midpoint S, and that S is the symmetry point for both pairs of opposite sides (see Major Exercise 12, Chapter 6 and Figure 6.27).

(d) Show that the diagonals are perpendicular if and only if all four sides are congruent, and in that case, □ABCD has an inscribed circle with center S.

(e) Show that the diagonals are congruent if and only if all four angles are congruent; however, in that case, show that all four sides need not be congruent.

(You can verify these assertions either by direct argument or by using the Klein model, placing S at the center O of the absolute and remarking that □ABCD is then a Euclidean parallelogram.)

K-21. It has been shown by Jenks that in hyperbolic geometry, "betweenness," "congruence," and "asymptotic parallelism" can all be defined in terms of incidence alone. (An important consequence of this observation is that every collineation of the hyperbolic plane is a motion; see Chapter 9). Here are his observations; draw diagrams in the Klein model to see what is going on. First, three distinct lines *a, b, c* form an asymptotic triangle *abc* if and only if for any point P on any one of them — say, on *a* — there exists a unique line *p ≠ a* through P which is parallel to both *b* and *c* (*p* is called an *asymptotic transversal* through P). Second, $a|b$ if and only if there exists a line c such that a, b, c form an asymptotic triangle. Third, given three points P, Q, R on a line m, P ∗ Q ∗ R if and only if given any $a \neq m$ through P, $b \neq m$ through R, and c such that a, b, c form an asymptotic triangle, every line through Q meets at least one of the sides of abc. Fourth, segment PQ on a is congruent to segment P′Q′ on a' if and only if either (1) $a|a'$ and both are asymptotically parallel to the join of the meets of the asymptotic transversals through P and P′ and through Q and Q′, or (2) both are asymptotically parallel to some line a'' on which lies a segment P″Q″ congruent with both PQ and P′Q′ in the sense of (1). Justify (1) by drawing the diagram in the Klein model and applying Lemma 7.5 and Theorem 7.4.

P-EXERCISES

P-1. Using the glossary for the Poincaré disk model, translate the following theorems in hyperbolic geometry into theorems in Euclidean geometry:

(a) If two triangles are similar, then they are congruent.

(b) If two lines are divergently parallel, then they have a common perpendicular and the latter is unique.

(c) The fourth angle of a Lambert quadrilateral is acute.

P-2. State and prove the analogue of Proposition 7.6 when O lies inside δ and the power p of O with respect to δ is negative.

P-3. Let δ be a circle with center C and α a circle not through C having center A. Let A′ be the inverse of A in δ and let circle α' be the image of α under inversion in δ. Prove that A′ is the inverse of C in α' and hence that A′ is not the center of α'. (Hint: Show that any circle β through A′ and C is orthogonal to α' by observing that the image β' of β under inversion in δ is a line orthogonal to α.)

P-4. Let l be a Poincaré line that is not a diameter of γ; l is then an arc of a circle δ orthogonal to γ. Prove that hyperbolic reflection across l is represented in the Poincaré model by inversion in δ. (Hint: Use Proposition 7.10 and the corollary to Proposition 7.6.)

P-5. Let C be a point in the Poincaré disk model. Prove that a circle centered at C in the sense of hyperbolic geometry is represented in the Poincaré model by a Euclidean circle whose center $\neq$ C unless C coincides with the center O of γ. (Hint: First take C = O and use Lemma 7.4. Then map the set of circles centered at O onto the set of circles with C as hyperbolic center by reflection in the Poincaré line that is the Poincaré perpendicular bisector of the Poincaré segment OC. Apply Exercises P-3 and P-4.)

P-6. In the hyperbolic plane with some given unit of length, the distance d for which the angle of parallelism $\Pi(d)° = 45°$ is called *Schweikart's constant.* Schweikart was the first to notice that if $\triangle ABC$ is an isosceles right triangle with base BC, then the length of the altitude from A to BC is bounded by this constant, which is the least upper bound of the lengths of all such altitudes. Prove that for the length function we have defined for the Poincaré disk model, Schweikart's constant equals $\log(1 + \sqrt{2})$ (see Figure 7.51). (Hint: Schweikart's constant is the Poincaré length d of segment OP in Figure 7.51. Show that the Euclidean length of OP is $\sqrt{2} - 1$ and apply Lemma 7.4 to solve for d.)

P-7. Let α be a circle with center A and radius of length r and β a circle with center B and radius of length s. Assume $A \neq B$ and let C be the unique point on $\overleftrightarrow{AB}$ such that $AC^2 - BC^2 = r^2 - s^2$. The line through C perpendicular to $\overleftrightarrow{AB}$ is called the *radical axis* of the two circles.

(a) Prove (e.g., by introducing coordinates) that C exists and is unique, and that for any point P different from A and B, P lies on the radical axis if and only if $\overline{PA}^2 - \overline{PB}^2 = r^2 - s^2$.

(b) For any point X outside both α and β, let T be a point of α such that $\overleftrightarrow{XT}$ is tangent to α at T; similarly let U on β be a point of tangency for $\overleftrightarrow{XU}$. Prove that $\overline{XT} = \overline{XU}$ if and only if X lies on the radical axis of α and β.

(c) Prove that if α and β intersect in two points P and Q, $\overleftrightarrow{PQ}$ is their radical axis.

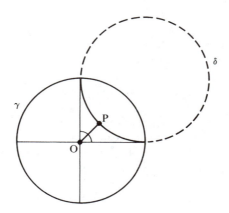

FIGURE 7.51

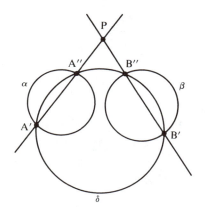

FIGURE 7.52

(d) Prove that if α and β are tangent at point C, the radical axis is the common tangent line through C.

(e) Let X be a point outside both α and β. Prove that X lies on the radical axis of α and β if and only if X has the same power with respect to α and β (see Lemma 7.1).

P-8. Given two nonintersecting, nonconcentric circles α and β with centers A and B, respectively. Justify the following straightedge-and-compass construction of the radical axis of α and β. Draw any circle δ that cuts α in two points A' and A" and cuts β in two points B' and B". If $\overleftrightarrow{A'A''}$ and $\overleftrightarrow{B'B''}$ intersect in a point P, then P lies on the radical axis; the latter is therefore the perpendicular to $\overleftrightarrow{AB}$ through P. (Hint: Draw tangents PS, PT, and PU from P to δ, α, and β and apply Exercises P-7(b) and P-7(c) to show that $\overline{PT} = \overline{PS} = \overline{PU}$. See Figure 7.52.)

P-9. Use Exercise P-7 to verify by a straightedge-and-compass construction that in the Poincaré model two divergently parallel Poincaré lines have a common perpendicular. (Hint: There are four cases to consider, depending on whether the Poincaré line is a diameter of γ or an arc of a circle α orthogonal to γ, and depending on whether radical axes intersect or not. One case is illustrated in Figure 7.53. In case the radical axes are parallel, use the fact that the perpendicular bisector of a chord of a circle passes through the center of the circle (Exercise 17, Chapter 4).)

P-10. Given any Poincaré line l and any Poincaré point P not on l. Construct the two rays from P in the Poincaré model that are limiting parallel to l. (If l is an arc of a circle α orthogonal to γ and intersecting γ at A_1 and A_2, then the problem amounts to constructing a circle β_i through P

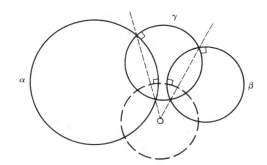

FIGURE 7.53

that is orthogonal to γ and tangent to α at A_i for each of $i = 1, 2$. See Figure 7.54, and use Proposition 7.5.)

P-11. Given an acute angle in the Poincaré model. Construct the unique Poincaré line that is perpendicular to a given side of this angle and limiting parallel to the other. This shows that the angle of parallelism can be any acute angle whatever. (Hint: If both Poincaré lines are arcs of orthogonal circles α and β, let P' be the intersection with γ of the part of α containing the given ray, and let P be the other intersection with γ of $\overleftrightarrow{P'B}$, B being the center of β; see Figure 7.55. Show that P and P' are inverse points in circle β, then find the point of intersection of the tangents to γ at P and P'. Compare with Major Exercise 9, Chapter 6.)

P-12. Given circle γ with center O. For any point $P \neq O$, if P' is the inverse of P in γ, then the line through P' that is perpendicular to $\overleftrightarrow{OP}$ is called the *polar* of P with respect to γ and will be denoted $p(P)$. When P lies outside γ, its polar joins the points of contact of the two tangents to γ from P (see Figure 7.22). When P lies on γ, its polar is the tangent to γ at P, and this is the only case in which P lies on $p(P)$. Prove the following duality property. B lies on $p(A)$ if and only if A lies on $p(B)$.

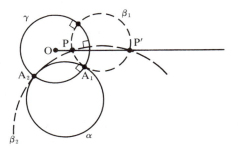

FIGURE 7.54

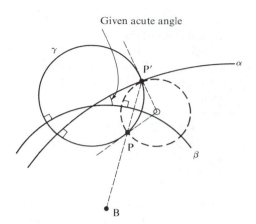

FIGURE 7.55

(Hint: If B lies on $p(A)$, let B′ be the foot of the perpendicular from A to $\overleftrightarrow{OB}$. See Figure 7.56. Show that △OAB′ is similar to △OBA′ and deduce that B′ is the inverse of B in γ. For the significance of this operation of polar reciprocation for the theory of conics, see Coxeter and Greitzer, 1967, Chapter 6.)

P-13. We define three types of *coaxal pencils of circles* as follows:

(1) Given a line t and a point C on t. The corresponding *tangent coaxal pencil* consists of all circles tangent to t at C.

(2) Given two points A and B. The corresponding *intersecting*

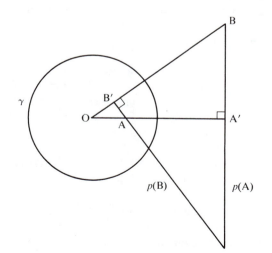

FIGURE 7.56

coaxal pencil consists of all the circles which pass through both A and B, and A and B are the *limiting points* of this pencil.

(3) Given a circle γ and a line t not meeting γ. The corresponding *nonintersecting coaxal pencil* consists of γ and all other circles δ such that t is the radical axis of γ and δ.

Prove the following:

(a) Any two nonconcentric circles belong to a unique coaxal pencil.

(b) Given a coaxal pencil C. All pairs of circles belonging to C have the same radical axis, and the centers of all circles in C lie on a line perpendicular to this radical axis called the *line of centers* of C. (Hint: See Exercise P-7).

P-14. Prove the following:

(a) The set of all circles orthogonal to two given circles γ and δ tangent at C is the tangent coaxal pencil through C whose line of centers is the common tangent t to γ and δ.

(b) The set of all circles orthogonal to two given nonintersecting nonconcentric circles γ and δ is the intersecting coaxal pencil whose line of centers is the radical axis t of γ and δ and whose limiting points are the two points at which every member of this pencil cuts the line joining the centers of γ and δ.

(c) The set of all circles orthogonal to two given circles γ and δ intersecting at A and B is the nonintersecting nonconcentric coaxal pencil whose line of centers is $\overleftrightarrow{AB}$ and whose radical axis is the perpendicular bisector of AB. (See Figure 7.57.)

P-15. Given three circles α, β, and γ. Is there always a fourth circle δ orthogonal to all three of them? If so, is δ unique? (Hint: Consider the radical axes of the three pairs of circles obtained from the three given circles; the center of δ must lie on all three radical axes and must lie outside the three circles.)

P-16. Given a circle γ with center O.

(a) Given $P \neq O$ and P' its inverse in γ. Prove that inversion in γ maps the pencil of lines through P' onto the intersecting coaxal pencil of circles through O and P and maps the orthogonal pencil of concentric circles centered at P' onto the nonintersecting coaxal pencil of circles whose radical axis is the perpendicular bisector of OP.

(b) Given a line l through O. Prove that inversion in γ maps the pencil of lines parallel to l onto the pencil of circles tangent to l at O.

P-17. *The inversive plane* is obtained from the Euclidean plane by adjoining a single point at infinity ∞, which by convention lies on every Euclidean line but does not lie on any Euclidean circle. By a "circle" we mean

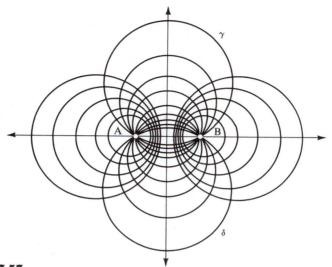

FIGURE 7.57

either an ordinary Euclidean circle or a line in the inversive plane. Two parallel Euclidean lines meet at ∞ when extended to inversive lines; as "circles" they will be considered to be tangent at ∞. Given an ordinary circle γ with center O, define the inverse of O in γ to be ∞. By inversion in a "circle" we mean either inversion in an ordinary circle or reflection across a line. Prove the following:

(a) Inversion in a given "circle" maps "circles" onto "circles."

(b) If A and B are inverse to each other in a "circle" α, and if under inversion in another "circle" β they map to A′, B′, α′, then A′ and B′ are inverse to each other in α′. (Hint for (b): Show that any "circle" γ′ through A′ and B′ is orthogonal to α′ by observing that inversion preserves orthogonality—use Propositions 7.5 and 7.9.)

P-18. In addition to the tangent, intersecting, and nonintersecting coaxal pencils of circles defined in Exercise P-13, define three further pencils of "circles" in the inversive plane as follows:

(4) All the circles having a given point as center

(5) All the lines passing through a given ordinary point

(6) A given line and all lines parallel to it

Furthermore, given a coaxal pencil of circles, we will consider its radical axis as one more "circle" belonging to the pencil. Prove the following:

(a) Two distinct "circles" belong to a unique pencil of "circles."

(b) A pencil of "circles" is invariant as a set under inversion in any "circle" in the pencil. (Hint for (b): The statement is obvious for the three new types of pencils just introduced. For the three coaxal types, use the two preceding exercises.)

P-19. Construct a regular 4-gon in the Poincaré disk model. (Hint: Choose a point A ≠ O on the line $y = x$; let B (respectively, D) be its reflection across the x axis (respectively, the y axis) and let C be obtained from A by 180° rotation about O. Show that □ABCD is a regular 4-gon. Note that as A approaches O, ⦨A approaches a right angle, while as A moves away toward the ideal end of ray $\overrightarrow{OA}$, ⦨A approaches the zero angle.)

P-20. Use the Poincaré model to show that in the hyperbolic plane, there exist two points A, B lying on the same side S of a line l such that no circle through A and B lies entirely within S. This shows that the result in Major Exercise 7, Chapter 5, is another statement equivalent to Euclid's parallel postulate. (Hint: Take l to be a diameter of the Poincaré disk and use Exercise P-5.)

H-EXERCISES

H-1. Let M be the midpoint of AB, $r = \overline{MA}$, and let C, D on $\overleftrightarrow{AB}$ lie on the same side of M, with A, B, C, D distinct. Then C and D are harmonic conjugates with respect to AB if and only if $r^2 = (\overline{MD})(\overline{MC})$.

H-2. If γ and δ are orthogonal circles, AB is a diameter of γ, and δ cuts $\overleftrightarrow{AB}$ in points C and D, then C and D are harmonic conjugates with respect to AB; conversely, if a diameter of one circle is cut harmonically by a second circle, then the two circles are orthogonal (see Figure 7.58). (Hint: If T is a point of intersection of γ and δ, use Lemma 7.1 to show

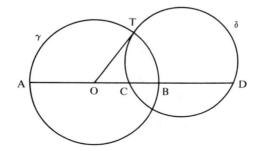

FIGURE 7.58

that the circles are orthogonal if and only if $(\overline{OT})^2 = (\overline{OC})(\overline{OD})$. Now apply Exercise H-1.)

H-3. Given three collinear points A, B, and C. Prove that the fourth harmonic point D is the inverse of C in the circle having AB as diameter. (Hint: Use Exercise H-2 and Proposition 7.5.)

H-4. *Sensed magnitudes.* Given two points A, B. Assign arbitrarily an order (i.e., a direction) to $\overset{\leftrightarrow}{AB}$. Then the length of AB will be considered positive or negative according to whether the direction from A to B is the positive or negative direction on the line. We will denote this signed length by AB, so that we have AB = −BA. If C is a third point on the directed line $\overset{\leftrightarrow}{AB}$, we define the signed ratio in which C divides AB to be AC/CB.

(a) Prove that this signed ratio is independent of the direction assigned to the line and that point C is uniquely determined by this ratio. (Note that C would not be uniquely determined by the unsigned ratio.)

(b) Given parallel lines *l* and *m*. Let transversals *t* and *t′* cut *l* and *m* in B, C and B′, C′, respectively, and let *t* meet *t′* at point A. Prove that AB/BC = AB′/B′C′ (see Exercise 18, Chapter 5).

H-5. *Theorem of Menelaus.* Given △ABC and points D on $\overset{\leftrightarrow}{BC}$, E on $\overset{\leftrightarrow}{CA}$, and F on $\overset{\leftrightarrow}{AB}$ that do not coincide with any of the vertices of the triangle. Define the *linearity number* by [ABC/DEF] = (AF/FB)(BD/DC)(CE/EA). Then a necessary and sufficient condition for D, E, and F to be collinear (Figure 7.59) is that [ABC/DEF] = −1. (Hint: If D, E, and F lie on a line *l*, let the parallel *m* to *l* through A cut $\overset{\leftrightarrow}{BC}$ at G. Use Exercise H-4 to get CE/EA = CD/DG and AF/FB = GD/DB and deduce that the linearity number is −1. Conversely, use Exercise H-4 to show that $\overset{\leftrightarrow}{EF}$ cannot be parallel to $\overset{\leftrightarrow}{BC}$. If these lines meet at D′, use the first part of the proof and the hypothesis to show that BD/DC = BD′/D′C and apply Exercise H-4(a).)

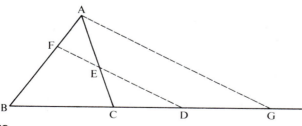

FIGURE 7.59

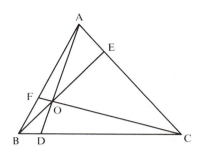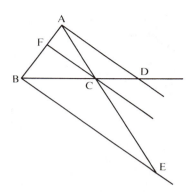

FIGURE 7.60

H-6. *Theorem of Ceva.* Given $\triangle ABC$ and a third point D (respectively E, F) on $\overleftrightarrow{BC}$ (respectively on $\overleftrightarrow{AC}$, $\overleftrightarrow{AB}$). Then the three lines $\overleftrightarrow{AD}$, $\overleftrightarrow{BE}$, and $\overleftrightarrow{CF}$ are either concurrent or parallel if and only if [ABC/DEF] $= +1$ (see Figure 7.60). (Hint: Suppose that the three lines meet at O; apply Menelaus' theorem to $\triangle ADB$ and $\triangle ADC$ to obtain two different expressions for $\underline{OD}/\underline{AO}$, then divide one expression by the other to see that the linearity number is $+1$. If the three lines are parallel, apply Exercise H-4(b). Conversely, if the linearity number is $+1$ and the three lines are not parallel, let $\overleftrightarrow{BE}$ and $\overleftrightarrow{CF}$, for example, meet at O, and let $\overleftrightarrow{AO}$ meet $\overleftrightarrow{BC}$ at D′. Use the first part of the proof and the hypothesis to show that $\underline{BD}/\underline{DC} = \underline{BD'}/\underline{D'C}$ and apply Exercise H-4(a).)

H-7. Given four collinear points A, B, C, and D. Define their *signed cross-ratio* $(\underline{AB}, \underline{CD})$ by $(\underline{AB}, \underline{CD}) = (\underline{AC}/\underline{CB})/(\underline{AD}/\underline{DB})$.
   (a) Prove that ABCD is a harmonic tetrad if and only if $(\underline{AB}, \underline{CD}) = -1$.
   (b) Prove that signed cross-ratios are preserved by perspectivities and parallel projections (see Lemma 7.5 and the parallel projection theorem preceding Exercise 18, Chapter 5).

H-8. Prove that ABCD is a harmonic tetrad if and only if $1/\underline{AB} = \frac{1}{2}(1/\underline{AC} + 1/\underline{AD})$.

H-9. Suppose the inscribed circle of $\triangle ABC$ touches sides BC, CA, and AB at D, E, and F, respectively. Prove that AD, BE, and CF are concurrent in a point G called the *Gergonne point* of $\triangle ABC$; see Figure 7.61. (Hint: By Exercise 16, Chapter 6, the center I of the inscribed circle lies on all three angle bisectors; this gives three pairs of congruent right triangles that can be used to verify the criterion of Ceva's theorem.)

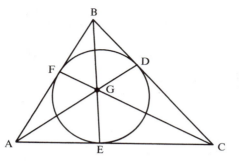

FIGURE 7.61

H-10. Use the theorem of Menelaus to prove Desargues's theorem as stated in Project 1, Chapter 2. (Hint: Referring to Figure 2.10, apply Menelaus' theorem to △BCP, △CAP, and △ABP, and then multiply the three equations to get [ABC/RST] $= -1$. Now apply Menelaus' theorem once more.)

H-11. The theorems of Menelaus and Ceva can be applied to prove famous theorems of Pappus and Pascal and to prove the existence of special points of a triangle. Report on these results, using Kay (1969) or Coxeter and Greitzer (1967) as references.

CHAPTER 8

PHILOSOPHICAL IMPLICATIONS

I have had my solutions for a long time, but I do not yet know how I am to arrive at them.

C. F. GAUSS

WHAT IS THE GEOMETRY OF PHYSICAL SPACE?

We have shown that if Euclidean geometry is consistent, so is hyperbolic geometry, since we can construct models for it within Euclidean geometry. Conversely, it can be proved that if hyperbolic geometry is consistent, so is Euclidean geometry, for the "horocycles" on the "horosphere" in hyperbolic space form a model of the lines on the Euclidean plane (see Kulczycki, 1961, §17). *Thus, the two geometries are equally consistent.*

You may grant now that, logically speaking, hyperbolic geometry deserves to be put on an equal footing with Euclidean geometry. But you may also feel that hyperbolic geometry is just an amusing intellectual pastime, whereas Euclidean geometry accurately represents the physical world we live in and is therefore far more important. Let's examine this idea a little more closely.

Certainly, engineering and architecture are evidence that Euclidean geometry is extremely useful for ordinary measurement of distances that are not too large. However, the representational accuracy of Euclidean geometry is less certain when we deal with larger distances. For example, let us interpret a "line" physically as the path traveled by a light ray. We could then consider three widely separated light sources forming a physical triangle. We would want to measure the angles of this physical triangle in order to verify whether the sum is 180° or not (such an experiment would presumably settle the question of whether space is Euclidean or hyperbolic).

F. W. Bessel, a friend of Gauss, performed such a measurement, using the angle of parallax of a distant star. The results were inconclusive. Why? Because any physical experiment involves experimental error. Our instruments are never completely accurate. Suppose the sum did turn out to be 180°. If the error in our measurement were at most 1/100 of a degree, we could conclude only that the sum was between 179.99° and 180.01°. We could never be sure that it actually was 180°.

Suppose, on the other hand, that measurement gave us a sum of 179°. Although we could conclude only that the sum was between 178.99° and 179.01°, we would be certain that the sum was less than 180°. In other words, the only conclusive result of such an experiment would be that space is hyperbolic![1] The inconclusiveness of Bessel's experiment shows only that if space is hyperbolic, the defects of terrestrial triangles are extremely small.

To repeat the point: Because of experimental error, a physical experiment can never prove conclusively that space is Euclidean — it can prove only that space is non-Euclidean.

The discussion can be made more subtle. We must question the nature of our instruments — aren't they designed on the basis of Euclidean assumptions? We must question our interpretation of "lines" — couldn't light rays travel on curved paths? We must question whether space, especially space of cosmic dimensions, cannot be described by geometries other than these two.

The latter is in fact our present scientific attitude. According to Einstein, space and time are inseparable and the geometry of space-time is affected by matter, so that light rays are indeed curved by the gravitational attraction of masses. Space is no longer conceived of as an empty Newtonian box whose contours are unaffected by the rocks put into it. The problem is much more complicated than Euclid or Loba-

---

[1] If the measurement gave us a sum of 181° with error at most .01°, we would conclude that space is elliptic.

Albert Einstein

chevsky ever imagined — neither of their geometries is adequate for our present conception of space. This does not diminish the historical importance of non-Euclidean geometry. Einstein said, "To this interpretation of geometry I attach great importance, for should I not have been acquainted with it, I never would have been able to develop the theory of relativity." [2]

Here is the famous response of Poincaré to the question of which geometry is true:

If geometry were an experimental science, it would not be an exact science. It would be subjected to continual revision. . . . *The geometrical axioms are therefore neither synthetic a priori intuitions nor experimental*

---

[2] See George Gamow (1956), which tells how Einstein developed a geometry appropriate to general relativity from the ideas of Georg Friedrich Bernhard Riemann (1826–1866).

facts. They are conventions. Our choice among all possible conventions is guided by experimental facts; but it remains free, and is only limited by the necessity of avoiding every contradiction, and thus it is that postulates may remain rigorously true even when the experimental laws which have determined their adoption are only approximate. In other words, *the axioms of geometry* (I do not speak of those of arithmetic) *are only definitions in disguise.* What then are we to think of the question: Is Euclidean Geometry true? It has no meaning. We might as well ask if the metric system is true and if the old weights and measures are false; if Cartesian coordinates are true and polar coordinates false. *One geometry cannot be more true than another: it can only be more convenient.* [italics added] [3]

You may think that Euclidean geometry is the most convenient — it is for ordinary engineering, but not for the theory of relativity. Moreover, R. K. Luneburg contends that visual space, the space mapped on our brains through our eyes, is most conveniently described by hyperbolic geometry. [4]

Philosophers are still arguing about Poincaré's philosophy of conventionalism. One school, which includes Newton, Helmholtz, Russell, and Whitehead, contends that space has an intrinsic metric or standard of measurement. The other school, which includes Riemann, Poincaré, Clifford, and Einstein, contends that a metric is stipulated by convention. The discussion can become very subtle (see Torretti, 1978, Chapter 4).

WHAT IS MATHEMATICS ABOUT?

The preceding discussion sheds new light on what geometry, and in general, mathematics, is about. Geometry is not about light rays, but the path of a light ray is one possible *physical interpretation* of the undefined geometric term "line." Bertrand Russell once said that "mathematics is the subject in which we do not know what we are

---

[3] H. Poincaré (1952), p. 50. See also Essay Topic 18 at the end of this chapter.

[4] R. K. Luneburg (1947), and his article in the *Optical Society of America Journal,* October, 1950, p. 629. See also the articles by O. Blank in that same journal, December, 1958, p. 911, and March, 1961, p. 335, and the explanation in Trudeau (1987), pp. 251–254.

talking about nor whether what we say is true." This is because certain primitive terms, such as "point," "line," and "plane," are undefined and could just as well be replaced with other terms without affecting the validity of results. Instead of saying "two points determine a unique line," we could just as well write "two alphas determine a unique beta." Despite this change in terms, the proofs of all our theorems would still be valid, because correct proofs do not depend on diagrams; they depend only on stated axioms and the rules of logic. Thus, geometry is a purely *formal* exercise in deducing certain conclusions from certain formal premises. Mathematics makes statements of the form "if . . . then"; it does not say anything about the meaning or truthfulness of the hypotheses. The primitive notions (such as "point" and "line") appearing in the hypotheses are implicitly defined by these axioms, by the rules as it were that tell us how to play the game.[5]

To illustrate how radically different is this view of mathematics, observe the following interaction (Torretti, 1987, p. 235). Gottlob Frege (1848 – 1925), who is considered the founder of modern mathematical logic, wrote to Hilbert:

> I give the name of axioms to propositions which are true, but which are not demonstrated because their knowledge proceeds from a source which is not logical, which we may call space intuition. The truth of the axioms implies of course that they do not contradict each other. That needs no further proof.

Frege has stated the traditional view. Hilbert replied:

> Since I began to think, to write and to lecture about these matters, I have always said exactly the contrary. If the arbitrarily posited axioms do not contradict one another or any of their consequences, they are true and the things defined by them exist. That is for me the criterion of truth and existence.

Hilbert knew that Euclidean and hyperbolic geometries were equally consistent, so it follows that for him they "exist" and are both "true." The discovery that Euclidean geometry was not "absolute truth" had a liberating effect on mathematicians, who now feel free to invent any

---

[5] For a clear exposition of this viewpoint, which is due to Hilbert, see Hempel (1945).

set of axioms they wish and deduce conclusions from them. In fact, this freedom may account for the great increase in the scope and generality of modern mathematics. In a 1961 address, Jean Dieudonné remarked on Gauss' discovery of non-Euclidean geometry:

> [It] was a turning point of capital significance in the history of mathematics, marking the first step in a new conception of the relation between the real world and the mathematical notions supposed to account for it; with Gauss' discovery, the rather naive point of view that mathematical objects were only "ideas" (in the Platonic sense) of sensory objects became untenable, and gradually gave way to a clearer comprehension of the much greater complexity of the question, wherein it seems to us today that mathematics and reality are almost completely independent, and their contacts more mysterious than ever.[6]

THE CONTROVERSY ABOUT THE FOUNDATIONS OF MATHEMATICS

It would be misleading to say that mathematics is just a formal game played with symbols and having no broader significance. Mathematicians do not arbitrarily make up axioms — it is unlikely that anyone would ever develop a geometry in which it was assumed that nonsupplementary right angles were never congruent to each other. Axioms must lead to interesting and fruitful results. Of course, some axioms that appear uninteresting may turn out to have surprising consequences — this was the case with the hyperbolic axiom, which was virtually ignored during the lifetimes of Gauss, Bolyai, and Lobachevsky. If, however, axiom systems do not bear interesting results, they become neglected and eventually forgotten.

Arguing against the description of mathematics as a "formal game," R. Courant and H. Robbins (in their fine book *What is Mathematics?*) insist that "a serious threat to the very life of science is implied in the assertion that mathematics is nothing but a system of conclusions drawn from definitions and postulates that must be con-

[6] J. Dieudonné, "L'Oeuvre Mathématique de C. F. Gauss," Poulet-Malassis Alençon: L'Imprimerie Alençonnaise, 1961.

sistent but otherwise may be created by the free will of the mathematician. If this description were accurate, mathematics could not attract any intelligent person. It would be a game with definitions, rules and syllogisms, without motivation or goal."

And Hermann Weyl has remarked: "The constructions of the mathematical mind are at the same time free and necessary. The individual mathematician feels free to define his notions and to set up his axioms as he pleases. But the question is, will he get his fellow mathematicians interested in the constructs of his imagination? We can not help feeling that certain mathematical structures which have evolved through the combined efforts of the mathematical community bear the stamp of a necessity not affected by the accidents of their historical birth." [7]

Axiom systems that are fruitful can also be controversial in the mathematical world, as are the axioms for infinite sets developed by Georg Cantor, E. Zermelo, and others. A controversy occurs because some outstanding mathematicians (such as Weyl, L. E. J. Brouwer, and Errett Bishop in the case of infinite sets) simply do not *believe* all these axioms. If axioms were truly meaningless formal statements, how could there be any controversy about them? Is there any controversy about the rules of chess? It would seem that the formalist viewpoint — the view that mathematics is just a formal game — is a dodge to avoid having to face the difficult philosophical and psychological problem of the nature of mathematical creations or discoveries. Just what is asserted when a mathematician claims that something exists? When the Pythagoreans discovered that the hypotenuse of an isosceles right triangle was not commensurable with the leg, they tried to keep this discovery secret, calling such lengths "irrational." Nowadays we aren't upset over numbers like $\sqrt{2}$. Similarly, mathematicians have accommodated themselves to "imaginary" numbers, such as $i = \sqrt{-1}$, exploited by J. Cardan. [8]

The most "fundamentalist" position on the philosophy of mathematics is that of Leopold Kronecker, who dominated the German

[7] From H. Weyl, "A Half-Century of Mathematics," *American Mathematical Monthly*, **58** (1951): 523–553.

[8] Jacques Hadamard has said about Cardan: "It would naturally be expected that the discovery of imaginaries, which seems nearer to madness than to logic and which, in fact, has illuminated the whole mathematical science, would come from such a man whose adventurous life was not always commendable from the moral point of view, and who from childhood suffered from fantastic hallucinations . . ." (Hadamard, 1945).

mathematical world in the late nineteenth century. According to Kronecker, "God created the whole numbers—all else is man-made." In particular, Kronecker repudiated Georg Cantor's theory of transfinite cardinal and ordinal numbers. Hilbert later defended Cantor, proclaiming that "no one shall expel us from the paradise which Cantor has created for us." Subsequently Kronecker was portrayed as the nasty reactionary whose rejection of Cantor's revolutionary new ideas drove Cantor to the insane asylum (see Bell, 1961); this is undoubtedly a myth, and the philosophical issues underlying the Kronecker-Cantor controversy are far from settled (see Fang, 1976).

In the twentieth century, Cantor's set theory, made precise by the Zermelo-Fraenkel (Z-F) axioms, became the new "absolute truth" that was the foundation for all of mathematics. However, there was some controversy about one axiom, the axiom of choice (AC), and there was so much uncertainty about another idea of Cantor's that it was called a "hypothesis"—the continuum hypothesis (CH). The first in Hilbert's famous 1900 list of 23 problems was to prove or disprove CH. Forty years later, Kurt Gödel created a model of the other Z-F axioms in which both AC and CH were true; that demonstrated the impossibility of disproving them. History repeated itself when, in 1963, models were created[9] in which either AC or CH or both were false. Thus AC and CH are independent of the other Z-F axioms and of each other. There exists an equally valid non-Cantorian set theory, just as there is an equally valid non-Euclidean geometry.

One mystery about mathematics is perhaps the most compelling of all. If mathematical creations are merely arbitrary fancies, how is it that some turn out to have physical applications, for example, applications that enable us to calculate orbits well enough to put men on the moon? When the Greeks developed the theory of ellipses they had no inkling that it would have applications to a "space race."[10]

These questions and viewpoints are not intended to confuse you, but to point up the fact that mathematics is alive, ever changing, and incomplete. Moreover, according to a metamathematical theorem of Kurt Gödel, mathematics is forever destined to remain incomplete. He proved that there will always be valid mathematical statements

[9] By Paul J. Cohen; see P. J. Cohen and R. Hersh, "Non-Cantorian Set Theory," *Scientific American*, **217** (December, 1967).

[10] See E. Wigner, "The Unreasonable Effectiveness of Mathematics," *Communications in Pure and Applied Mathematics*, **13** (1960): 1 ff.

Kurt Gödel

that cannot be demonstrated from systems of axioms that are broad enough to include arithmetic (see DeLong, 1970). In other words, Gödel provided a formal demonstration of the inadequacy of formal demonstrations!

Perhaps the following remarks by René Thom are an appropriate reaction to Gödel's incompleteness theorem:

> The mathematician should have the courage of his private convictions; he would then affirm that mathematical structures have an existence independent of the human mind that thinks about them. The form of this existence is undoubtedly different from the concrete and material existence of the external world, but it is nevertheless subtly and profoundly linked to objective existence. For how else explain — if mathematics is merely a gratuitous game, the random product of our cerebral activities — its indisputable success in describing the universe? Mathematics is encountered — not only in the rigid and mysterious

laws of physics — but also, in a more hidden but still indubitable manner, in the infinitely playful succession of forms of the animate and inanimate world, in the appearance and destruction of their symmetries. That's why the Platonic hypothesis of Ideas informing the universe is — despite appearances — the most natural and philosophically the most economical. But, at any instant, mathematicians have only an incomplete and fragmentary vision of this world of Ideas . . . , we have to recreate it in our consciousness by a ceaseless and permanent reconstruction. . . . With this confidence in the existence of an ideal universe, the mathematician will not overly worry about the limits of formal procedures, he will be able to forget the problem of consistency. For the world of Ideas infinitely exceeds our operational possibilities, and the ultima ratio of our faith in the truth of a theorem resides in our intuition — a theorem being above all, according to a long-forgotten etymology, the object of a vision.[11]

THE MESS

In the first edition of this book, I ended this chapter with that inspiring quote from Thom (the founder of "catastrophe theory"). Further inquiry into these questions prompts me to a more somber conclusion. Namely, there is at present no intelligible account of what the statements of pure mathematics are about. The philosophy of mathematics is in a mess!

My claim that the formalist viewpoint is a dodge is substantiated by the following revealing admission by Jean Dieudonné:[12]

On foundations we believe in the reality of mathematics, but of course when philosophers attack us with their paradoxes we rush to hide behind formalism and say "Mathematics is just a combination of meaningless symbols," and then we bring out Chapters 1 and 2 on set theory. Finally we are left in peace to go back to our mathematics and do it as we have always done, with the feeling each mathematician has that he is working with something real. This sensation is probably an

---

[11] R. Thom, " 'Modern' Mathematics: An Educational and Philosophic Error?," *American Scientist*, November, 1971, p. 695 ff. The translation here is my own from the original (in *L'Age de Science*, III (3): 225).

[12] "The work of Nicholas Bourbaki," *Amer. Math. Monthly*, **77** (1970): 134–145.

illusion, but is very convenient. That is Bourbaki's attitude toward foundations.

An article by Reuben Hersh[13] forcefully demonstrates the philosophical plight of the working mathematician, who is "a Platonist on weekdays and a formalist on Sundays." Hersh contends that the tension caused by holding contradictory views on the nature of his work must affect the self-confidence of a person who is supposed, above all things, to hate contradiction.

Dieudonné admits that the Platonic view is probably an illusion. In a very interesting essay, Gabriel Stolzenberg[14] argues that the illusion consists in being taken in by a present tense *language of objects and their properties*, a language that has the appearance — but only that — of being meaningful. The psychological act of accepting this appearance produces a notion of "reality" so strong that it becomes very difficult to step aside and question it.

We have already seen examples of such illusion. If one believes that points and lines in the plane are "real objects," then they either satisfy Euclid's postulate or they don't (with the corollary belief that Euclidean geometry is either "true"or "false"). Similarly, if sets are "real objects," then they either satisfy Cantor's continuum hypothesis or they don't (Gödel believed that they don't).

The fundamental illusion, according to Stolzenberg (and Brouwer before him), is the belief that a mathematical statement can be "true" without anyone being able to know it. This belief is so strong that only the few constructivist mathematicians have been willing to give it up. They contend that "$\mathscr{S}$ is true" is a signal to announce a state of knowing which one has attained by means of an act of proof. Stolzenberg (1978) claims (p. 265):

> What one "sees" or "discovers" at the conclusion of an act of proof is that a certain structure (which is constructed in the course of the proof) displays a certain form: a form of the type that, according to the conventions of mathematical language use that have been established,

---

[13] "Some Proposals for Reviving the Philosophy of Mathematics," *Advances in Math.*, **31** (1979): 31–50.

[14] "Can an Inquiry into the Foundations of Mathematics Tell Us Anything Interesting about Mind? " in *Psychology and Biology of Language and Thought*, Essays in Honor of Eric Lenneberg, G. Miller and Elizabeth Lenneberg, eds., New York: Academic Press (1978): 221–269.

entitles anyone who observes it to say "$\mathscr{S}$ is true." But "$\mathscr{S}$ is true" is merely what one *says*, not what one *sees;* the expression itself is merely the "brand name" for the type of thing that one sees at the conclusion of the proof. And it is a type of thing that may be seen only by constructing a proof — not because we need to use the proof as "a ladder" to get ourselves into a position to see it but rather because what one sees is "in" the structure that is created by the act "of making the proof."

An interesting consequence of this position is that "the knower" is brought into the philosophy of mathematics (just as "the observer" has been brought into the philosophy of physics by Heisenberg's uncertainty principle).

If indeed the philosophical mess is the result of a linguistic illusion, then deep insights are needed to develop a new language system. This system would not be a mere rephrasing of current usage (if it were, it wouldn't be worth the bother). It would be a tool to gain higher levels of understanding.

On the other hand, the Platonic "illusion" has shown itself to be very valuable heuristically (e.g., Gödel credited the Platonic viewpoint for his insights). An intelligible justification for Platonic heuristics may someday be found (just as one was found in the twentieth century by the logician Abraham Robinson for the "illusory" infinitesimals used in the seventeenth century by the founders of the calculus). Physics has continued to advance despite the even worse mess in its philosophical foundations, so the proverbial "working mathematicians" will have no trouble continuing to ignore the irritating question of the *meaning* of their theorems.

REVIEW EXERCISE

Which of the following statements are correct?

(1) It is impossible to verify by physical experiments whether hyperbolic geometry is true because hyperbolic geometry is not about physical entities.

(2) If we interpret the undefined terms of geometry physically, e.g., by interpreting "line" as "path of a light ray in empty space," then it

makes sense to ask whether this interpretation is a model of Euclidean geometry; however, due to experimental error, physical experiments could never prove conclusively that it is a model.

(3) Hyperbolic geometry is consistent if and only if Euclidean geometry is consistent.

(4) Poincaré maintained that it was meaningless to ask which geometry is "true," and that it only made sense to ask which geometry is more "convenient" for physics.

(5) The most convenient geometry for astrophysics is neither Euclidean nor hyperbolic geometry but a more complicated geometry of space-time developed by Einstein out of ideas from Riemann.

(6) The Klein and Poincaré models, although they appear to be different, are actually isomorphic to each other.

(7) Hyperbolic geometry, although equally as consistent as Euclidean geometry, has no application to other branches of mathematics or to other sciences.

SOME TOPICS FOR ESSAYS

1. Comment on this quotation from Albert Einstein: "As far as the mathematical theorems refer to reality, they are not sure, and as far as they are sure, they do not refer to reality." (See Hempel, 1945, for a development of this theme.)

2. Report on the debate about the philosophy of *conventionalism,* using Grünbaum, (1968), Poincaré (1952), and Nagel (1939) as sources.

3. Report on the use of hyperbolic geometry to describe binocular vision, referring to Luneburg and Blank (see note 4 in this chapter).

4. It can be said that the discovery of non-Euclidean geometry led to the extensive modern development of mathematical logic. Elaborate on this statement, using DeLong (1970), Chapters 1 and 2, as a source.

5. Jacques Hadamard said: "Practical application is found by not looking for it, and one can say that the whole progress of civilization rests on that principle. . . . It seldom happens that important mathematical researches are *directly* undertaken in view of a given practical

use: they are inspired by the desire which is the common motive of every scientific work, the desire to know and understand."[15]

Along the same lines, David Hilbert maintained that in spite of the importance of the applications of mathematics, these must never be made the measure of its value. And the mathematician Jacobi said that "the glory of the human spirit is the sole aim of all science."

Nevertheless, Lobachevsky believed that "there is no branch of mathematics, however abstract, they may not someday be applied to phenomena of the real world."

Comment on these viewpoints.

6. Read the "Socratic Dialogue on Mathematics" in Renyi (1967), and discuss the following questions therein:

(a) "Is it not mysterious that one can know more about things which do not exist than about things which do exist?"

(b) "How do you explain that, as often happens, mathematicians living far from each other and having no contact independently discover the same truths?"

7. Comment on the following statement by Michael Polanyi (1964; see especially Chapter 6, Sections 9–11):

We can now turn to the paradox of a mathematics based on a system of axioms which are not regarded as self-evident and indeed cannot be known to be mutually consistent. To apply the utmost ingenuity and the most rigorous care to prove the theorems of logic or mathematics while the premises of these inferences are cheerfully accepted, without any grounds being given for doing so . . . might seem altogether absurd. It reminds one of the clown who solemnly sets up in the middle of the arena two gateposts with a securely locked gate between them, pulls out a large bunch of keys, and laboriously selects one which opens the lock, then passes through the gate and carefully locks it after himself—while all the while the whole arena lies open on either side of the gateposts where he could go round them unhindered.

8. Comment on the following statements:

There is a scientific taste just as there is a literary or artistic one. . . . Concerning the fruitfulness of the future result—about which, strictly

---

[15] Hadamard (1945); see especially Chapter 9.

speaking, we most often do not known anything in advance — [the] sense of beauty can inform us and I cannot see anything else allowing us to foresee. . . . Without knowing anything further we *feel* that such a direction of investigation is worth following. . . . Everybody is free to call or not to call that a feeling of beauty. This is undoubtedly the way the Greek geometers thought when they investigated the ellipse, because there is no other conceivable way. (Hadamard, 1945.)

We dwell on mathematics and affirm its statements for the sake of its intellectual beauty. . . . For if this passion were extinct, we would cease to understand mathematics; its conceptions would dissolve and its proofs carry no conviction. Mathematics would become pointless and lose itself in a welter of insignificant tautologies. . . . (Polanyi, 1964.)

We all believe that mathematics is an art. The author of a book or the lecturer in a classroom tries to convey the structural beauty of mathematics to his readers, to his listeners. In this attempt he must always fail. Mathematics is logical, to be sure; each conclusion is drawn from previously derived statements. Yet the whole of it, the real piece of art, is not linear; worse than that, its perception should be instantaneous.[16]

9. Comment on the following statements. G. H. Hardy (1940) said:

For me, and I suppose for most mathematicians, there is another reality, which I will call "mathematical reality"; and there is no sort of argument about the nature of mathematical reality among either mathematicians or philosophers. . . . A man who could give a convincing account of mathematical reality would have solved very many of the most difficult problems of metaphysics. . . . I believe that mathematical reality lies outside us, that our function is to discover or *observe* it, and that the theorems which we prove, and which we describe grandiloquently as our "creations," are simply the notes of our observations. This view has been held, in one form or another, by many philosophers of high reputation from Plato onwards. . . .

Heinrich Hertz, the discoverer or radio waves, said:

One cannot escape the feeling that these mathematical formulas have an independent existence and an intelligence of their own, that they

[16] Emil Artin, "Review of *Algèbre* by N. Bourbaki," *Bulletin American Mathematical Society,* **59** (1953): 474.

are wiser than we are, wiser even than their discoverers, that we get more out of them than was originally put into them.

10. Comment on the following remarks by Kurt Gödel:

I don't see any reason why we should have less confidence in this kind of perception, i.e., in mathematical intuition, than in sense perception, which induces us to build up physical theories and to expect that future sense perceptions will agree with them and, moreover, to believe that a question not decidable now has meaning and may be decided in the future. The set theoretical paradoxes are hardly any more troublesome for mathematics than deceptions of the senses are for physics. . . . Evidently the "given" underlying mathematics is closely related to the abstract elements contained in our empirical ideas. It by no means follows, however, that the data of this second kind [mathematical intuitions], because they cannot be associated with actions of certain things upon our sense organs, are something purely subjective, as Kant asserted. Rather, they, too, may represent an aspect of objective reality. But as opposed to the sensations, their presence in us may be due to another kind of relationship between ourselves and reality.[17]

Gödel in this passage speaks primarily of *set theoretical intuition*. As far as geometrical intuition is concerned, the following, according to Gödel, would have to be added:

Geometrical intuition, strictly speaking, is not mathematical, but rather a priori physical intuition. In its purely mathematical aspect our Euclidean space intuition is perfectly correct, namely, it represents correctly a certain structure existing in the realm of mathematical objects. Even physically it is correct "in the small."[18]

11. Comment on the following quotation from Rolf R. Loehrich:

The communication of a new mathematical system or game meets with peculiar obstacles. Each mathematician has a preferred game. A new game may not capture his interest if it is significantly different from those he has been accustomed to play. . . .

[17] K. Gödel, "What Is Cantor's Continuum Problem?" in Benacerraf and Putnam's *Philosophy of Mathematics*, 2nd ed. (Englewood Cliffs, N.J.: Prentice-Hall, 1964), p. 271.
[18] Private communication to the author, October, 1973.

A mathematical system is hardly ever presented axiomatized at its inception. Successful axiomatization is a fruition of an *exercitium cogitandi.* Once a system is axiomatized, mathematical activity can be played as a game, as a manipulation of symbols by virtue of rule-systems thought of as invented, but this does not assert that the mathematician who invented or presumably discovered the system meant to play a game. . . . Roberts and I are convinced that there is what might be adequately referred to as a mathematical universe. We believe that, with the complex instrumentations and empirical data set forth in *Exercitium Cogitandi,* the ontological value of confrontations belonging to this universe can be determined with a high degree of accuracy (such confrontations are to be thought [of as] sign-values of signs, and these signs are the symbol systems as known and/or as to be invented by virtue of new conceptual systems with ever increasing ranges). . . . If this is true, then indeed a mathematician may think of himself as an explorer of the mathematical universe, and any new mathematical system functions as the inception of a possible creation of a universe which comprehends any of the other universes.[19]

12. Write an essay on the development of geometry in ancient Greece, using the resources of your school library. You may be particularly interested in the female mathematician Hypatia.

13. Comment on the following remarks about the true role of logic in mathematics:

If logic is the hygiene of the mathematician, it is not his source of food; the great problems furnish the daily bread on which he thrives. We have learned to trace our entire science back to a single source, constituted by a few signs and by a few rules for their use; this is an unquestionable stronghold, inside which we could hardly confine ourselves without risk of famine, but to which we are always free to retire in case of uncertainty or external danger. (A. Weil, "The Future of Mathematics," *American Mathematical Monthly,* **57** (1950); 295 – 306.)

All physicists, and a good many quite respectable mathematicians, are contemptuous about proof. (G. H. Hardy, *Ramanujan,* Cambridge University Press, New York, 1940, p. 15.)

Discovery, after all, is more important in science than strict deductive proof. Without discovery there is nothing for deduction to attack

---

[19] R. R. Loehrich (with L. G. Roberts), *Exercitium Cogitandi,* vol. II (Center for Medieval and Renaissance Studies, Oxford, 1978).

and reduce to order. (E. T. Bell, *Development of Mathematics,* 2nd ed. McGraw-Hill, New York, 1945, p. 83.)

14. Report on Imre Lakatos' critique of the formalist philosophy of mathematics and his ideas on how mathematics is discovered, as presented in his book *Proofs and Refutations: The Logic of Mathematical Discovery* (Cambridge University Press, 1976). Here are some pertinent Lakatos quotes:

> Euclid has been the evil genius particularly for the history of mathematics and for the teaching of mathematics, both on the introductory and the creative levels. . . . The two activities of guessing and proving are rigidly separated in the Euclidean tradition. . . . It was the infallibilist philosophical background of Euclidean method that bred the authoritarian traditional patterns in mathematics, that prevented publication and discussion of conjectures, that made impossible the rise of mathematical criticism. . . . The discovery of non-Euclidean geometries (by Lobatschewsky in 1829 and Bolyai in 1831) shattered infallibilist conceit. . . . There is no infallibilist logic of scientific discovery, one which would infallibly lead to results; there is a fallibilist logic of discovery which is the logic of scientific process.

15. Write a detailed report on the theory of area in hyperbolic geometry using Moise (1990), Chapter 24, as a reference.

16. Report on Bertrand Russell's doctoral dissertation *An Essay on the Foundations of Geometry* (Dover reprint, 1956). Show how Russell very capably refutes theories of geometry due to Kant and other philosophers, but then proclaims his own incorrect notion of space (that was later refuted by Einstein). See also the critique in Torretti (1978), Chapter 4.

17. Report on Chapter 3 of Roberto Torretti's sublime treatise *Philosophy of Geometry from Riemann to Poincaré* (1978). This chapter is on the foundations of geometry. Here is one important quote:

> The fact that these semi-circles [in the Poincaré upper half-plane model] behave exactly like Euclidean lines with regard to every logical consequence of Hilbert's axioms [for neutral geometry] bespeaks a deep analogy between them, which can come as a shock only to the mathematically uneducated. To maintain that *line* means something entirely different in Bolyai-Lobachevsky geometry and in Euclidean geometry, is not more reasonable than to say that *heart* has a completely

different meaning in the anatomy and physiology of elephants and in that of frogs.

18. To further illustrate his contention that it is meaningless to ask which geometry is "true," Poincaré invented a "universe" U occupying the interior of a sphere S of radius R in Euclidean space, in which the following physical laws hold:

(a) At any point P inside S, the absolute temperature T is directly proportional to $R^2 - r^2$, where r is the distance from P to the center of S.

(b) The length, width, and height of an object vary directly with the absolute temperature of the object.

(c) All objects in U instantaneously take on the temperatures of their locations.

(d) Light travels along the shortest path from one point to another.

Show that an inhabitant of U could not detect his change in temperature and size as he moves about with a thermometer or a tape measure, and that he could never reach the boundary S of his universe, so would consider it infinitely far away. Poincaré showed that the shortest path in U joining point A to point B is the smaller arc of the circle through A and B that cuts S orthogonally. Hence, if an inhabitant interprets "straight line segment" in his universe to be the path of a light ray, he would conclude that the "true" geometry of his world was hyperbolic. In other words, this is a region of Euclidean space which because of different and undetectable physical laws appears to its inhabitants to be non-Euclidean. Comment, using Poincaré (1952) as a reference, as well as Torretti (1978) and Grünbaum (1968).

19. Write an essay on a topic of your own.

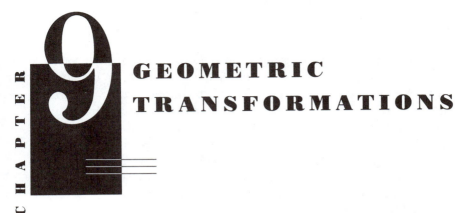

GEOMETRIC TRANSFORMATIONS

CHAPTER 9

I have spent a lifetime applying Klein's program to differential geometry.

W. BLASCHKE

KLEIN'S *ERLANGER PROGRAMME*

In 1872, a year after his decisive publication of the projective models for non-Euclidean geometries, Felix Klein was appointed (at age 23) to a chair at the University of Erlangen. He delivered an inaugural address proposing a new unifying principle for classifying the various geometries that were rapidly being developed, and for discovering relationships between them. This *Erlanger Programme* has had an enormous impact on all of mathematics to the present day.[1]

The key notion, according to Klein, involves the group of all automorphisms of a mathematical structure. In Chapter 2 we defined the concept of an isomorphism of one model onto another, and in Chapter 7 we used a specific isomorphism to relate the Klein and Poincaré models of the hyperbolic plane. An isomorphism mapping a given model onto itself is called an *automorphism* of that model; thus, an automorphism is a one-to-one mapping (or transformation) of each basic set of objects in the model onto itself which preserves the basic relations among the objects.

---

[1] For an English translation of Klein's lecture, see the *Bulletin of the New York Mathematical Society*, **2** (1893): 215–249.

The importance of the group of automorphisms was first recognized in connection with the problem of solving an algebraic equation by radicals. Évariste Galois (1811 – 1832) showed that a solution by radicals was possible if and only if the group of automorphisms of the field extension generated by the roots of the equation is a solvable group. This implies Abel's particular discovery that the general equation of degree 5 cannot be solved by radicals. Klein later discovered a relation between the group of rotations of a dodecahedron and the roots of the quintic equation that explained why the latter can be solved by elliptic functions.

Here is an example of the simplest type of geometric automorphism.

Example 1. Consider models of incidence geometry (Chapter 2). The basic sets of objects are the sets of points and lines, and the only basic relation is incidence of a point and line. An automorphism T will therefore map each point P and each line l onto a point P$'$ and a line l' such that P lies on l if and only if P$'$ lies on l'. By Axiom I-1, a line is determined by any two points lying on it, so T is determined as a mapping of lines once its effect on the points is known — namely

$$T(\overleftrightarrow{PQ}) = \overleftrightarrow{P'Q'}.$$

Since T preserves incidence and is one-to-one on the set of lines, it has the property that three points O, P, Q are collinear if and only if their images O$'$, P$'$, Q$'$ are collinear. Hence an automorphism of a model of incidence geometry is called a *collineation*.

For example, in the 3-point model, every permutation of the three noncollinear points is a collineation. However, for the 7-point projective plane (Figure 9.1), you can show that, of the 7! $= 5040$ permutations of the points, only 168 are collineations (Exercise 1).

It is important to note that an automorphism not only preserves the basic relations, but also *all* the relations that can be defined from them. For example, a collineation of an incidence plane preserves parallelism ($l\|m \Rightarrow l'\|m'$).

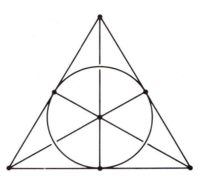

FIGURE 9.1

GROUPS

Transformations of a set onto itself can be multiplied by first applying one transformation T and then another transformation S; thus the composite transformation ST is defined by the equation

$$(0) \qquad\qquad ST(x) = S(T(x))$$

for all x in the set.

With this multiplication, the set $\mathcal{G}$ of all automorphisms of a structure has itself the structure of a *group*, which means that the following properties hold:

1. $S, T \in \mathcal{G} \Rightarrow ST \in \mathcal{G}$.
2. $I \in \mathcal{G}$ (where I is the *identity* transformation that leaves all the objects fixed; the identity transformation satisfies $IT = T = TI$ for all $T \in \mathcal{G}$).
3. $T \in \mathcal{G} \Rightarrow T^{-1} \in \mathcal{G}$ (where the *inverse* T^{-1} of T is characterized by the equations $TT^{-1} = I = T^{-1}T$).
4. $S(TU) = (ST)U$ for all $S, T, U \in \mathcal{G}$ (this *associative law* is an immediate consequence of the definition (0) of multiplication).

To illustrate these properties, let us consider rotations about a point O, which will be rigorously defined later but can now be thought of as transformations that turn the entire plane through a certain angle about O. If T is the rotation through $t°$ clockwise and S the rotation through $s°$ clockwise, then ST is the rotation through $(s + t)°$ clockwise. T^{-1} is the rotation through $t°$ counterclockwise. I can be thought of as the rotation through $0°$.

Warning. The product ST is not, in general, equal to the product TS in the opposite order, as the next example shows.

Example 2. Consider the equilateral triangle $\triangle ABC$ situated symmetrically about the point O in Figure 9.2. If we let T be the rotation through $120°$ counterclockwise about O and let S be the reflection across the vertical line $\overleftrightarrow{AO}$, then TS leaves C fixed and interchanges A and B (in fact, TS is the reflection across $\overleftrightarrow{CO}$); whereas ST leaves B fixed and interchanges A and C (ST is the reflection across $\overleftrightarrow{BO}$).

If two transformations S, T happen to have the property $ST = TS$, we say that they *commute*, and a collection of transformations in which every pair commute is called *commutative* (or Abelian, after the great Norwegian mathematician N. H. Abel). For instance, any two rotations about the same point O commute.

The more structure a geometry has, the smaller is its group of automorphisms. Neutral geometry is incidence geometry with the additional relations of betweenness and congruence; hence the group of automorphisms of a neutral geometry is the *subgroup* of those collineations T for which betweenness and congruence are *invariant;* i.e., for which

$$A * B * C \Rightarrow A' * B' * C'$$
$$AB \cong CD \Rightarrow A'B' \cong C'D'$$

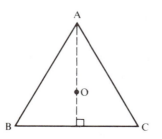

FIGURE 9.2

(*we will systematically use* X′ *to denote the image of any object* X — *point, line, circle, etc.* — *under a transformation denoted* T). We have not assumed T preserves congruence of angles because this can be proved: If ∢ABC ≅ ∢DEF, we can assume by Axiom C-1 that AB ≅ DE and BC ≅ EF, so that AC ≅ DF (SAS); since T preserves congruence of segments, △A′B′C′ ≅ △D′E′F′ (SSS), hence ∢A′B′C′ ≅ ∢D′E′F′. Notice also that if a transformation preserves betweenness it must be a collineation (by Axioms B-1 and B-3).

The principal objective of this chapter will be to explicitly determine all the automorphisms of Euclidean and hyperbolic planes and to classify them according to their geometric properties, particularly their invariants.

We say that a property or relation is "invariant" under a transformation or group of transformations if the property or relation still holds after the transformations are applied; a geometric figure is "invariant" if it is mapped onto itself by the transformations.

"Invariance" and "group" are the unifying concepts in Klein's *Erlanger Programme*. Groups of transformations had been used in geometry for many years, but *Klein's originality consisted in reversing the roles, in making the group the primary object of interest and letting it operate on various geometries, looking for invariants.* For example, the group $PSL(2, \mathbb{R})$ of 2-by-2 projective transformations with real coefficients (see Proposition 9.26) operates on both the hyperbolic plane and the real projective line; for the latter operation, the cross-ratio of four points is the fundamental invariant, whereas for the former operation, the length of a segment (which is calculated by means of cross-ratios in the Klein and Poincaré models) is the fundamental invariant.

Klein classified the following geometries as subgeometries of real plane projective geometry:

1. Affine geometry is the study of invariants of the subgroup of those projective transformations (called *affine transformations*) which leave the line at infinity invariant.
2. Hyperbolic geometry is the study of invariants of the subgroup of those projective transformations which leave a given real conic ("the absolute") invariant.
3. Elliptic geometry is the study of invariants of the subgroup of those projective transformations which leave a given imaginary conic invariant.
4. Parabolic geometry is the study of invariants of the subgroup of

those affine transformations (called *similarities*) which leave invariant the two imaginary circular points at infinity (see p. 362 Coxeter, 1960).
5. Euclidean geometry is the study of invariants of the subgroup of those similarities (called *motions*) which preserve length (which is defined in terms of an arbitrarily chosen unit segment).

During the two decades preceding Klein's address, Cayley and Sylvester had developed a general theory of algebraic invariants together with a systematic procedure for determining generators and relations for them (see J. Dieudonné and J. Carrell, *Invariant Theory, Old and New*, Academic Press, 1971). Klein proposed to translate geometric problems in projective geometry into algebraic problems in invariant theory, where such problems could be solved by the known algebraic methods (for a readable explanation of this program, see Part Three of Klein's *Geometry*, which is Part 2 of his *Elementary Mathematics from an Advanced Standpoint*, Dover, 1948).

Klein's idea of looking for various actions or representations of a group and their invariants has proved to be fruitful in many branches of mathematics and physics, not just in geometry.

In physics, for example, the invariance of Maxwell's equations for electromagnetism under Lorentz transformations suggested to Minkowski a new geometry of space-time whose group of automorphisms is the Lorentz group; this was the beginning of relativity theory, for which Einstein at one point considered the name "Invariantentheorie." In atomic physics, the regularities revealed in the periodic table are a direct consequence of invariance under rotations. In elementary particle physics, considerations of invariance and symmetry have led to several nontrivial predictions. E. Wigner has said that in the future we may well "derive the laws of nature and try to test their validity by means of the laws of invariance rather than to try to derive the laws of invariance from what we believe to be the laws of nature." [2]

In this chapter we will explore the insights Klein's point of view gives to plane Euclidean and hyperbolic geometries. From our axioms we will deduce a description of all possible motions, showing how they are built up from reflections (see Table 9.1, p. 343). Then we will show how to calculate using these transformations in terms of the

[2] E. Wigner, "Invariance in Physical Theory," *Proceedings of the American Philosophical Society*, **93** (1949): 521–526.

coordinates in our models. We will implement Klein's program by replacing congruence axioms with group axioms. Finally, we will apply group-theoretic methods to questions of symmetry.

APPLICATIONS TO GEOMETRIC PROBLEMS

Here are some examples[3] of geometric problems that can easily be solved using transformations; the solutions will use certain properties of reflections, rotations, translations, and dilations which will be demonstrated in the following sections. The purpose in discussing these problems at this time is to illustrate concretely the power of transformation techniques. *You will better comprehend the solutions after you study the theory that follows,* and I suggest that you then reread these solutions and then test your understanding with Exercises 69–77.

Problem 1.   Given two points A, B on the same side of line l. Find the point C on l such that $\overrightarrow{CA}$ and $\overrightarrow{CB}$ make congruent angles with l (if l were a mirror, ACB is the path of a ray of light traveling from A to B by reflecting in l).

Solution.   (See Figure 9.3.) Let B′ be the reflection of B across l. Then C is the intersection of AB′ with l.

FIGURE 9.3

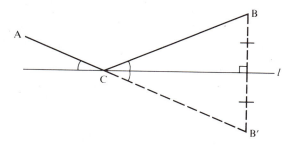

[3] Several hundred more examples will be found in the monumental three volume treatise by I. M. Yaglom, *Geometric Transformations,* Mathematical Association of America, 1962.

Problem 2. Point Q is called a *center of symmetry* for figure F if whenever AA′ is a segment having Q as midpoint and A is in F, then A′ also belongs to F. Show that a figure can only have zero, one, or infinitely many centers of symmetry.

Solution. Q is a center of symmetry if and only if the figure is invariant under the half-turn (180° rotation) H_Q about Q. A triangle has zero, a circle has one, and a line has infinitely many centers of symmetry. Suppose figure F has at least two centers Q and Q′. Then $H_Q(Q') = Q''$ is a third center, $H_{Q'}(Q'')$ is a fourth center, etc.

Note. The preceding problems were stated and solved in neutral geometry. For the remaining problems we will assume the geometry to be Euclidean.

Problem 3. Let L, M, N be the respective midpoints of sides AB, BC, CA of $\triangle$ABC. Let O_1, O_2, O_3 be the *circumcenters* (i.e., the centers of the circumscribed circles) of triangles $\triangle$ALN, $\triangle$BLM, $\triangle$CMN respectively, and let P_1, P_2, P_3 be the *incenters* (i.e., the centers of the inscribed circles) of these same triangles. Show that $\triangle O_1 O_2 O_3 \cong \triangle P_1 P_2 P_3$.

Solution. (See Figure 9.4.) Observe that each of the three triangles is obtained from each of the others by a translation — e.g., translating $\triangle$ALN in direction $\overrightarrow{AB}$ through distance $\overline{AL} = \overline{LB}$ gives $\triangle$LBM. This translation carries the circumscribed circle (and its center) of one triangle onto the circumscribed circle (and its center) of

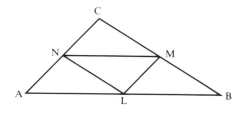

FIGURE 9.4

the other; similarly for the inscribed circles. Hence we not only have $O_1O_2 \cong AL \cong P_1P_2$, etc., giving $\triangle O_1O_2O_3 \cong \triangle P_1P_2P_3$, but we also see that corresponding sides of these two triangles are parallel.

Problem 4. Given an acute-angled triangle, find the inscribed triangle of minimum perimeter (Fagnano's problem).

Solution. Consider $\triangle XYZ$ inscribed as in Figure 9.5(a). Reflect X across $\overleftrightarrow{AB}$ to point X_1 and across $\overleftrightarrow{AC}$ to point X_2. Then the perimeter of $\triangle XYZ$ is equal to the length of the polygonal path X_1ZYX_2. If we fix X, this length will be minimized when Z and Y are chosen to lie on $\overline{X_1X_2}$, and then $\overline{X_1X_2}$ equals the perimeter of $\triangle XYZ$. We have $AX_1 \cong AX \cong AX_2$ and $\sphericalangle X_1AX_2 \cong 2 \sphericalangle A$. If we now vary X, the summit angle of isosceles triangle $\triangle X_1AX_2$ remains constant in measure and the base $\overline{X_1X_2}$ varies in direct proportion to $\overline{AX}$ (in fact, trigonometry gives us $\overline{X_1X_2} = 2\overline{AX}\sin \sphericalangle A$). Hence the minimum perimeter is achieved when $\overline{AX}$ is a minimum, and that occurs when X is the foot of the altitude from A (Figure 9.5(b)). We leave for Exercise 74 the verification that Y and Z must then also be the feet of the altitudes from B and C. Hence the unique inscribed triangle of minimum perimeter is the *orthic* or *pedal triangle* formed by the feet of the altitudes of $\triangle ABC$.

Problem 5. Given three parallel lines, find an equilateral triangle whose vertices lie on them.

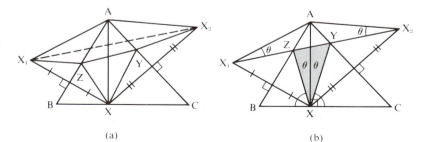

(a)   (b)

FIGURE 9.5

Solution. Choose any point A on the first line l. Rotate the second line m about A through $60°$ to a new line m'. Let C be the intersection of m' with the third line n, and let B be the point on m obtained by rotating C about A through $60°$ in the opposite direction. Then $\triangle ABC$ is a solution.

Problem 6. For any triangle $\triangle ABC$, construct equilateral triangles on the sides of $\triangle ABC$, exterior to it. Show that the centers of these triangles also form the vertices of an equilateral triangle.

Solution. Call the centers O_1, O_2, and O_3, and consider the rotations R_1, R_2, and R_3 through $120°$ counterclockwise about O_1, O_2, and O_3 respectively; then $R_1(A) = B, R_2(B) = C$, and $R_3(C) = A$. Now $R_2 R_1$ is the clockwise rotation through $120°$ about the point O_3' of the intersection of two lines, one through O_1 and the other through O_2, each making an angle of $60°$ with $\overleftrightarrow{O_1 O_2}$, so that $\triangle O_1 O_2 O_3'$ is equilateral. Since R_3^{-1} is also a clockwise rotation through $120°$ taking A into C, we must have $R_3^{-1} = R_2 R_1$ and $O_3' = O_3$.

Problem 7. Given a circle κ and a point P on κ. Find the locus κ' of midpoints M of all chords PA of κ through P.

Solution. (See Figure 9.6.) Since κ' is obtained from κ by dilation of center P and ratio $\frac{1}{2}$, κ' is the circle with diameter OP, O being the center of κ.

Problem 8. Given any triangle $\triangle ABC$, consider its *circumcenter* O (point of concurrence of the perpendicular bisectors of the sides), its *centroid* G (point of concurrence of the medians), and its *orthocenter* H (point of concurrence of the altitudes). You showed O exists in Exercise 12, Chapter 6. An easy argument using analytic geometry (Exercise 69) shows that G exists and lies $\frac{2}{3}$ of the distance from each vertex to the midpoint of the opposite side; thus the dilation T of center G and ratio $-\frac{1}{2}$ maps $\triangle ABC$ onto the *medial triangle* $\triangle A'B'C'$.

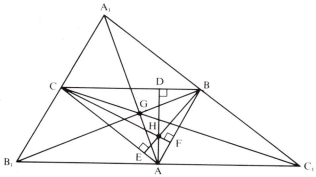

FIGURE 9.6

The problem we pose now is to show that H exists, that O, G, and H lie on a line (called the *Euler line* of △ABC), and that G lies $\frac{2}{3}$ of the distance from H to O.

Solution. Dilation T^{-1} maps △ABC onto △$A_1B_1C_1$ having sides parallel to the respective sides of △ABC and twice as long (Figure 9.7). △ABC is then the medial triangle of △$A_1B_1C_1$, and the altitudes of △ABC are the perpendicular bisectors of △$A_1B_1C_1$, hence are concurrent in a point H.

The original dilation T, being a similarity, preserves perpendicularity, hence maps the orthocenter H of △ABC onto the orthocenter of the medial triangle △A′B′C′, which is O; since G is the center of T and $-\frac{1}{2}$ the ratio, the conclusion follows from the definition of dilation.

FIGURE 9.7

Problem 9. Let H be the orthocenter, O the circumcenter, L, M, N the midpoints of the sides, D, E, F the feet of the altitudes of △ABC. Show that L, M, N, D, E, F and the midpoints of segments HA, HB, HC all lie on a circle whose center U lies on the Euler line and is the midpoint of HO (*the 9-point circle* of △ABC).

Solution. Consider the dilation T of center H and ratio 2. If we show that T maps all nine points onto the circumscribed circle κ of △ABC, the conclusion will follow from Lemma 7.2, Chapter 7, applied to dilation T^{-1} of ratio $\frac{1}{2}$ (T^{-1} maps κ onto a circle of half the radius and center the midpoint of OH). Clearly T maps the midpoints of HA, HB, HC onto A, B, C on κ.

Let P be the point on κ diametrically opposite to A (see Figure 9.8). Since ⦡ACP is inscribed in a semicircle, $\overleftrightarrow{PC} \perp \overleftrightarrow{AC}$, hence $\overleftrightarrow{PC}$ is parallel to altitude $\overleftrightarrow{BH}$. Similarly $\overleftrightarrow{PB} \parallel \overleftrightarrow{CH}$. Thus □PCHB is a parallelogram, hence the midpoint of diagonal HP coincides with the midpoint L of BC. This shows $T(L) = P$ on κ (and similarly for $T(M)$ and $T(N)$).

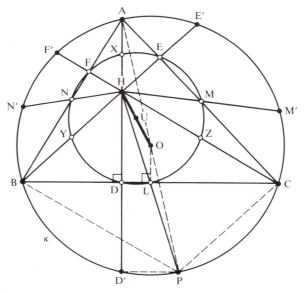

FIGURE 9.8

Let ray $\overleftrightarrow{HD}$ meet κ at D'. Since $\angle AD'P$ is inscribed in a semicircle of κ, $\overrightarrow{D'P} \perp \overleftrightarrow{AD'} = \overleftrightarrow{AD} \perp \overleftrightarrow{DL}$; i.e., $\overrightarrow{D'P} \parallel \overleftrightarrow{DL}$, which implies that D is the midpoint of HD' (since L is the midpoint of HP). Thus $T(D) = $ D' on κ (and similarly for $T(E)$ and $T(F)$).

MOTIONS AND SIMILARITIES

Henceforth the word "automorphism" will be used only for an automorphism of a neutral geometry, i.e., for a transformation that preserves incidence, betweenness, and congruence.

DEFINITION. A transformation T of the entire plane onto itself is called a *motion*[4] or an *isometry* if length is invariant under T, i.e., if for every segment AB, $\overline{AB} = \overline{A'B'}$.

PROPOSITION 9.1. (a) Every motion is an automorphism. (b) The motions form a subgroup of the group of automorphisms.

Proof:
(a) Let T be a motion. If AB $\cong$ CD, then

$$\overline{A'B'} = \overline{AB} = \overline{CD} = \overline{C'D'}$$

so that A'B' $\cong$ C'D'. That T also preserves betweenness follows from Theorem 4.3(9), which says that A * B * C if and only if $\overline{AC} = \overline{AB} + \overline{BC}$.
(b) You must verify properties 1 through 3 in the definition of a group, which is an easy exercise. ∎

PROPOSITION 9.2. Automorphisms preserve angle measure.

Proof:
Given an automorphism T, define a possibly new measure of $\angle A$ to be the measure $(\angle A')°$ of its image under T. You will show in

---

[4] Some authors call these transformations *rigid motions*. The term "motion" as we used it here does not mean continuous movement of a physical body as in common usage, although it is suggested by the latter.

Exercise 3 that this new measure satisfies all the basic properties 1 through 6 in Theorem 4.3. But that theorem says there is a *unique* degree measure with these properties. Hence $(\sphericalangle A)° = (\sphericalangle A')°$. ■

COROLLARY 1. If $\triangle A'B'C'$ is the image of $\triangle ABC$ under an automorphism, then $\triangle A'B'C'$ is similar to $\triangle ABC$.

Proof:
Corresponding angles are congruent. ■

COROLLARY 2. In a hyperbolic plane, every automorphism is a motion.

Proof:
Theorem 6.2 says $\triangle ABC \cong \triangle A'B'C'$, hence $\overline{AB} = \overline{A'B'}$. ■

In fact, as we have already observed using the Klein model (Exercise K-21, Chapter 7), every collineation of the hyperbolic plane *onto* itself is a motion.

Because of Corollary 1, an automorphism of a Euclidean plane is called a *similarity;* by definition, it is a collineation that preserves angle measure. An example of a similarity that is not a motion is the *dilation*[5] with center O and ratio $k \neq 0$: if $k > 0$ (respectively, $k < 0$) this transformation T fixes O and maps any other point P onto the unique point P' on ray $\overrightarrow{OP}$ (respectively, on the ray opposite to $\overrightarrow{OP}$) such that

$$\overline{OP'} = |k|\overline{OP}.$$

If we introduce Cartesian coordinates with origin O, this transformation is represented by

$$(x, y) \longrightarrow (kx, ky).$$

Hence, if A, B have coordinates (a_1, a_2), (b_1, b_2), we have

$$(\overline{A'B'})^2 = (ka_1 - kb_1)^2 + (ka_2 - kb_2)^2 = k^2(\overline{AB})^2;$$

i.e., $\overline{A'B'} = |k|\overline{AB}$. From this you can show that T preserves betweenness and congruence (Exercise 4), which is the "if" part of the next proposition.

---

[5] This notion of dilation is more general than the one on p. 250, where only the case $k > 0$ was considered. A dilation is also called a *homothety* or *central similarity*. Note that similarities are characterized as collineations that preserve circles (see p. 362).

PROPOSITION 9.3. A transformation T of a Euclidean plane is a similarity if and only if there is a positive constant k such that

$$\overline{A'B'} = k\,\overline{AB}$$

for all segments AB.

Proof:
Given similarity T and segment AB, choose any point C not collinear with A, B, and consider $\triangle A'B'C'$ similar to $\triangle ABC$. By the fundamental theorem on similar triangles (Exercise 18, Chapter 5), there is a positive constant k such that the ratio of corresponding sides of these triangles is equal to k. If D is any other point, the same argument applied to $\triangle ACD$ (or $\triangle BCD$ if A, C, D are collinear) gives $\overline{C'D'} = k\overline{CD}$. And if D, E lie on $\overleftrightarrow{AB}$, the same argument applied to $\triangle CDE$ gives $\overline{D'E'} = k\overline{DE}$. Thus k is the proportionality constant for all segments. ∎

The proof just given, together with Exercise 4, shows the following.

COROLLARY. A one-to-one transformation T of a Euclidean plane onto itself is an automorphism if and only if for every triangle $\triangle ABC$, we have $\triangle ABC \sim \triangle A'B'C'$.

We can conclude from these results that hyperbolic planes have *invariant* distance functions AB $\rightarrow \overline{AB}$, whereas Euclidean planes do not. According to Klein's viewpoint, any function or relation that is not invariant under the group of automorphisms of a structure is not an intrinsic part of the theory of that structure; it is only part of the theory of the new structure described by those transformations that do leave it invariant. So if we want distance to be a part of Euclidean geometry, we would have to redefine Euclidean geometry as the study of invariants of the group of Euclidean *motions* only. Klein suggested the name *parabolic geometry* for the study of invariants of the full group of Euclidean similarities. We will not adopt this terminology, but you should be aware that other authors do.[6]

---

[6] Hermann Weyl contends that (for three-dimensional geometry) the group of motions is the group of *physical* automorphisms of space, because the mass and charge of an electron supply us with an absolute standard of length.

REFLECTIONS

The most fundamental type of motion from which we will generate all others is the *reflection* R_m across line m, its *axis* (see p. 111). We will denote the image of a point A under R_m by A^m. Reflecting across m twice sends every point back where it came from, so $R_m R_m = I$ or $R_m = (R_m)^{-1}$. A transformation that is equal to its own inverse and that is not the identity is called an *involution*. The 180° rotation about a point is another example of an involution. (You will show in Exercise 9 that there are no other involutions.)

A *fixed point* of a transformation T is a point A such that $A' = A$. The fixed points of a reflection R_m are the points lying on m. We will use fixed points to classify motions.

LEMMA 9.1. If an automorphism T fixes two points A, B, then it is a motion and it fixes every point on line $\overleftrightarrow{AB}$.

Proof:
Since $AB = A'B'$, the constant k in the corollary to Proposition 9.2 is equal to 1. Let C be a third point on $\overleftrightarrow{AB}$. Consider the case $A * B * C$ (the other two cases are treated similarly). Then $A * B * C'$ and $\overline{AC} = \overline{AC'}$. By Axiom C-1, $C = C'$. ∎

LEMMA 9.2. If an automorphism fixes three noncollinear points, then it is the identity.

Proof:
If A, B, C are fixed, then by Lemma 9.1 so is every point on the lines joining these three points. If D is not on those three lines, choose any E between A and B. By Pasch's theorem, line $\overleftrightarrow{DE}$ meets another side of $\triangle ABC$ in a point F. Since E and F are fixed, Lemma 9.1 tells us D is fixed. ∎

PROPOSITION 9.4. If an automorphism fixes two points A, B and is not the identity, then it is the reflection across line $\overleftrightarrow{AB}$.

Proof:
Lemma 9.1 ensures that every point of $\overleftrightarrow{AB}$ is fixed. Let C be any point off AB and let F be the foot of the perpendicular from C to $\overleftrightarrow{AB}$.

Since automorphisms preserve angle measure they preserve perpendicularity, so C′ must lie on $\overleftrightarrow{CF}$. Lemma 9.2 ensures that C′ ≠ C, and since $\overline{CF} = \overline{C'F}$, C′ is the reflection of C across $\overleftrightarrow{AB}$. ∎

The next result shows that "motion" is the precise concept that justifies Euclid's idea of superimposing one triangle on another.

PROPOSITION 9.5. $\triangle ABC \cong \triangle A'B'C'$ if and only if there is a motion sending A, B, C respectively onto A′, B′, C′ and that motion is unique.

Proof:
Uniqueness follows from Lemma 9.2, for if T and T' had the same effect on A, B, C, then $T^{-1}T'$ would fix these points, hence $T^{-1}T' = I$ and $T = T'$. It's clear that a motion maps $\triangle ABC$ onto a congruent triangle (SSS). So we will assume conversely that $\triangle ABC \cong \triangle A'B'C'$ and construct the motion. We may assume A ≠ A′ and let t be the perpendicular bisector of AA′. Then reflection across t sends A to A′ and B, C to points B′, C′. If the latter are B′, C′, we're done, so assume B′ ≠ B′. We have

$$A'B' \cong AB \cong A'B'.$$

Let u be the perpendicular bisector of B′B′, so that R_u sends B′ to B′ (Figure 9.9). This reflection fixes A′, because if A′, B′, B′ are collinear, A′ is the midpoint of B′B′ and lies on u, whereas if they are not collinear, u is the perpendicular bisector of the base of isosceles triangle $\triangle B'A'B'$ and u passes through the vertex A′.

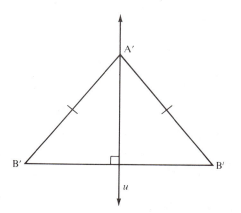

FIGURE 9.9

Thus, the composite $R_u R_t$ sends the pair (A, B) to the pair (A′, B′). If it also sends C to C′ we're done; otherwise let C″ be its effect on C. Then

$$A'C' \cong AC \cong A'C''$$
$$B'C' \cong BC \cong B'C''$$

so that $\triangle A'B'C' \cong \triangle A'B'C''$. An easy argument with congruent triangles (see Figure 9.10) shows that C′ is the reflection of C″ across $v = \overleftrightarrow{A'B'}$. Thus $R_v R_u R_t$ is the motion we seek. ∎

COROLLARY. Every motion is a product of at most three reflections.

This was shown in the course of the proof (where we consider the identity a "product of zero reflections" and a reflection a "product of one reflection").

We are next going to examine products of two reflections $T = R_l R_m$. If l meets m at a point A, T is called a *rotation* about A. If l and m have a common perpendicular t, T is called a *translation along t*. Finally, in the hyperbolic plane only, if l and m are asymptotically parallel in the direction of an ideal point Ω, T is called a *parallel displacement* about Ω. These cases are mutually exclusive, but by convention the identity motion will be considered to be a rotation, translation, and parallel displacement (this is the case $l = m$).

Proposition 9.4 showed the importance of fixed points in describing motions. Another important tool is invariant lines: we say that line l is *invariant* under T if $l' = l$. This does not imply that all the points on l are fixed; it only implies that if a point on l is moved by T, it is moved to another point on l. For example, the only lines besides m that are invariant under a reflection R_m are the lines perpendicular to m (Exercise 7).

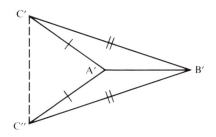

FIGURE 9.10

ROTATIONS

PROPOSITION 9.6. Let $l \perp m$, let A be the point of intersection of l and m, and let $T = R_l R_m$. Then for any point B $\neq$ A, A is the midpoint of BB′.

Proof:
The assertion is clear if B lies on either l or m, so assume it does not (Figure 9.11).

Let C be the foot of the perpendicular from B to m. Then B′ is on the opposite side of both l and m from B, and C′ is on the opposite side of l from C. From the congruence $\sphericalangle$BAC $\cong$ $\sphericalangle$B′AC′ we deduce that these must be vertical angles, hence A, B, B′ are collinear. Since AB $\cong$ AB′, A is the midpoint. ■

The motion T in Proposition 9.6 can be described as the 180° rotation about A; we will call it the *half-turn* about A and denote it H_A. The image of a point P under H_A will be denoted P^A.

COROLLARY. H_A is an involution and its invariant lines are the lines through A.

PROPOSITION 9.7. A motion $T \neq I$ is a rotation if and only if T has exactly one fixed point.

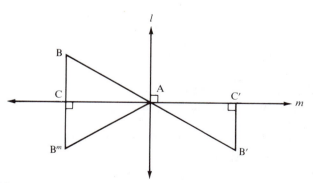

FIGURE 9.11

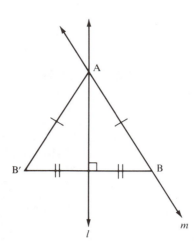

FIGURE 9.12

Proof:

Suppose T has only one fixed point, A, and choose B ≠ A. Let l be the perpendicular bisector of BB'. Since AB ≅ AB', A lies on l, and the motion $R_l T$ fixes both A and B. If $R_l T = I$, then $T = R_l$, which contradicts the hypothesis that T has only one fixed point. Hence if $m = \overleftrightarrow{AB}$, Proposition 9.4 implies $R_l T = R_m$, so that $T = R_l R_m$ and T is a rotation about A (see Figure 9.12).

Conversely, given rotation $T = T_l R_m$ about A, assume on the contrary that point B ≠ A is fixed. Then $B^l = B^m$, so that joining this point to B gives a line perpendicular to both l and m, which is impossible. ■

Note. This last argument breaks down in an elliptic plane, because, there, it is possible for intersecting lines to have a common perpendicular. In fact, each point P has a line l called its *polar* such that l is perpendicular to every line through P (see Figure 9.13).

In the elliptic plane, the half-turn H_p about P is the same as the reflection R_l across l. (Lemmas 9.1 and 9.2 are also false in the elliptic plane.) It can be shown that rotations are the only motions of an elliptic plane (see Ewald, 1971, p. 50). In the sphere model, with antipodal points identified, the motions are represented by Euclidean rotations about lines through the center of the sphere (Artzy, 1965, p. 181).

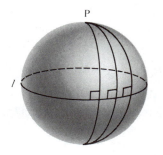

FIGURE 9.13

If you reread the first part of the proof of Proposition 9.7 and refer to Figure 9.12, you will see that we have also proved the following proposition, which is the first case of the fundamental result Proposition 9.19 on three reflections.

PROPOSITION 9.8. If T is a rotation about A and m is any line through A, then there is a unique line l through A such that $T = R_l R_m$. If l is not perpendicular to m, then for any point B $\neq$ A,

$$(\sphericalangle BAB')° = 2d°.$$

(So, for example, to express a 90° rotation about A in the form $R_l R_m$, you must choose lines l and m through A that make a 45° angle.)

Warning. The rotation $R_l R_m$ is not the same as the rotation $R_m R_l$ unless $l \perp m$. Intuitively, one of these rotations is the "clockwise rotation" through $2d°$ about A while the other is the "counterclockwise rotation" through $2d°$. For more rigorous argument, note that

$$(R_m R_l)(R_l R_m) = R_m(R_l^2) R_m = R_m I R_m = R_m^2 = I,$$

so that $R_m R_l$ is the inverse of $R_l R_m$. In Exercise 9 you will show that the only rotation equal to its inverse is the half-turn.

PROPOSITION 9.9. Given a point A, the set or rotations about A is a commutative group.

Proof:
The identity is a rotation about A by definition, and we have just shown that the inverse of a rotation about A is a rotation about A.

We must show that the product TT' of rotations about A is a rotation about A. Let $T' = R_l R_m$. By Proposition 9.8, there is a unique line k through A such that $T = R_k R_l$. Then

$$TT' = (R_k R_l)(R_l R_m) = R_k(R_l^2)R_m = R_k I R_m = R_k R_m,$$

which is a rotation about A. To prove commutativity, apply Proposition 9.8 again to get a unique line n such that $T^{-1} = R_n R_m$. Then $T = R_m R_n$ and $T'T = (R_l R_m)(R_m R_n) = R_l(R_m^2)R_n = R_l R_n$. Since $TT' = R_k R_m$, $R_k(TT')R_m = R_k^2 R_m^2 = I$. But we also have $R_k(T'T)R_m = R_k(R_l R_n)R_m = (R_k R_l)(R_n R_m) = TT^{-1} = I$. Hence, $TT' = T'T$ by canceling on the right and the left. ∎

Warning. Rotations about different points never commute (unless at least one rotation is the identity). For if T is a rotation about A and T' is a rotation about B, $T'T$ sends A to A″, whereas TT' sends A to (A″)′. Furthermore, the product of such rotations may or may not be a rotation (Exercise 10).

TRANSLATIONS

We turn next to translations $T = R_l R_m$, where l and m have a common perpendicular t. The geometric properties of translations are different in hyperbolic planes from those in Euclidean planes (unlike rotations, which behave the same in both geometries).

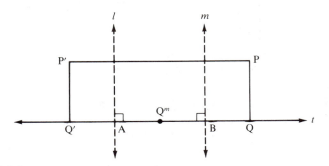

FIGURE 9.14

PROPOSITION 9.10. Let $l \perp t$ at A, $m \perp t$ at B, $T = R_l R_m$. If Q lies on t, then $\overline{QQ'} = 2(\overline{AB})$. If P does not lie on t, then P' lies on the same side of t as P, and $\overline{PP'} = 2(\overline{AB})$ if the plane is Euclidean, $\overline{PP'} > 2(\overline{AB})$ if the plane is hyperbolic.

Proof:
(See Figure 9.14.) We will prove the assertion about $\overline{QQ'}$ when A * B * Q, leaving the other cases as an exercise. If $\overline{BQ} < \overline{AB}$, then A * Q^m * B and

$$\begin{aligned} \overline{QQ'} &= \overline{Q'A} + \overline{AB} + \overline{BQ} \\ &= \overline{Q^mA} + \overline{AB} + \overline{BQ^m} \\ &= 2\overline{AB}. \end{aligned}$$

If $\overline{BQ} = \overline{AB}$, then $Q' = A$ and $\overline{QQ'} = 2(\overline{AB})$. If $\overline{BQ} > \overline{AB}$, then Q^m * A * B and $\overline{Q^mA} = \overline{Q^mB} - \overline{AB} = \overline{BQ} - \overline{AB}$. Hence, Q^m * A * Q' * Q and we have

$$\begin{aligned} 2\overline{BQ} = \overline{QQ^m} &= \overline{QQ'} + 2\overline{Q'A} \\ &= \overline{QQ'} + 2(\overline{BQ} - \overline{AB}), \end{aligned}$$

which gives $\overline{QQ'} = 2\overline{AB}$.

If P does not lie on t, then P and P^m lie on a line perpendicular to m hence parallel to t, and thus are on the same side of t; similarly, P^m and P' are on the same side of t; so, by Axiom B-4, P and P' are on the same side of t. Let Q be the foot of the perpendicular from P to t. Since T preserves perpendicularity, Q' is the foot of the perpendicular from P' to t, and since T is a motion, $\overline{P'Q'} \cong \overline{PQ}$. Thus, $\square PQQ'P'$ is a Saccheri quadrilateral. In Euclidean geometry it's a rectangle and its opposite sides are congruent, so $\overline{PP'} = \overline{QQ'} = 2(\overline{AB})$; in hyperbolic geometry, the summit is larger than the base (Exercise 1, Chapter 6), so $\overline{PP'} > \overline{QQ'} = 2(\overline{AB})$. ∎

COROLLARY. If a translation has a fixed point, then it is the identity motion.

PROPOSITION 9.11. If T is a translation along t and m is any line perpendicular to t, then there is a unique line $l \perp t$ such that $T = R_l R_m$.

Proof:

Let m cut t at Q and let l be the perpendicular bisector of QQ'. Then $R_l T$ fixes Q. Let P be any other point on m, so that as before, $\square$PQQ'P' is a Saccheri quadrilateral. Since l is perpendicular to the base QQ' at its midpoint, l is also perpendicular to the summit PP' at its midpoint (Lemma 6.2, p. 193), so that P is the reflection of P' across l. Thus, $R_l T$ fixes every point on m, whence $R_l T = R_m$, and

$$T = (R_l)^2 T = R_l R_m.$$

As for uniqueness, if $T = R_k R_m$, then

$$R_l = T R_m = R_k$$

and $l = k$. ∎

PROPOSITION 9.12.   Given a line t, the set of translations along t is a commutative group.

The proof is the same as the proof of Proposition 9.9, using Proposition 9.11 in place of Proposition 9.8. ∎

PROPOSITION 9.13.   Let $T \neq I$ be a translation along t. If the plane is Euclidean, the invariant lines of T are t and all lines parallel to t. If the plane is hyperbolic, t is the only invariant line.

Proof:

It's clear that t is invariant. In the Euclidean case, If $u \| t$, then T is also a translation along u (Proposition 4.9), so u is invariant. In both cases, if u meets t at A, then A' lies on t and A' $\neq$ A, so A' does not lie on u and u is not invariant. Suppose in the hyperbolic case u is invariant and parallel to t. Choose any P on u; then $u = \overleftrightarrow{PP'}$. But we have already seen that P and P' are equidistant from t, whence u and t have a common perpendicular m (Theorem 6.4). We've shown that m is not invariant, and since T preserves perpendicularity, m' is also perpendicular to $t = t'$ and $u = u'$. This contradicts the uniqueness of the common perpendicular in hyperbolic geometry (Theorem 6.5). ∎

PROPOSITION 9.14.   Given a motion T, a line t, and a point B on t. Then T is a translation along t if and only if there is a unique point A on t such that T is the product of half-turns $H_A H_B$.

Proof:

Let m be the perpendicular to t through B. If T is a translation along t, then by Proposition 9.11 there is a unique line $l \perp t$ such that $T = R_l R_m$. If l meets t at A, then $H_A H_B = (R_l R_t)(R_t R_m) = R_l (R_t^2) R_m = R_l R_m = T$. Reverse the argument to obtain the converse. ■

HALF-TURNS

Having shown that the product of two half-turns in a translation, we now naturally ask: What is the product of three half-turns? Once again, the answer depends on whether the geometry is Euclidean or hyperbolic.

PROPOSITION 9.15. In a Euclidean plane, the product $H_A H_B H_C$ of three half-turns is a half-turn. In a hyperbolic plane, the product is only a half-turn when A, B, C are collinear, and if they are not, the product could be either a rotation, a translation, or a parallel displacement.

Proof:

Suppose that A, B, C are collinear, lying on t, and that l, m, n are the respective perpendiculars to t through these points. Then

$$\begin{aligned}
H_A H_B H_C &= (R_l R_t)(R_t R_m)(R_n R_t) \\
&= R_l (R_t^2) R_m R_n R_t \\
&= (R_l R_m R_n) R_t \\
&= R_k R_t,
\end{aligned}$$

where the line $k \perp t$ such that $R_k R_n = R_l R_m$ is furnished by Proposition 9.11. If k meets t at D, we have shown $H_A H_B H_C = H_D$.

Suppose that A, B, C are not collinear, that $t = \overleftrightarrow{AB}$, that $l \perp t$ at A, and $m \perp t$ at B. We may assume C lies on m (otherwise replace B by the foot of the perpendicular from C to t and replace A by the point furnished by Proposition 9.14). Let u be the perpendicular to m through C (Figure 9.15). Then

$$H_A H_B H_C = (R_l R_t)(R_t R_m)(R_m R_u) = R_l (R_t^2)(R_m^2) R_u = R_l R_u.$$

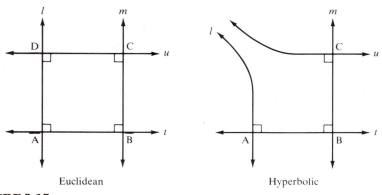

Euclidean   Hyperbolic

FIGURE 9.15

In the Euclidean case, l meets u at a point D and $l \perp u$ (Propositions 4.7 and 4.9), so $H_A H_B H_C = H_D$. In the hyperbolic case, l and u may meet, be divergently parallel, or be asymptotically parallel (Major Exercise 10, Chapter 6); if they do meet at point D, then $H_A H_B H_C$ is a rotation about D, but it is not a half-turn, because $\sphericalangle$D is the fourth angle of a Lambert quadrilateral. ■

COROLLARY. In a Euclidean plane, the product of two translations along different lines is again a translation, and the set of all translations along all lines is a commutative group (proof left for Exercise 13).

IDEAL POINTS IN THE HYPERBOLIC PLANE

We next study the effect of motions in the hyperbolic plane on ideal points. An *ideal point* Ω is by definition an equivalence class of rays, where rays are in the same class if one is contained in the other, or if they are limiting parallel to each other (Major Exercises 2 and 3 of Chapter 6 ensure that this situation does define an equivalence relation).

Now limiting parallelism is defined in terms of incidence and betweenness (see p. 196), hence motions preserve the relation of limiting parallelism. Thus, it makes sense to choose any ray r from the class Ω, consider its image r' under T, and define the image Ω' to be the class of r'. If $\Omega = \Omega'$ (which means that either r' is limiting parallel to r or one ray contains the other), we say Ω is an *ideal fixed point* of T.

Given a line t containing a ray r, the class of r and the class of the opposite ray are called the two *ends* of t and are said to *lie on* t. Two ideal points Ω, Σ lie on a unique line $\Omega\Sigma$ (namely, if rays $r \in \Omega$, $s \in \Sigma$ emanate from the same point and are not opposite, then $\Omega\Sigma$ is the line of enclosure of the angle formed by r and s — see Major Exercise 8, Chapter 6). We say that Ω and Σ are *on the same side* of line t if neither of them is an end of t and if line $\Omega\Sigma$ is parallel to t. This defines a transitive relation on the set of ideal points off t.

PROPOSITION 9.16

(a) The ends of m are the only ideal fixed points of the reflection R_m and any translation along m $(\neq I)$.

(b) If a rotation has an ideal fixed point, then it is the identity.

(c) If (Φ, Σ, Λ) and $(\Omega', \Sigma', \Lambda')$ are any triples of ideal points, then there is a unique motion sending one triple onto the other.

Proof:

(a) It is clear that R_m and any translation $T \neq I$ along m fix the ends Σ, Ω of m. If any other ideal point Λ were fixed, then the line $\Sigma\Lambda$ would be invariant; but T has no other invariant lines than m (Proposition 9.13), and the only other invariant lines of R_m are the perpendiculars to m, whose ends are interchanged by R_m.

(b) If a rotation about A fixes Ω, then ray $A\Omega$ would be invariant; but Propositions 9.6 and 9.8 imply that only the identity rotation has an invariant ray emanating from A.

(c) There is a unique point B on $\Sigma\Omega$ such that $\sphericalangle\Lambda B\Omega$ is a right angle (Major Exercise 10, Chapter 6). Let B′ be the point on $\Sigma'\Omega'$ such that $\sphericalangle\Lambda'B'\Omega'$ is a right angle. Let A be any point $\neq$ B on B Λ, and let C be any point $\neq$ B on BΩ. By Axiom C-1, there are unique points A′ on B′Λ', and C′ on B′Ω', such that

$$AB \cong A'B' \qquad \text{and} \qquad CB \cong C'B'.$$

Then $\triangle ABC \cong \triangle A'B'C'$ (SAS), so by Proposition 9.5, there is a unique motion T effecting this congruence. Clearly T sends $(\Omega, \Sigma, \Lambda)$ to $(\Omega', \Sigma', \Lambda')$. Conversely, any such motion must send (A, B, C) onto (A′, B′, C′), so by Proposition 9.5, T is unique. ∎

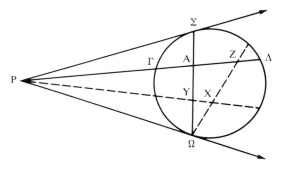

FIGURE 9.16

Note. Part (c) of this proposition can be visualized nicely in terms of the Klein model. Ideal points are represented by points on the absolute (the unit circle). There is a unique motion T mapping any triple of points $(\Sigma, \Omega, \Lambda)$ on the absolute onto any other. The effect of T on the other points can be described as follows (see Figure 9.16).

If P is the pole of chord $\Omega\Sigma$, then line $P\Lambda$ is Klein-perpendicular to $\Omega\Sigma$ at some point A. Then the image A' of A must be the intersection of $\Omega'\Sigma'$ with $\Lambda'P'$, where P' is the pole of $\Omega'\Sigma'$. Take any other point B on $\Omega\Sigma$; say $\Omega * B * A$. By Theorem 7.4, the image B' of B is the unique point between Ω' and A' such that cross-ratios are preserved:[7]

$$(AB, \Omega\Sigma) = (A'B', \Omega'\Sigma').$$

Let Γ be the other intersection of $P\Lambda$ with the absolute — its image Γ' is the other intersection of $P'\Lambda'$ with the absolute. As previously, we can use cross-ratios to determine the image of any point on $\Gamma\Lambda$. Finally, given any other point X (ideal or ordinary), represent it as the intersection of two lines, one being XP, which cuts $\Omega\Sigma$ at some point Y, and the other being $X\Omega$ (or $X\Sigma$), which cuts $\Gamma\Lambda$ at some point Z. Then X' is the intersection of P'Y' and $Z'\Omega'$ (or $Z'\Sigma'$).

This construction describes the motion T in terms of incidence alone. It suggests the conjecture that *every collineation of the hyperbolic plane is a motion;* this conjecture was demonstrated by Karl Menger

---

[7] The mapping of $\Omega\Sigma$ on to $\Omega'\Sigma'$ given by this equality of cross-ratios is called a *projectivity*. It can be described more geometrically by a sequence of at most three perspectivities (see p. 266); this is essentially the "fundamental theorem of projective geometry" (Ewald, 1971, Theorem 5.9.5, p. 226).

and his students.[8] In the Euclidean plane there are lots of collineations that are not motions or similarities (see Exercise 34 on affine transformations).

PARALLEL DISPLACEMENTS

We next study parallel displacements about an ideal point Σ.

PROPOSITION 9.17. Given a parallel displacement $T = R_l R_m$, where l and m are asymptotically parallel in the direction of ideal point Σ. Then

(a) T has no ordinary fixed points.
(b) Let k be any line through Σ and A any point on k. Then Σ lies on the perpendicular bisector h of AA' and $T = R_h R_k$.
(c) T has no invariant lines.
(d) The only ideal fixed point of T is Σ.
(e) The set of parallel displacements about Σ is a commutative group.
(f) A motion with exactly one ideal fixed point is a parallel displacement.

Proof:

(a) Assume A is fixed. Then the line joining A to $A^m = A'$ is perpendicular to both l and m, contradicting the hypothesis.

(b) Σ lies on two perpendicular bisectors l and m of $\triangle AA^mA'$, so by Major Exercise 7, Chapter 6, Σ also lies on the third perpendicular bisector h. Then $R_h T$ fixes A and Σ. By Proposition 9.16(b), $R_h T$ cannot be a rotation about A. By Proposition 9.4, it must be a reflection, and by Proposition 9.16(a), it has to be the reflection across the line k joining A to Σ. (See Figure 9.17.)

(c) Suppose line t were invariant under T. Choose any point A lying on t and let h, k be as in (b). Then $h \perp t = \overleftrightarrow{AA'}$, so t is invariant under R_h too. Hence t is invariant under $R_k = R_h T$, which means either $t \perp k$ or $t = k$. But the asymptotically parallel lines h and k cannot have a common (or be) perpendicular.

[8] See L. Blumenthal and K. Menger (1970, p. 220). See also K. Menger, "The New Foundation of Hyperbolic Geometry," in J. C. Butcher (ed.), *A Spectrum of Mathematics* (Auckland and Oxford University Presses, 1971), p.86. The idea of the proof is given in Exercise K-21, Chapter 7.

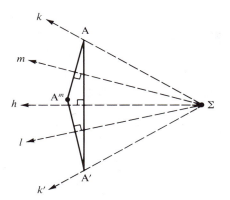

FIGURE 9.17

   (d) If T had another ideal fixed point Ω, then line $\Sigma\Omega$ would be invariant, contradicting part (c).

   (e) The proof is the same as the proof of Proposition 9.9, using part (b) instead of Proposition 9.8.

   (f) This follows from the classification of motions (Theorem 9.1, later in this chapter), and is inserted here for convenience. ■

GLIDES

We come now to our final type of motion, a *glide* alone a line t, defined as a product $T' = R_t T$ of a nonidentity translation T along t followed by reflection across t. (If you walk straight through the snow, your consecutive footprints are related by a glide.)

PROPOSITION 9.18.   (See Figure 9.18.) Given $l \perp t$ at A, $m \perp t$ at B, $T = R_l R_m$, $T' = R_t T$. Then

   (a) $TR_t = T'$.
   (b) $H_A R_m = T' = R_l H_B$.
   (c) T' maps each side of t onto the opposite side.
   (d) T' has no fixed points.
   (e) The only invariant line of T' is t.

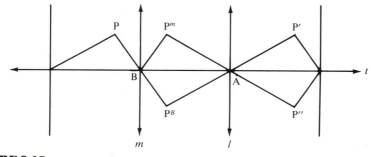

FIGURE 9.18

(f) Conversely, given point B and line l, let t be the perpendicular to l through B. Then $R_l H_B$ is a glide along t if B does not lie on l, and is R_t if B does lie on l.

Proof:
(a) and (b) follow from the formulas $H_A = R_t R_l = R_l R_t$ and $H_B = R_t R_m = R_m R_t$. (c) is clear, and (d) follows from it. (e) follows from (c) and (d). As for (f), if B lies on l, then $H_B = R_l R_t$, so $R_l H_B = R_l(R_l R_t) = (R_l^2)R_t = R_t$. If B does not lie on l, let $m \neq l$ be the perpendicular to t through B; then $T = R_l R_m \neq I$ and $R_l H_B = T R_t$. ■

Glides are characterized in Euclidean geometry by having only one invariant line. In hyperbolic geometry this characteristic does not distinguish them from translations, so we must add the condition that the two sides of the invariant line are interchanged. The invariant line is called the *axis* of the glide.

HJELMSLEV'S LEMMA. Let G be a glide, l a line not invariant under G and l'' the image of l under G. As point P varies on l and its image P' varies on l'', the midpoints of the segments PP' all lie on the axis t of G. Furthermore, those midpoints are all distinct, except in the case of $G = H_M R_l$, where the midpoints all coincide with M; that case occurs if and only if the axis t of G is perpendicular to both l and l''.

The proof will be left for Exercise 21. The lemma gives a method of locating the axis of a glide.

CLASSIFICATION OF MOTIONS

Our next objective is to show that every motion is either a reflection, a rotation, a translation, a parallel displacement, or a glide. The first step is to describe products of three reflections. Toward that end, we introduce three types of *pencils of lines:*

1. The pencil of all lines through a given point P.
2. The pencil of all lines perpendicular to a given line *t*.
3. The pencil of all lines through a given ideal point Σ (hyperbolic plane only).

Clearly, two lines *l* and *m* determine a unique pencil (if *l* and *m* are divergently parallel in the hyperbolic plane, they have a common perpendicular *t* by Theorem 6.6, Chapter 6). Moreover, if A is any point, there is a line *n* in that pencil through A. For the three types of pencils, *n* is:

1. The line $\overleftrightarrow{AP}$ if $A \neq P$.
2. The perpendicular to *t* through A.
3. The line $A\Sigma$.

The first part of the next proposition is sometimes called "the theorem on three reflections." F. Bachmann takes it as an axiom for his development of geometry without continuity or betweenness axioms.[9]

PROPOSITION 9.19.   Let $T = R_l R_m R_n$. (1) If *l*, *m*, and *n* belong to a pencil, then *T* is a reflection in a line of that pencil. (2) If *l*, *m*, and *n* do not belong to a pencil, then *T* is a glide.

This proposition is wonderful. It leads to the complete classification of motions. You may wonder how the axis of glide *T* is related to the three given lines; Exercises 55–57 answer this question in the Euclidean case.

---

[9] Bachmann introduces ultra-ideal and ideal points as pencils of the second and third types, and, using a technique developed by the Danish geometer J. Hjelmslev, is able to prove that the plane so extended is a projective plane coordinatized by a commutative field. See Appendix B.

Proof:

Part 1 of Proposition 9.19 follows from Propositions 9.8, 9.11, and 9.17(b), so assume the lines do not belong to a pencil. Choose any point A on l. Let m' be the line through A belonging to the pencil determined by m and n (Figure 9.19). Then line n' exists such that

$$R_{m'} R_m R_n = R_{n'}.$$

Let B be the foot of the perpendicular k from A to n'. Since l, m', and k pass through A, line h exists such that

$$R_l R_{m'} R_k = R_h.$$

Then B does not lie on h (by assumption on l, m, n) so by Proposition 9.18(f), $R_h H_B$ is a glide along the perpendicular to h through B. But

$$R_h H_B = R_h(R_k R_{n'}) = R_l R_{m'} R_k R_{m'} R_m R_n = T. \quad \blacksquare$$

COROLLARY. Every product $R_l R_m H_A$ equals a product $R_h R_k$.

Proof:

Let n be a line through A in the pencil determined by l and m. Let h be the line such that

$$R_l R_m R_n = R_h$$

and let k be the perpendicular to n through A. Then

$$R_l R_m H_A = R_l R_m R_n R_k = R_h R_k. \quad \blacksquare$$

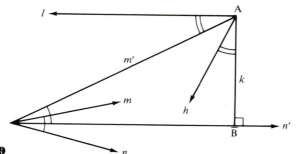

FIGURE 9.19

DEFINITION. A motion is called *direct* (or *proper* or *orientation-preserving*) if it is a product of two reflections or else is the identity. It is called *opposite* (or *improper* or *orientation-reversing*) if it is a reflection or a glide.

THEOREM 9.1. Every motion is either direct or opposite and not both. The set of direct motions is a group. The product of two opposite motions is direct, whereas the product of a direct and an opposite motion is opposite.

Proof:

We know that every motion is a product of at most three reflections (corollary to Proposition 9.5), so by Proposition 9.19, every motion is either direct or opposite. *The opposite motions are characterized by having an invariant line whose sides are interchanged.*

Given a product $(R_k R_l)(R_m R_n)$ of direct motions. If l, m, n belong to a pencil, $R_l R_m R_n = R_h$ and the product reduces to $R_k R_h$, which is direct. Otherwise $R_l R_m R_n = R_h H_B$ (Propositions 9.19 and 9.18(b)), and the corollary tells us that $R_k(R_h H_B)$ is direct. It follows that the direct motions form a group.

The product of a reflection and a direct motion is opposite by Proposition 9.19. The product of a glide and a direct motion is a product of five reflections, which reduces to a product of three reflections by the previous paragraph, hence is opposite by Proposition 9.19. Similarly, a product of four to six reflections reduces to a product of two. ■

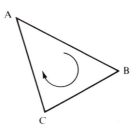

FIGURE 9.20

Note. The intuitive idea behind our classification of motions is that the plane can be given two distinct "orientations," so that, for example, the vertices of $\triangle ABC$ can be ordered in a "clockwise direction" (Figure 9.20). When the triangle is moved by a rotation, translation, or parallel displacement, the orientation of $\triangle A'B'C'$ will remain clockwise, whereas under a reflection or a glide, the orientation becomes counterclockwise. (See Exercise 23 for further discussion of orientation.)

Note. In the elliptic plane, no such invariant exists that is preserved by rotations and reversed by reflections, since every reflection is a 180° rotation; there is no distinction between direct and opposite motions in the elliptic plane. The only motions are rotations.

TABLE 9.1   Table of motions.

	Orientation	Fixed Points	Invariant Lines	Ideal Fixed Points (Hyperbolic Plane)
Identity	direct	all	all	all
Reflection R_l	opposite	points on l	l and all $m \perp l$	the two ends of l
Half-turn H_A	direct	A	all lines thru A	none
Rotation that is not involutory	direct	one	none	none
Euclidean translation along t	direct	none	t and all $u \| t$	—
Hyperbolic translation along t	direct	none	only t	the two ends of t
Parallel displacement	direct	none	none	one
Glide along t	opposite	none	t	the two ends of t

AUTOMORPHISMS OF THE CARTESIAN MODEL

Our next objective is to *rapidly* describe the groups of motions explicitly in terms of coordinates in models of our geometries. We begin with the Cartesian model of the Euclidean plane, and we assume in both this section and the next that the reader has some familiarity with vectors, matrices, and complex numbers.

The easiest transformations to describe are the *translations*. As the proof of Proposition 9.10 showed, a translation moves each point a fixed distance and in a fixed direction (in Figure 9.8, it moves the distance $2\overline{AB}$ in the direction $\overrightarrow{BA}$). This can be represented by a vector emanating from the origin of our coordinate system of length $2\overline{AB}$ and pointing in the given direction. If the coordinates of the endpoint of this vector are (e, f), then, by definition of vector addition, the translation is given by

$$T(x, y) = (x, y) + (e, f) = (x + e, y + f)$$

(Figure 9.21) or

$$x' = x + e$$
$$y' = y + f.$$

If we apply a second translation T' corresponding to the vector with endpoint (e', f'), then the image (x'', y'') of (x, y) under $T'T$ is given by

$$(x'', y'') = (x', y') + (e', f') = (x + e + e', y + f + f').$$

Thus $T'T$ is the translation by the sum of the vectors determining T and T'. This proves the next proposition.

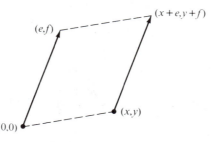

FIGURE 9.21

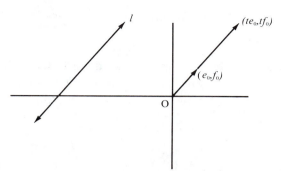

FIGURE 9.22

PROPOSITION 9.20. In the Cartesian model of the Euclidean plane, the translations form a commutative group isomorphic to the group of vectors under addition.

(According to our general definition of isomorphic models, two groups are *isomorphic* if there is a one-to-one correspondence between them and that correspondence preserves the group laws; here the two groups are considered as models of the system of axioms 1 through 4 on p. 311.)

We say that the translations form a *two-parameter group* since they depend on two real variables (e, f).[10]

PROPOSITION 9.21. In the Cartesian model of the Euclidean plane, the translations along a fixed line form a one-parameter group isomorphic to the group of real numbers under addition.

Proof:

Let (e_0, f_0) be a unit vector parallel to the fixed line l (Figure 9.22). Then the vector corresponding to a translation T along l has the form $t(e_0, f_0) = (te_0, tf_0)$, where $|t|$ is the distance translated and t is positive or negative according to whether the direction of translation is the same as (e_0, f_0) or opposite. If T' corresponds to vector $t'(e_0, f_0)$, then $T'T$ corresponds to vector $t(e_0, f_0) + t'(e_0, f_0) =$

---

[10] The theory of groups of transformations that depend continuously on real parameters was first developed by the great Norwegian mathematician Sophus Lie in the late nineteenth century, and has become one of the most fruitful ideas in twentieth-century mathematics and physics. (For example, this theory was used to predict the existence of certain subatomic particles—see F. S. Dyson, "Mathematics in the Physical Sciences," *Scientific American,* September 1964.)

$(t + t')(e_0, f_0)$. Thus assigning the parameter t to T gives the isomorphism. ∎

We next discuss rotations about a fixed point A. Our first step is to reduce to the case where A is the origin O: if not, let T be the translation along $\overleftrightarrow{\text{AO}}$ taking A to O. Then (by Proposition 9.11)

$$T = R_m R_l = R_{l^*} R_m,$$

where m is the perpendicular bisector of AO and l, l^* are the perpendiculars to $\overleftrightarrow{\text{AO}}$ through A, O. The given rotation R about A can be written (by Proposition 9.8) as

$$R = R_l R_k,$$

where k passes through A. Let k^* be the reflection of k across m (see Figure 9.23). Then $R^* = R_{k^*} R_{l^*}$ is a rotation about O and

$$\begin{aligned}
T^{-1} R^* T &= (R_l R_m)(R_{k^*} R_{l^*})(R_{l^*} R_m) \\
&= R_l R_m R_{k^*} (R_{l^*})^2 R_m \\
&= R_l (R_m R_{k^*} R_m) \\
&= R_l R_k \\
&= R.
\end{aligned}$$

This shows that the rotation R about A is uniquely determined by the rotation R^* about O. Moreover, the mapping $R^* \to T^{-1} R^* T$ is an isomorphism of the group of rotations about O onto the group of rotations about A, as you can easily verify. Thus we may assume A = O.

By Proposition 9.8, the given rotation about O can be written as $R = R_l R_m$, where m is the x axis. If $l \perp m$, then R is represented in complex coordinates as

$$z \to -z$$

(Proposition 9.6). Otherwise, if the acute angle from m to l has radian measure $\theta/2$, $0 < \theta/2 < \pi/2$, then R is represented in complex coordinates[11] as

$$\begin{aligned}
z &\to e^{i\theta} z \quad &&\text{(if } l \text{ has positive slope)} \\
z &\to e^{-i\theta} z \quad &&\text{(if } l \text{ has negative slope)}
\end{aligned}$$

---

[11] Recall that $e^{i\theta} = \cos\theta + i \sin\theta$.

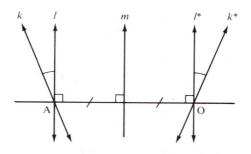

FIGURE 9.23

(see Figure 9.6 and Proposition 9.8). Combining these cases, we see that rotations about O are uniquely represented by the transformations

$$z \to e^{i\theta}z \qquad (-\pi < \theta \le \pi).$$

Since $e^{i\phi}(e^{i\theta}z) = (e^{i\phi}e^{i\theta})z$, the product of two rotations about O corresponds to the product $e^{i\phi}e^{i\theta}$ of complex numbers of absolute value 1. This proves the following proposition.

PROPOSITION 9.22.   In the Cartesian model of the Euclidean plane, the group of rotations about a fixed point is isomorphic to the one-parameter multiplicative group S^1 of complex numbers $e^{i\theta}$ of absolute value 1 (θ is the real parameter).

Let us combine our results, using complex coordinates. If a point has complex coordinate z, translating it by a vector (e_0, f_0) is the same as adding to z the complex number $z_0 = e_0 + if_0$, since addition of complex numbers is the same as vector addition. Now if T is any *direct* motion and T moves the origin O to the point O′ with complex coordinate z_0, follow T by the translation by $-z_0$ to obtain a direct motion fixing O. This motion is a rotation about O by our previous results, hence has the form $z \to e^{i\theta}z$. Therefore our original motion is equal to this rotation followed by translation by z_0. We have proved the following proposition.

PROPOSITION 9.23.   The group of direct motions of the Cartesian model of the Euclidean plane is isomorphic to the three-parameter group given in complex coordinates by

$$z \to e^{i\theta}z + z_0.$$

Let us be more explicit on the multiplication law for this group. Let T have complex parameters $(e^{i\theta}, z_0)$ and T' complex parameters $(e^{i\theta'}, z_0')$. Then the image of z under $T'T$ is

$$e^{i\theta'}(e^{i\theta}z + z_0) + z_0' = e^{i(\theta+\theta')}z + (e^{i\theta'}z_0 + z_0')$$

so the complex parameters are $(e^{i(\theta+\theta')}, e^{i\theta'}z_0 + z_0')$. In other words, the rotation parameters multiply, but the translation parameters do not add — there is a "twist" involved in multiplying z_0 by $e^{i\theta'}$. This accounts for the noncommutativity of the group (which technically is a "semidirect product" of S^1 with the additive group $\mathbb{C}$ of complex numbers).

Put another way, let $T = T_1 R_1$ and $T' = T_2 R_2$, where the T_i are translations and the R_i rotations about O, $i = 1, 2$. Then

$$T'T = T_2 R_2 T_1 R_1 = T_2 R_2 T_1 (R_2^{-1}R_2)R_1 = T_2(R_2 T_1 R_2^{-1})(R_2 R_1).$$

In this last expression, the factor on the right is the product $R_2 R_1$ of the two rotations about O, and the factor T_2 on the left is the second translation. The middle factor reveals the "twist," because $R_2 T_1 R_2^{-1}$ is the translation by $e^{i\theta'}z_0$:

$$\begin{aligned}
R_2 T_1 R_2^{-1}(z) &= R_2 T_1(e^{-i\theta'}z) \\
&= R_2(e^{-i\theta'}z + z_0) \\
&= e^{i\theta'}(e^{-i\theta'}z + z_0) \\
&= z + e^{i\theta'}z_0.
\end{aligned}$$

COROLLARY. Opposite motions of the Cartesian plane have a unique representation in the form

$$z \to e^{i\theta}\bar{z} + z_0.$$

Proof:
By Theorem 9.1, all opposite motions are obtained by following all the direct motions with one particular opposite motion, which we can choose to be reflection across the X axis $z \to \bar{z}$. Since $e^{\overline{i\theta}} = e^{-i\theta}$, the complex conjugate of $e^{i\theta}z + z_0$ is $e^{-i\theta}\bar{z} + \bar{z}_0$, and relabeling $-\theta$ for θ, $\bar{z}_0$ for z_0 gives the result. ∎

These results easily generalize to similarities.

PROPOSITION 9.24. In the Cartesian model of the Euclidean plane, a similarity is represented in complex coordinates either in the form

$$z \longrightarrow w_0 z + z_0 \qquad w_0 \neq 0$$

(in which case it is called *direct*), or in the form

$$z \longrightarrow w_0 \bar{z} + z_0 \qquad w_0 \neq 0$$

(in which case it is called *opposite*). The direct similarities form a 4-parameter group.

Here w_0 ranges through the multiplicative group $\mathbb{C}^*$ of nonzero complex numbers while z_0 ranges through the additive group $\mathbb{C}$ of all complex numbers; the group of direct similarities is the "semidirect product" of $\mathbb{C}^*$ with $\mathbb{C}$. The modulus $k = |w_0|$ is the constant of proportionality for the similarity. Geometrically, this representation means that a direct similarity is equal to a dilation centered at the origin followed by a rotation about the origin followed by a translation; an opposite similarity is equal to the reflection in the x axis followed by a direct similarity. The proof is left for Exercise 24.

MOTIONS IN THE POINCARÉ MODEL

We turn next to the coordinate description of hyperbolic motions, and for this purpose the most convenient representation is the Poincaré upper half-plane model (see p. 237). Recall that the Poincaré lines are either vertical rays emanating from points on the x axis or semicircles with center on the x axis. The ideal points are represented in this model by the points on the x axis and a point at infinity ∞ which is the other end of every vertical ray (Figure 9.24).

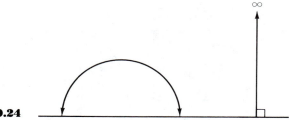

FIGURE 9.24

We have seen (Exercise P-4, Chapter 7) that hyperbolic reflections are represented in the Poincaré *disk* model by either Euclidean reflections in diameters of the absolute circle γ or by inversions in circles δ orthogonal to γ. Let us show that *in the upper half-plane model, hyperbolic reflections are represented by either Euclidean reflections in the vertical lines or by inversions in circles δ orthogonal to the x axis,* i.e., circles δ with center on the x axis.

In Exercise 38, you will show that the mapping

$$E : z \longrightarrow i \, \frac{i+z}{i-z}$$

sends the unit disk one-to-one onto the upper half-plane, sends i to ∞, and all other points of the unit circle onto the x axis.

The Poincaré lines of the disk model are mapped onto the Poincaré lines of the upper half-plane model — in fact, all Euclidean circles and lines are mapped into either Euclidean circles or lines by E, and orthogonality is preserved (E is *conformal*); see Figure 9.25. So we can use E as the isomorphism which defines congruence in the upper half-plane interpretation (just as we previously established congruence in the Klein model via the isomorphism F — see p. 258).

For simplicity, let us agree to also call the Euclidean reflection in a Euclidean line "inversion"; this will enable us to avoid discussing this special case separately. Figure 9.26 shows a hyperbolic reflection R_l in the disk model represented as inversion. For any point A, drop hyperbolic perpendicular t from A to l, and let M be the foot on l of this perpendicular. Let α be the *hyperbolic* circle through A with hyperbolic center M.

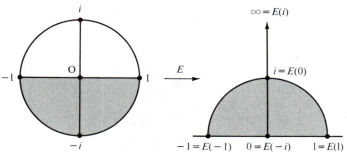

FIGURE 9.25

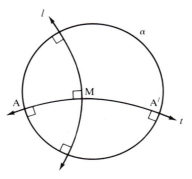

FIGURE 9.26

In Exercise P-5 of Chapter 7 you showed that α is also a Euclidean circle (with a different Euclidean center). The reflection A′ of A across l is then the other intersection of α with t. Now α is orthogonal to l since l is the extension of a hyperbolic diameter of α. Hence, α is mapped onto itself by inversion in l, and so is t (Proposition 7.11), so A′ must be the inverse of A in l.

If we apply the mapping E, this entire figure is transformed onto an isomorphic figure in the upper half-plane. Thus, the argument just given shows that R_l is also represented in the upper half-plane model by inversion. ■

We next calculate the formulas for these inversions. For a vertical line $x = k$, the inversion is given by

$$(x, y) \rightarrow (2k - x, y).$$

In terms of the single complex coordinate $z = x + iy$, this becomes

$$z \rightarrow 2k - \bar{z}.$$

For consistent notation later on, set $b = 2k$ and write this as

$$z \rightarrow -\bar{z} + b.$$

For a circle centered at $(k, 0)$ of radius r, make a change of coordinates $x' = x - k$, $y' = y$ (i.e., translate the center to the origin); the complex coordinate change is $z' = z - k$. Then, by definition of inversion, the image w' of z' is determined by the two equations

$$|z'||w'| = r^2$$
$$\frac{z'}{|z'|} = \frac{w'}{|w'|},$$

whose solution is

$$w' = |w'| \frac{z'}{|z'|} = \frac{r^2}{|z'|} \frac{z'}{|z'|} = r^2/\bar{z}'$$

since $|z'|^2 = z'\bar{z}'$. So in the original coordinate system with $w = w' + k$, we get

$$w = \frac{r^2}{\bar{z} - k} + k = \frac{k\bar{z} + r^2 - k^2}{\bar{z} - k}.$$

For convenience, we set $c = 1/r$, $a = kc$, and $b = r(1 - a^2)$. The inversion then takes the form

$$z \longrightarrow \frac{a\bar{z} + b}{c\bar{z} - a},$$

which includes the previous case when we set $c = 0$ and $a = -1$. We have shown

PROPOSITION 9.25. In the Poincaré upper half-plane model of the hyperbolic plane, reflections are represented in complex coordinates by

$$z \longrightarrow \frac{a\bar{z} + b}{c\bar{z} - a} \qquad a^2 + bc = 1$$

(where a, b, c are real numbers).

We can next determine the representation of all the *direct* hyperbolic motions, since they are products of two reflections. The calculation is simplified by the following general observations.

For any coefficient field K (such as the field $\mathbb{R}$ of all real numbers or the field $\mathbb{C}$ of all complex numbers), we define *the projective line* $\mathcal{P}^1(K)$ over K to be $K \cup \{\infty\}$, where "∞" just means another point not in K. Each point on this "line" will be assigned homogeneous coordinates $[x_1, x_2]$, where $x_1, x_2 \in K$ and are not both zero. These coordinates will only be determined up to multiplication by a nonzero scalar λ; that is,

$$[x_1, x_2] = [\lambda x_1, \lambda x_2] \qquad \lambda \neq 0.$$

Specifically, the point $x \in K$ is assigned the homogeneous coordinates

$$[x, 1] = [\lambda x, \lambda] \qquad \lambda \neq 0$$

while the point ∞ is assigned the homogeneous coordinates

$$[1, 0] = [\lambda, 0] \qquad \lambda \neq 0.$$

We will operate on the points of $\mathscr{P}^1(K)$ with nonsingular 2×2 matrices with coefficients in K, in the usual way that matrices operate on vectors:

$$\begin{bmatrix} a_{11} & a_{12} \\ a_{21} & a_{22} \end{bmatrix} \begin{bmatrix} x_1 \\ x_2 \end{bmatrix} = \begin{bmatrix} a_{11}x_1 + a_{12}x_2 \\ a_{21}x_1 + a_{22}x_2 \end{bmatrix},$$

where the brackets around the matrix again mean that its entries are only determined up to multiplication by a nonzero scalar. These operators are called *projective transformations,* and they form a group under matrix multiplication denoted $PGL(2, K)$. Now, a *linear fractional transformation*

$$x \rightarrow \frac{ax + b}{cx + d}$$

defined on K can be obtained by operating on the homogeneous coordinates $[x, 1]$ of x with the projective transformation $\begin{bmatrix} a & b \\ c & d \end{bmatrix}$, obtaining $[ax + b, cx + d]$, and then dehomogenizing the coordinates to get $[(ax + b)/(cx + d), 1]$. Viewed thusly it becomes clear that the composite of two such linear fractional transformations can be calculated by multiplying the two matrices, so the composite is again a linear fractional transformation.

Returning to our representation of reflections, we have the added complication of the complex conjugate $\bar{z}$ occurring in the formula of Proposition 9.25; but it is clear that for a product of two reflections, the two conjugations cancel each other, with the coefficients being unaffected because they are real numbers. Furthermore, the condition $a^2 + bc = 1$ in Proposition 9.25 means that the matrix

$$\begin{bmatrix} a & b \\ c & -a \end{bmatrix}$$

of the transformation has determinant -1. By the formula

$$\det (AB) = (\det A)(\det B)$$

for the determinant of a product of matrices, we see that the product will have determinant $+1$.

We claim that conversely, every linear fractional transformation with real coefficients and determinant $+1$ is a product of two hyperbolic reflections, i.e., the matrix equation

$$\begin{bmatrix} a' & b' \\ c' & -a' \end{bmatrix}\begin{bmatrix} a & b \\ c & -a \end{bmatrix} = \begin{bmatrix} x & y \\ u & v \end{bmatrix}$$

can be solved for the eight unknowns on the left, given the four real numbers on the right:

Case 1. $u \neq 0$. Then a solution is $c' = 0$, $a' = -1$, $c = u$, $a = -v$, $b' = (x - v)/u$, $b = vb' - y$.

Case 2. $u = 0$ and $y = 0$. We may assume $x > 0$. Then a solution is $a = 0 = a'$, $c = \sqrt{x} = b'$, $c^{-1} = b = c'$.

Case 3. $u = 0$ and $x = v = 1$. Then a solution is $c = 0 = c'$, $a = -1 = a'$, $b' = 0$ and $b = -y$.

Case 4. $u = 0$. This follows from the preceding cases because

$$\begin{bmatrix} x & y \\ 0 & x^{-1} \end{bmatrix} = \begin{bmatrix} x & 0 \\ 0 & x^{-1} \end{bmatrix}\begin{bmatrix} 1 & y/x \\ 0 & 1 \end{bmatrix}$$

and we know that a product of four reflections reduces to a product of two (Theorem 9.1.).

We have proved the next proposition.

PROPOSITION 9.26. In the Poincaré upper half-plane model of the hyperbolic plane, the direct hyperbolic motions are represented by all the linear fractional transformations

$$z \to \frac{ax + b}{cz + d} \qquad ad - bc = 1$$

$(a, b, c, d, \text{real})$.

This group is denoted $PSL(2, \mathbb{R})$, and is called the *projective special linear group* over the real field. It is a 3-parameter group (one of the four parameters being eliminated by the condition that the determinant be $+1$).[12]

We can next obtain all opposite hyperbolic motions by multiplying all the direct motions by one fixed opposite motion. For the latter, let's use the reflection in the y axis, $z \rightarrow -\bar{z}$. The result is

$$z \rightarrow \frac{-a\bar{z} + b}{-c\bar{z} + d},$$

which after relabeling gives the next proposition.

PROPOSITION 9.27. In the Poincaré upper half-plane model of the hyperbolic plane, the opposite hyperbolic motions are represented by all the mappings

$$z \rightarrow \frac{a\bar{z} + b}{c\bar{z} + d} \qquad ad - bc = -1$$

$(a, b, c, d$ real$)$.

We can combine the direct and opposite motions and represent them by all real projective transformations of the real projective line $\mathscr{P}^1(\mathbb{R})$, with the matrices representing direct or opposite motions according to whether the determinant D is positive or negative, since

$$\begin{bmatrix} a & b \\ c & d \end{bmatrix} = \begin{bmatrix} a/\sqrt{|D|} & b/\sqrt{|D|} \\ c/\sqrt{|D|} & d/\sqrt{|D|} \end{bmatrix}$$

and the matrix on the right has determinant ± 1.

COROLLARY. The group of all hyperbolic motions is isomorphic to $PGL(2, \mathbb{R})$.

This isomorphism suggests analogies between one-dimensional real projective geometry and two-dimensional hyperbolic geometry.[13] For example, Proposition 9.16(c) corresponds to the theorem in projective geometry that for any two triples of points on the projective

[12] $SL(2, \mathbb{R})$ is the group of all 2×2 real matrices of determinant $+1$. It is extremely important in analytic number theory—see the book by S. Lang devoted entirely to this group (Addison Wesley, Reading, Mass., 1975).

[13] Klein's *Erlanger Programme* pointed out many other analogies between geometries whose groups are isomorphic, e.g., the analogies among *the inversive plane* (Exercise P-17, Chapter 7), the one-dimensional complex projective geometry $\mathscr{P}^1(\mathbb{C})$ (which from the real point of view is a geometry on a sphere), and three-dimensional hyperbolic geometry.

line, there is a unique projective transformation mapping one triple onto the other (see Exercise 65). Also, projective transformations are classified by their fixed points. The equation for a fixed point is

$$x = \frac{ax + b}{cx + d}$$

or

$$cx^2 + (d - a)x - b = 0$$

showing that the number of finite fixed points is 0, 1, or 2. Since

$$\begin{bmatrix} a & b \\ c & d \end{bmatrix} \begin{bmatrix} 1 \\ 0 \end{bmatrix} = \begin{bmatrix} a \\ c \end{bmatrix},$$

we see that ∞ is a fixed point if and only if $c = 0$; in that case, the quadratic equation above becomes linear, and if the transformation is not the identity, it has a finite fixed point only when $a \neq d$.

Our classification of hyperbolic motions showed that the number of ideal fixed points is 0 for rotations, 1 for parallel displacements, and 2 for reflections, translations, and glides.

Example 3. Let us determine the group of all parallel displacements about ∞. We just showed that these are represented by matrices with $c = 0$ and $a = d$; they form the group of mappings

$$z \longrightarrow z + b,$$

which is isomorphic to the one-parameter additive group of all real numbers (these are *Euclidean* translations along the x axis).

Example 4. Consider next the two ideal fixed points 0 and ∞ and the group of hyperbolic translations along the Poincaré line joining them (the upper half of the y axis). They are represented by matrices with $c = b = 0$ and $ad = 1$; they form the group of mappings

$$z \longrightarrow a^2 z$$

and since a^2 can be any positive number, this group is isomorphic to the multiplicative group of all positive real numbers. By taking the logarithm, this group in turn is isomorphic to the additive group of all real numbers, just as in the Euclidean case of translations along a fixed line — Proposition 9.21. (The mappings are *Euclidean* dilations centered at 0.)

Example 5. Finally, let us determine the group of all hyperbolic rotations about the point i in the upper half-plane. They are the direct motions that fix i:

$$i = \frac{ai + b}{ci + d}$$

so $a = d$ and $b = -c$, with $1 = ad - bc = a^2 + b^2$. If we set $a = \cos\theta$, $b = -(\sin\theta)$, rotations about i are represented by

$$z \longrightarrow \frac{(\cos\theta)z - \sin\theta}{(\sin\theta)z + \cos\theta}.$$

But the matrix

$$\begin{bmatrix} \cos\theta & -\sin\theta \\ \sin\theta & \cos\theta \end{bmatrix}$$

is just the matrix of Euclidean rotation through θ about the origin in the Cartesian model of the Euclidean plane. Thus these two groups are isomorphic to each other and to the multiplicative group S^1 of all complex numbers of modulus 1.

Note. In the representation of hyperbolic rotations in Example 5, when $\theta = \pi$, z is mapped to z and we get the identity rotation, whereas the Euclidean rotation through π is a half-turn. So in order to have a one-to-one (instead of two-to-one) mapping of the group of hyperbolic rotations about i onto the group of Euclidean rotations about 0, we must represent the hyperbolic rotation through θ by

$$z \longrightarrow \frac{\left(\cos\dfrac{\theta}{2}\right)z - \sin\dfrac{\theta}{2}}{\left(\sin\dfrac{\theta}{2}\right)z + \cos\dfrac{\theta}{2}}.$$

CONGRUENCE DESCRIBED BY MOTIONS

In neutral geometry, motions can be used to define congruence of arbitrary figures, namely S is *congruent* to S' if there is a motion T mapping S onto S'. By Proposition 9.5, this definition gives the same congruence relation for triangles as before (hence, the same congruence relation for segments and for angles).

A particularly important figure is a *flag*, which is defined to consist of a point A, a ray $\overrightarrow{AB}$ emanating from A, and a side S of line $\overleftrightarrow{AB}$ (Figure 9.27).

LEMMA 9.3. Any two flags are congruent by a unique motion.

Proof
Choose B' so that AB $\cong$ A'B'. Choose any point C $\in S$, and let C' be the unique point in side S' of $\overleftrightarrow{A'B'}$ such that $\triangle$ABC $\cong \triangle$A'B'C' (Corollary to SAS, Chapter 3). By Proposition 9.5, there is a unique motion T taking A, B, C respectively into A', B', C', and this maps flag $\overrightarrow{AB} \cup S$ onto $\overrightarrow{A'B'} \cup S'$. ∎

We say that the group of motions *operates simply transitively* on the flags. This property expresses the homogeneity of the plane, and corresponds to our physical intuition of performing measurements by moving a rigid ruler around. This property is crucial in our next theorem, which answers the converse question: Given a model $\mathcal{M}$ for our incidence and betweenness axioms, and given a group $\mathcal{G}$ of betweenness-preserving collineations of $\mathcal{M}$. Define congruence by the action of $\mathcal{G}$; e.g., *define* AB $\cong$ A'B' to mean that some transformation T in $\mathcal{G}$ maps segment AB onto segment A'B' (similarly for angles). What additional assumptions on $\mathcal{G}$ guarantee that with this definition, our Congruence Axioms C-1 through C-6 hold in $\mathcal{M}$?

FIGURE 9.27

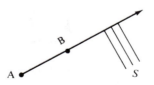

THEOREM 9.2. Assume the group $\mathcal{G}$ of betweenness-preserving collineations satisfies the following conditions:

(i) $\mathcal{G}$ operates simply transitively on the flags.
(ii) For any two points A, B, there is at least one transformation $T \in \mathcal{G}$ which interchanges A and B.
(iii) For any two rays r, s emanating from the same vertex, there is at least one transformation $T \in \mathcal{G}$ which interchanges r and s.

Then, with congruence defined in terms of the action of $\mathcal{G}$, Axioms C-1 through C-6 hold, and $\mathcal{G}$ is the group of motions.

We know that these conditions are necessary for the group of motions — condition (i) by our lemma, condition (ii) using the reflection across the perpendicular bisector of AB, and condition (iii) using the reflection across the bisector of the angle $r \cup s$.

Proof:
The proof that the conditions are sufficient proceeds in 10 steps.

Step 1. Congruence is reflexive, symmetric, and transitive (in particular, C-2 and C-5 hold). $\mathcal{G}$ is a group of automorphisms.

For if T maps S onto S', then T^{-1} maps S' onto S; and if T' maps S' onto S'', then $T'T$ maps S onto S''. Obviously, I maps S onto S. (We use here the condition that $\mathcal{G}$ is a group.) If $S \cong S'$, i.e., $S' = TS$, and $U \in \mathcal{G}$, then $US' = (UTU^{-1})US$, so $US \cong US'$; this shows that U is an automorphism.

Step 2. Existence of reflections.

Let point P lie on line l. Let S_1, S_2 be the two sides of l, and let r_1, r_2 be the two rays of l emanating from P. By condition (i), there is a unique $T \in \mathcal{G}$ which interchanges S_1 and S_2 and leaves r_1 invariant. Since T^2 leaves r_1 and S_1 invariant, $T^2 = I$ (by (i)). We claim T fixes every point of l: By definition of T, P is fixed. Since T is betweenness-preserving, r_2 is also invariant under T. Suppose point A on l is moved to A', and say P * A * A'. Then $(A')' = A$ since $T^2 = I$, so P' * (A')' * A', contradicting the assumption that T preserves betweenness. Since the reflection R_l is the involutory automorphism that leaves each point of l fixed, we may write $T = R_l$.

Step 3. Let r be a ray of line l, and let r' be any ray. Then there are exactly two transformations in $\mathscr{G}$ that map r onto r', and they both have the same effect on the points of l.

For by condition (i), given a side S of l, the two transformations are uniquely determined by which side of l' $(r' \subset l')$ S is mapped onto. If T is one such transformation, the other is TR_l, and they both agree on l.

Step 4. If AB $\cong$ CD, then there are exactly two transformations in $\mathscr{G}$ sending A to C and B to D, and they agree on $\overleftrightarrow{AB}$.

By definition of $\cong$, there is a transformation T in $\mathscr{G}$, mapping segment AB onto segment CD. If T sends A to D and B to C, follow T by a transformation in $\mathscr{G}$ that interchanges C and D (condition (ii)). We can therefore assume A goes to C and B to D. Since betweenness is preserved, ray $\overrightarrow{AB}$ is mapped onto ray $\overrightarrow{CD}$. Hence, step 3 applies.

Step 5. Congruence Axiom C-1 holds. (This follows from steps 3 and 4.)

Step 6. Congruence Axiom C-3 holds.

Let T send A to A' and B to B'. If A * B * C, then T maps ray $\overrightarrow{BC}$ onto ray $\overrightarrow{B'C'}$, where A' * B' * C'. If BC $\cong$ B'C', then there is a motion T' sending B to B' and C to C'. By step 3, T and T' agree at every point on the line through A, B, C. Hence they both send A to A' and C to C', so AC $\cong$ A'C'.

Step 7. Congruence Axiom C-4 holds.

Given $\sphericalangle$BAC, ray $\overrightarrow{A'C'}$, and side S' of $\overleftrightarrow{A'C'}$. Let S be the side of $\overleftrightarrow{AC}$ on which B lies. Let T in $\mathscr{G}$ be the unique transformation which, according to condition (i), maps $(\overrightarrow{AC}, S)$ onto $(\overrightarrow{A'C'}, S')$. If B' is the image of B under T then $\sphericalangle$BAC $\cong$ $\sphericalangle$B'A'C'. Conversely, if this congruence is effected by a transformation T', where B' lies in S', then T' maps $(\overrightarrow{AC}, S)$ onto $(\overrightarrow{A'C'}, S')$, so by the uniqueness part of condition (i), $T = T'$, and ray $\overrightarrow{A'B'}$ in S' is uniquely determined.

Step 8. If $\sphericalangle BAC \cong \sphericalangle B'A'C'$, there is a unique transformation in $\mathcal{G}$ sending $\overrightarrow{AB}$ to $\overrightarrow{A'B'}$, and $\overrightarrow{AC}$ to $\overrightarrow{A'C'}$.

By definition of congruence, there is a $T \in \mathcal{G}$ mapping $\sphericalangle BAC$ onto $\sphericalangle B'A'C'$. If T maps $\overrightarrow{AC}$ onto $\overrightarrow{A'B'}$ and $\overrightarrow{AB}$ onto $\overrightarrow{A'C'}$, then condition (iii) allows us to follow T with a transformation in $\mathcal{G}$ interchanging the two sides of $\sphericalangle B'A'C'$; so we can assume T sends $\overrightarrow{AC}$ to $\overrightarrow{A'C'}$ and $\overrightarrow{AB}$ to $\overrightarrow{A'B'}$. Then uniqueness was shown in the proof of step 7.

Step 9. Congruence Axiom C-6 (SAS) holds.

Given $AB \cong A'B'$, $\sphericalangle BAC \cong \sphericalangle B'A'C'$, and $AC \cong A'C'$. Let these congruences be effected by transformations $T_1, T_2, T_3 \in \mathcal{G}$, where by steps 4 and 8, we may assume T_1 sends A to A' and B to B', T_2 sends $\overrightarrow{AB}$ to $\overrightarrow{A'B'}$ and $\overrightarrow{AC}$ to $\overrightarrow{A'C'}$, and T_3 sends A to A' and C to C'. By step 3, T_2 agrees with T_1 on $\overrightarrow{AB}$, and T_2 agrees with T_3 on $\overrightarrow{AC}$. Hence, T_2 sends B to B' and C to C', so that via T_2 we have $BC \cong B'C'$, $\sphericalangle ABC \cong \sphericalangle A'B'C'$, and $\sphericalangle ACB \cong \sphericalangle A'C'B'$.

Step 10. $\mathcal{G}$ is the group of all motions.

Choose a flag F. A motion T transforms F into F', and by condition (i), an automorphism $T' \in \mathcal{G}$ has the same effect; by Lemma 9.2, $T = T'$, so every motion belongs to $\mathcal{G}$. By Lemma 9.3 and the same argument, every member of $\mathcal{G}$ is a motion. ∎

Theorem 9.2 is one step in Klein's program to describe the geometry in terms of action of a group. F. Bachmann (1973) carries the program further by describing points, lines, and incidence in terms of involutions in the group—see Exercise 50 and Ewald (1971).

Note on Similarities. A *pointed flag* is a figure consisting of an ordered pair of distinct points A, B together with a specific side S of line $\overleftrightarrow{AB}$. Every flag with vertex A supports infinitely many pointed flags corresponding to the various choices of point B on the ray of the flag. It is easy to show that for any two pointed flags in the Euclidean plane, there is a unique similarity mapping one onto the other (namely, follow the motion given by Lemma 9.3 with a dilation centered at the vertex). This leads to the following nice characterization: *similarities are collineations which map circles onto circles.*

Proof:

Let T be such a collineation. In Major Exercise 7, Chapter 5, you showed that given any line l, two points off l are on the same side S of l if and only if they lie on a circle contained in S. Hence T maps S onto a side S' of l'. Also, if T maps circle γ onto circle γ', then T maps the interior of γ onto the interior of γ' — because a point P not on γ lies in the interior of γ if and only if every line through P intersects γ. Next we claim that T maps perpendicular lines onto perpendicular lines. This is because a collineation maps parallelograms onto parallelograms, a parallelogram in the Euclidean plane has a circumscribed circle iff it is a rectangle (Major Exercise 1, Chapter 5), and since T preserves circles, it must map rectangles onto rectangles. Moreover, since a square is characterized as a rectangle with perpendicular diagonals, T maps squares onto squares. T also maps the center of a square onto the center of the image square (since it is the intersection of the diagonals) and the midpoints of the sides onto the midpoints of the sides (since they form squares with the center).

Now let us work with Cartesian coordinates. Consider the basic pointed flag F consisting of the origin $(0, 0)$, the unit point $(1, 0)$ on the x axis, and the side S of the x axis containing the unit point $(0, 1)$ of the y axis. Our transformation T maps F onto some flag F'; let U be the unique similarity which maps F' back onto F. We will show that UT is the identity, so that T is equal to the similarity inverse to U: UT fixes $(0, 0)$ and $(1, 0)$ and maps each side of the x axis onto itself. By the remark above about squares, the points $(0, 1)$, $(1, 1)$, $(0, -1)$, and $(1, -1)$ must also be fixed, as must the midpoints and centers of these squares. We can infer successively that all points whose coordinates are integers, half-integers, or dyadic rationals are fixed by UT. Since each point in the plane is interior to an arbitrarily small circle through three fixed points, it must be fixed. (This neat proof is due to Werner Fenchel, 1989.) ∎

SYMMETRY

We conclude this chapter with a brief discussion of symmetry, which is one of the main applications of the transformation approach to geometry.

Given a plane figure S, the motions that leave S invariant (i.e., that map S onto itself) are called *symmetries of S;* clearly the symmetries of S form a group. Intuitively, the larger this group is, the more symmetrical is the figure.

For example, a circle γ is highly symmetric. Its symmetry group consists of all rotations about the center O of γ and all reflections across lines through O; this group has the cardinality of the continuum.

A square seems to be fairly symmetric, yet we will show it only has eight symmetries (see Example 6 later in this section).

The frieze pattern shown in Figure 9.28 has a countably infinite group of symmetries: it consists of all integer powers T^n of a fixed translation T which shifts the pattern one unit to the right.

The first problem is to find a *minimal set of generators* of the group of symmetries. This means finding as small a collection of symmetries as possible with the property that all other symmetries can be expressed as products of the symmetries in this collection and their inverses. For the frieze pattern in Figure 9.28, there is a single generator T (or T^{-1}).

A second problem is to describe the basic *relations* among the generators. For the generator T above, there is no relation: all the powers of T are distinct. Consider, however, Figure 9.29.

The only symmetries of these figures are the identity I, the rotation R about the center O through 120° clockwise, and the rotation R^2 about O through 240° clockwise. R is a generator of this group and satisfies the relation $R^3 = I$. We say this group is cyclic of order 3.

More generally, a group is called *cyclic of order n* if it has a single generator and n elements; it is called *infinite cyclic* if it has a single generator and infinitely many elements (the group of symmetries of

FIGURE 9.28

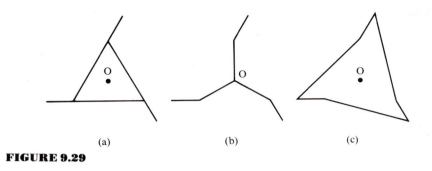

(a)   (b)   (c)

FIGURE 9.29

the frieze pattern in Figure 9.28 is infinite cyclic). Let us denote by C_n any cyclic group of order n generated by a rotation through $(360/n)°$ about some point. The constructions in Figure 9.29 can be generalized from 3 to n to obtain a figure having C_n as its symmetry group. The graph in Figure 9.29(b) is called a *triquetrum;* its generalization to $n = 4$ is a swastika. The $2n$-sided convex polygons obtained from generalizing Figure 9.29(c) are called *ratchet polygons.*

A third basic problem is to describe the structure of the symmetry group, showing if possible that it is isomorphic to some familiar group.

Example 6. We will solve these problems for the group of symmetries of a square.

Any symmetry must leave the center O fixed (since, for example, O is the intersection of the diagonals, and each diagonal must be mapped onto itself or the other diagonal). Hence, the symmetries must either be rotations about O or reflections across lines through O. The only rotations about O that are symmetries are I, R, R^2, R^3, where R can be taken to be the counterclockwise rotation through 90°; these form a cyclic subgroup of order 4. There are also four reflections that are symmetries: the two reflections across diagonals c, d and the two reflections across the perpendicular bisectors a, b of the sides (see Figure 9.30). Let T be any one of these reflections; e.g., $T = R_c$. Then $\{R, T\}$ is a minimal set of generators of the group.

For R can be written in four ways as a product of reflections,

$$R = R_b R_d = R_c R_b = R_a R_c = R_d R_a,$$

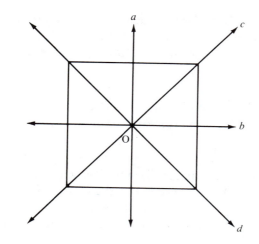

FIGURE 9.30

so that

$$TR = R_c(R_cR_b) = (R_c)^2R_b = R_b$$
$$TR^2 = (TR)R = R_b(R_bR_d) = R_d$$
$$TR^3 = (TR^2)R = R_d(R_dR_a) = R_a.$$

The basic relations between these generators are

$$R^4 = I$$
$$T^2 = I$$
$$RT = TR^3$$

(the last because $RT = (R_aR_c)R_c = R_a = TR^3$). This last relation shows that the group is noncommutative. It is denoted D_4, and is called the *dihedral group* of order 8.

More generally, if $n \geq 3$, D_n denotes the group of symmetries of a regular n-gon. It has $2n$ symmetries and is generated by two elements $\{R, T\}$, where R is rotation about the center of the n-gon through $(360/n)°$, and T is reflection across a line joining a vertex to the center. For $n = 2$, D_2 denotes the group generated by a half-turn H_P and a reflection across a line through P, while for $n = 1$, D_1 denotes a cyclic group of order 2 generated by a reflection.

The following remarkable theorem has been attributed to Leonardo da Vinci.

THEOREM 9.3. In both the Euclidean and hyperbolic planes, the only *finite* groups of motions are the groups C_n and D_n ($n \geqq 1$).

The proof will be based on the following series of lemmas, Lemmas 9.4 through 9.9.

LEMMA 9.4. A finite group of motions cannot contain nonidentity translations, parallel displacements, or glides.

Proof:
The point is to show that if T is any of these three types of motions, no power T^n, $n \neq 0$, of T is equal to the identity I. We will show this in case $T = R_l R_m$, $l \| m$, leaving the case of a glide for Exercise 18. By Exercise 7, we can also write $T = R_k R_l$, where k is the reflection of m across l, so that

$$T^2 = (R_k R_l)(R_l R_m) = R_k R_m.$$

Repeating this argument, we can show by induction that for any positive integer n, we can write

$$T^n = R_h R_m,$$

where h lies in the half-plane bounded by m and containing l—in particular, $h \| m$, so $T^n \neq I$. Applying this result to T^{-1} we get $T^n \neq I$ for negative n as well.

Now if T belonged to a finite group, we would have $T^n = T^m$ for distinct integers n, m; hence, $T^{n-m} = I$, a contradiction. ∎

LEMMA 9.5. If a finite group of motions contains rotations, all those rotations have the same center.

Proof:
Let T be a rotation about A, U a rotation about B $\neq$ A, and let $l = \overleftrightarrow{AB}$. By Proposition 9.8, there is a unique line m through A (respectively, n through B) such that $T = R_l R_m$ (respectively, $U = R_l R_n$). Then $U^{-1} T^{-1} U T$, which belongs to the finite group, is equal to $(R_n R_l R_m)^2$, which is a translation (see Exercise 15). This contradicts Lemma 9.4 unless the translation is the identity, but in that case $UT = TU$, which can only happen when at least one of U, T is the identity (see warning after Proposition 9.9). ∎

LEMMA 9.6. If a finite group of motions contains reflections, the axes of those reflections are concurrent.

Proof:
Otherwise we would obtain a contradiction of the previous lemmas, since the group must contain the products $R_l R_m$. ∎

LEMMA 9.7. If a finite group of motions contains a rotation and a reflection, then the center of the rotation lies on the axis of the reflection.

Proof:
Otherwise the product of the rotation and the reflection would be a glide (Proposition 9.19), contradicting Lemma 9.4. ∎

COROLLARY. A finite group of motions of order > 2 has a unique fixed point.

Proof:
Since the order is > 2, the group must contain a nonidentity rotation (by Lemmas 9.4 and 9.6), and every symmetry in the group fixes the center of that rotation, which is uniquely determined (by Lemmas 9.4, 9.5, and 9.7). ∎

LEMMA 9.8. If a finite group of motions of order n contains only rotations, then it is cyclic or order n.

Proof:
The cases $n = 1$ or 2 are trivial, so assume $n > 2$. Let O be the center of all the rotations (Lemma 9.5). Choose any point $P_1 \neq O$. The images $P_1, \ldots, P_n$ of P_1 under the rotations in our group are all distinct, since the rotations are distinct (Exercise 19). We assume these images numbered so that $(\sphericalangle P_1 O P_2)°$ is the minimum of all the degrees $(\sphericalangle P_i O P_j)°$. Let R be the rotation taking P_1 to P_2. We claim that R generates the group.

Let Q_i be the image of P_1 under R^{i-1}, $i = 1, 2 \ldots$ (so that $Q_1 = P_1$, $Q_2 = P_2$). Then

$$(\sphericalangle Q_i O Q_{i+1})° = (\sphericalangle P_1 O P_2)°$$

for each i. If some P_j were not among the Q_i's, ray $\overrightarrow{OP}_j$ would lie between some $\overrightarrow{OQ}_i$ and $\overrightarrow{OQ}_{i+1}$, hence $(\sphericalangle P_j O Q_i)°$ would be smaller than $(\sphericalangle P_1 O P_2)°$, contradicting our choice of P_2. Therefore, every rotation in our group is equal to a power of R. ∎

LEMMA 9.9. If a finite group of motions contains a reflection, then it is a dihedral group D_n.

Proof:
Partition the group $\mathscr{G}$ into its set $\mathscr{D}$ of direct motions and its set $\mathscr{E}$ of opposite motions. Let n be the number of elements in $\mathscr{D}$. By Lemma 9.8, $\mathscr{D}$ is a cyclic subgroup of order n generated by a rotation R. Let T be any reflection. Then the opposite motions in $\mathscr{E}$ are also reflections and since the product of two of them is a direct motion, they can be written uniquely in the form

$$TR^i, \qquad i = 0, 1, \ldots, n.$$

This is the group D_n. ∎

Thus, Theorem 9.3 is proved.

Much is known about particular types of infinite groups of motions. For instance, a *frieze group* is a group of motions that has an invariant line t and whose translations form an infinite cyclic group $\langle T \rangle$ generated by one particular translation T along t. It is not difficult to prove that *there are exactly seven frieze groups* in Euclidean or hyperbolic planes:

1. $\langle T \rangle$.
2. The group $\langle T, R_t \rangle$, generated by T and the reflection across t.
3. The group $\langle T, R_u \rangle$, generated by T and a reflection across a perpendicular u to t.
4. The infinite cyclic group $\langle G \rangle$ generated by the unique glide G such that $G^2 = T$.
5. The group $\langle T, H_A \rangle$ generated by T and a half-turn about a point A on t.
6. The group $\langle T, H_A, R_t \rangle$.
7. The group $\langle G, H_A \rangle$.

(For the proof, see Martin, 1982, and Exercise 60.)

Another type of infinite group is called a *wallpaper group*, whose subgroup of translations is generated by two translations along distinct intersecting lines. In the Euclidean plane, *there are exactly 17 wallpaper groups.* The ornamental patterns designed on the walls of the Alhambra in Spain by the Moors illustrate these 17 groups.

The classic treatise on symmetry is Hermann Weyl's *Symmetry.*[14] You will find therein a discussion of the 17 wallpaper groups, plus an analysis of three-dimensional symmetry, including the generalization of our Theorem 9.3 to three dimensions. Most important, you will find a fascinating treatment in words and pictures of how these purely mathematical abstractions relate to the physical universe in the form of crystals, biological specimens, and works of art throughout the ages.[15]

REVIEW EXERCISES

Which of the following statements are correct?

(1) In the Cartesian model, the equations for reflection across the y axis are $y' = y$ and $x' = -x$.

(2) In the Cartesian model, the equations for the $90°$ clockwise rotation about the origin are $y' = x$ and $x' = -y$.

(3) In the Euclidean plane, a similarity that is not a motion must be a dilation.

(4) In the Cartesian model, the equations for the translation moving the origin to $(1, 1)$ are $y' = y + 1$ and $x' = x + 1$.

(5) An involution is a transformation equal to its own inverse but not equal to the identity.

(6) If a motion leaves a circle invariant, it must be a rotation about the center of the circle.

(7) In the Cartesian model, if k is the x axis, l the line $y = x$, m the y axis, and n the line $y = -x$, then $R_k R_l R_m = R_n$.

(8) In the Cartesian model, the equations for the half-turn about the point $(1, 0)$ are $x' = -x + 2$ and $y' = -y$.

---

[14] Hermann Weyl, *Symmetry* (Princeton University Press, 1952).

[15] Other excellent references on this subject are: J. N. Kapur, *Transformation Geometry* (Affiliated East-West Press, 1976); Joe Rosen, *Symmetry Discovered* (Cambridge University Press, 1975); E. H. Lockwood and R. H. Macmillan, *Geometric Symmetry* (Cambridge University Press, 1978); and I. Stewart and M. Golubitsky, *Fearful Symmetry: Is God a Geometer?* (Blackwell, 1992).

(9) In the Cartesian model, the glide along the x axis mapping $(0, 1)$ to $(1, -1)$ is given by the equations $x' = x + 1$ and $y' = -y$.

(10) If A, A′ are distinct points in the Euclidean plane, there is a unique translation T such that $T(A) = A'$, i.e., the group of translations operates simply transitively on the points.

(11) If A, A′ are distinct points in the hyperbolic plane, there are infinitely many translations T such that $T(A) = A'$.

(12) In neutral geometry, if A, A′ are distinct points and O is any point on the perpendicular bisector of AA′, then there is a unique rotation T with center O such that $T(A) = A'$.

(13) In neutral geometry, the set of all translations is a group.

(14) In neutral geometry, if the product of two rotations is a nonidentity translation, then the rotations are half-turns about distinct points.

(15) In hyperbolic geometry, for any two distinct points A, A′, there are exactly two parallel displacements T such that $T(A) = A'$.

(16) In Euclidean geometry, the product of a non identity rotation and a translation is a rotation.

(17) Reflections in distinct lines never commute.

(18) In hyperbolic geometry, the product of two rotations about the same point could be any of the three types of direct motion.

(19) In Euclidean geometry, the set of all half-turns and all translations is a group.

(20) If $R_l R_m R_n$ is a glide, then lines l, m, and n do not lie in a pencil.

(21) In the Cartesian model, the equations for a direct similarity have the form $x' = ax - by + e$, $y' = bx + ay + f$, where $a^2 + b^2 \neq 0$, and all lengths are multiplied by $k = \sqrt{a^2 + b^2}$ under this transformation.

(22) A rotation through angle θ can be written as a product of reflections in two lines which meet and form an angle θ.

(23) The group of motions operates simply transitively on the set of all equilateral triangles whose sides have length 1.

(24) In hyperbolic geometry, the group of motions operates simply transitively on the set of all ordered triples of distinct ideal points.

(25) A finite group of motions cannot contain more than one half-turn.

(26) In the Poincaré upper half-plane model, the linear fractional transformation

$$z \to \frac{3z + 4}{z + 1}$$

represents a direct hyperbolic motion.

(27) In the Poincaré upper half-plane model, the transformation $z \to z - 1$ represents a parallel displacement about ∞.

(28) In the Cartesian model, the product of a translation along the x axis with the reflection across the y axis is a reflection.

(29) The group of symmetries of a regular pentagon is cyclic of order 5.

(30) No opposite motion commutes with a nonidentity direct motion.

(31) In Euclidean geometry, two figures are congruent if and only if there is an automorphism mapping one onto the other.

(32) Given a triangle, if there exists a reflection leaving the triangle invariant, then the triangle is isosceles.

(33) If A and A′ are any two points on opposite sides of line t, then there exists a glide T along t such that $T(A) = A′$.

(34) In the Cartesian model, the equations for rotation through angle θ about the origin are $x' = x \cos \theta + y \sin \theta$ and $y' = x \sin \theta - y \cos \theta$.

(35) In neutral geometry, if a motion has a unique invariant line, then it is a glide.

(36) A motion that is a product of an odd number of reflections is opposite.

(37) In the Cartesian model, the equations for reflection across the line $y = x$ are $y' = x$ and $x' = y$.

(38) In Euclidean geometry, the group of symmetries of a quadrilateral has order $\geqq 4$ if and only if the quadrilateral is a rectangle.

(39) In hyperbolic geometry, the group of symmetries of a quadrilateral must have order < 4.

(40) In Euclidean geometry, if the group of symmetries of a convex quadrilateral has order 2, then the quadrilateral must be an isosceles trapezoid.

(41) In Euclidean geometry, the group of symmetries of every triangle has order $\geqq 3$.

(42) In neutral geometry, if there exists an automorphism that is not a motion, then the geometry is Euclidean.

(43) In neutral geometry, the product of reflections in two parallel lines is a translation.

(44) In neutral geometry, an automorphism with exactly one fixed point must be a rotation.

(45) In the Cartesian model, an opposite similarity that fixes the origin O is equal to a product DR, where D is a dilation centered at O, R is a reflection in a line through O, and D and R commute.

(46) In the Poincaré upper half-plane model, the transformation

$$z \longrightarrow \frac{6\bar{z} - 4}{8\bar{z} - 6}$$

represents a reflection.

(47) In the Cartesian model, for any complex number z_0, the transformation $z \longrightarrow \bar{z} + z_0$ represents a reflection.

(48) In Euclidean geometry, a betweenness-preserving transformation that doubles the length of every segment must be a similarity.

(49) In hyperbolic geometry, there is no collineation that doubles the length of every segment.

(50) In neutral geometry, every motion either has an invariant line or a fixed point or both.

EXERCISES

Outline: Exercises 2–22 consist of supplementary results and proofs left to the reader on the classification of motions; Exercise 23 is an essay question on orientation; Exercises 24–33 give more information on similarities; Exercises 34–35 are about linear and affine transformations; Exercises 36–38 discuss Möbius transformations; Exercises 39–49 deal with orbits of groups of transformations (particularly with horocycles and equidistant curves in the Poincaré model); Exercise 50 exhibits algebraic equations in the group of motions and their geometric meaning; Exercise 51 refers to invariant sets of transformations; Exercise 52 presents another attempt to prove the parallel postulate using rotations, and Exercise 53 shows what happens when translations are used; Exercises 54–57 determine the axis of a glide in the Euclidean plane; Exercise 58 presents a natural hyperbolic transformation that is not a collineation; Exercises 59–61 are about symmetry; Exercises 62–63 give some unusual models; Exercise 64 raises the question of two-dimensionality in incidence planes; Exercises 65–68 are fundamental for one-dimensional projective geometry; and Exercises 69–77 are applications to the 9-point circle and other topics in Euclidean geometry.

1. Show that there are 168 collineations of the 7-point projective plane (Figure 9.1). (Hint: A collineation is uniquely determined by its effect on four points, no three of which are collinear.)
2. Prove that the set of all automorphisms of a model of neutral geometry is a group, and that the set of motions is a subgroup.
3. Finish the proof of Proposition 9.2.
4. Prove that a transformation of the plane that multiplies all lengths by a constant $k > 0$ preserves betweenness and congruence of segments.
5. Prove that a reflection is a motion.

6. In a Euclidean plane, prove that $\triangle ABC \sim \triangle A'B'C'$ if and only if there is a similarity sending A, B, C respectively onto A', B', C', and that similarity is unique. (Hint: Use Lemma 9.2, Proposition 9.5, and Exercise 20 of Chapter 5.)

7. Prove that the invariant lines of a reflection R_m are m and all lines perpendicular to m. Prove also that $R_m R_k R_m = R_{k^*}$, where k^* is the reflection of k across m.

8. Prove the corollary to Proposition 9.6.

9. Prove that if an automorphism is an involution then it is either a reflection or a half-turn. (Hint: If A and A' are interchanged, show that the midpoint of AA' is fixed, and apply Propositions 9.4 and 9.6.)

10. Show that the product $T'T$ of rotations about distinct points can be any of the three types of direct motions. (Hint: Apply Proposition 9.8 to the line joining the centers of rotation.)

11. Prove that a nonidentity rotation that is not a half-turn has no invariant lines. (Hint: Apply Proposition 9.8.)

12. Let T be a translation along t, l a line that is not invariant under T. If l meets t (automatic in case the plane is Euclidean, by Proposition 9.13), prove that $l \| l'$. Suppose now the plane is hyperbolic. If l is asymptotically parallel to t in direction Ω, prove that l' is also (in particular, $l \| l'$). If l is divergently parallel to t, show by diagrams from a Poincaré model that l' could either meet l, be divergently parallel to l, or be asymptotically parallel to l.

13. Prove the corollary to Proposition 9.15.

14. Show that in the hyperbolic plane, the product of translations along distinct lines could be any of the three types of direct motions, and that two such translations do not commute unless at least one of them is the identity. (Hint: Apply the warning after Proposition 9.9 to the invariant lines.)

15. If R_1, R_2, R_3 are reflections, prove that $(R_1 R_2 R_3)^2$ is a translation. (Hint: Use Propositions 9.18 and 9.19.)

16. Let T, T' be glides along perpendicular lines. Prove that TT' is a half-turn if and only if the plane is Euclidean. (Hint: Use Propositions 9.15 and 9.18.)

17. In the hyperbolic plane, prove that every direct motion can be expressed as a product of three half-turns. (Hint: Refer back to Figure 9.9. Show that for any two lines l, u, there is a perpendicular m to u that is divergently parallel to l.)

18. If T is a glide and $n \neq 0$, prove that $T^n \neq I$. (Hint: This has already been proved for translations.)

19. If rotations T, T' about O have the same effect on a point $P \neq O$, then $T = T'$.

20. If T is any motion, A any point, and l any line, prove that $TH_A T^{-1}$ is a half-turn and that $TR_l T^{-1}$ is a reflection.

21. Prove Hjelmslev's lemma (p. 339). (Hint: See Figure 9.31.)

22. (Hjelmslev's theorem) Let l and l' be distinct lines and let T be a motion transforming l onto l'. As point P varies on l and its image P' under T varies on l', the midpoints of the segments PP' are either distinct and collinear or else they all coincide. (Hint: Show that you can assume that T is an opposite motion and apply Exercise 21 if T is a glide.)

23. How would you go about precisely defining the notion of an "orientation" of the (neutral) plane? The requirements are that you must be able to show there are exactly two "orientations," which are interchanged by opposite motions and preserved by direct motions. (If your definition uses words such as "clockwise" or "counterclockwise," you must define them precisely.) Making this notion precise is surprisingly tricky. It can be done in several ways, all of which appear artificial — see Ewald (1971), p. 65, for one. Perhaps the reason for the artificiality is, as Hermann Weyl says, that "to the scientific mind there is no difference, no polarity between left and right. . . . It requires an arbitrary act of choice to determine what is left and what is right; . . . in all physics nothing has shown up indicating an intrinsic difference of left and right." (See Weyl's *Symmetry,* pp. 16–38, for a fascinating discussion of this problem, illustrated with examples from physics, biology, and art. The Nobel Prize–winning work of C. N. Yang and T. D. Lee, done after Weyl's death, does indicate a physical difference between left and right at the subatomic level.)

24. Prove Proposition 9.24. (Hint: Follow the given similarity by a translation, a rotation about 0, and a dilation centered at 0 — if necessary — to obtain a similarity fixing 0 and 1; then apply Lemma 9.1.)

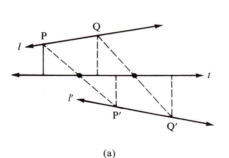
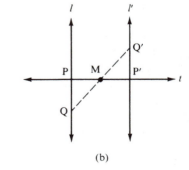

(a)   (b)

FIGURE 9.31

25. Prove that every similarity that is not a motion has a unique fixed point. (Hint: Use Proposition 9.24. For a synthetic proof of this, see Coxeter, 1969, p. 72.)

26. Prove that the reflections in the Cartesian model are characterized among all opposite motions $z \to e^{i\theta}\bar{z} + z_0$ by the equation

$$e^{i\theta}\bar{z_0} + z_0 = 0$$

and that the axis of such a reflection is the line

$$2(yy_0 + xx_0) = x_0^2 + y_0^2$$

if $z_0 = x_0 + iy_0 \neq 0$, whereas if $z_0 = 0$, it is the line

$$y \cos(\theta/2) - x \sin(\theta/2) = 0.$$

(Hint: If $z_0 \neq 0$, the axis must pass through $z_0/2$ and be perpendicular to the line joining 0 to z_0; if $z_0 = 0$, the axis passes through $e^{i(\theta/2)}$.) Conversely, given a line $ax + by = c$, find z_0 and θ for the reflection across this line.

27. What is the representation in complex coordinates of a dilation of ratio k whose center has complex coordinate z_1?

28. The definition of direct and opposite similarities given in Proposition 9.24 depended on the representation in complex coordinates. Prove that a coordinate-free description is given as follows: a similarity is direct (respectively, opposite) if and only if it is the product of a dilation with a direct (respectively, opposite) motion. Prove also that if AB and A'B' are any two segments, there is a unique *direct* similarity taking A to A', and B to B'.

29. Let γ_1 and γ_2 be circles in the Euclidean plane with distinct centers O_1, O_2 and distinct radii r_1, r_2. Prove that there are two dilations D_1, D_2 which transform γ_1 onto γ_2. (Hint: Choose any point A on γ_1, and let A_1, A_2 be the ends of the diameter of γ_2, which is parallel to O_1A; then O_2A_1 (respectively, O_2A_2) will be the image of O_1A under D_1 (respectively, D_2).) The centers of D_1 and D_2 are called the *centers of similitude* of the two circles. (See Coxeter, 1969, p. 71, for an application to the famous 9-point circle.)

30. Given any $\triangle ABC$ in the Euclidean plane, let A', B', C' be the midpoints of its sides, labeled so that the medians are AA', BB', CC'. Show that there is a unique dilation of ratio $-\frac{1}{2}$ taking $\triangle ABC$ onto $\triangle A'B'C'$. (Hint: See Problem 8, p. 318, and Coxeter, 1969, p. 10.)

31. Show that the set of all dilations (with all possible centers and ratios) is not a group, whereas the set of all translations and dilations (of a Euclidean plane) is a group, and that this group is noncommutative.

32. Prove that dilations are geometrically characterized among all similari-

ties by the two properties of (i) mapping each line *l* onto a line equal to or parallel to *l*; and (ii) having a fixed point.

33. A *point at infinity* for the Cartesian plane is an equivalence class of all lines equal to or parallel to some given line, and the *line at infinity* is the set of all points at infinity (compare Chapter 2). Since automorphisms (in fact, collineations) of the Cartesian plane preserve parallelism, they induce transformations of the line at infinity onto itself, and we can investigate the fixed points at infinity of these transformations. Show that an automorphism which (i) fixes every point at infinity is either a dilation or a translation; (ii) fixes two points at infinity is an opposite similarity (if *l*, *m* determine the two fixed points, then $l \perp m$); (iii) has no fixed points at infinity is either a rotation that is not the identity and not a half-turn, or the product of such a rotation with a dilation.

   (Hint: In the representation Proposition 9.24 gives for similarities, let $w_0 = \alpha + \beta i$; if the similarity is direct, it takes lines of slope *m* onto lines of slope $(\beta + \alpha m)/(\alpha - \beta m)$, whereas if the similarity is opposite, it takes lines of slope *m* onto lines of slope $(\beta - \alpha m)/(\alpha + \beta m)$.)

34. A *linear transformation* of the Cartesian plane is a transformation *T* given in coordinates by

$$x' = a_{11}x + a_{12}y$$
$$y' = a_{21}x + a_{22}y,$$

where the matrix

$$A = \begin{bmatrix} a_{11} & a_{12} \\ a_{21} & a_{22} \end{bmatrix}$$

has nonzero determinant. In vector notation, the transformation has the form

$$z' = Az.$$

An *affine transformation* is a linear transformation followed by a translation:

$$z \rightarrow Az + z_0.$$

Prove that an affine transformation is a collineation and that the set of all affine transformations is a group (called *the affine group*). It can be shown that, conversely, every collineation of the Cartesian plane is an affine transformation — see Artzy (1965), p. 155.

35. Give an example of a linear transformation of the Cartesian plane that fixes exactly one point at infinity.

36. Linear fractional transformations

$$T: z \rightarrow \frac{az + b}{cz + d}$$

with complex coefficients and nonzero determinant

$$\delta = ad - bc$$

are called *Möbius transformations* or *homographies*. Show that such a transformation with $c \neq 0$ can be factored into a composite

$$T = T_4 T_3 T_2 T_1,$$

where T_1 is the translation $z \rightarrow z + (a/c)$, T_2 is the similarity $z \rightarrow (-\delta/c^2)z$, T_3 is mapping $z \rightarrow z^{-1}$, and T_4 is translation $z \rightarrow z + (d/c)$.

37. Show that a Möbius transformation maps the set of all circles and lines in the Cartesian model onto itself, preserving orthogonality. (Hint: In case $c \neq 0$, use the factorization in the previous exercise, observing that T_3 is the composite of inversion in the unit circle with reflection across the x axis and using Exercise P-17, Chapter 7. In case $c = 0$, use the fact that T is an automorphism of the Cartesian model.)

38. Show that the mapping

$$E : z \rightarrow i \frac{i + z}{i - z}$$

has all the properties claimed on p. 350. (Hint: Calculate the real and imaginary parts of $E(z)$ and use the previous exercise.)

39. Let $\mathscr{G}$ be a subgroup of the group of motions. For any point P, the *orbit* $\mathscr{G}$P of P under $\mathscr{G}$ is defined to be the set of all images of P under motions in $\mathscr{G}$. For example, if $\mathscr{G}$ is the entire group of motions, then $\mathscr{G}$P is the entire plane. Let $\mathscr{G}$ be the group of all rotations about a point O; if P $\neq$ O, prove that the orbit $\mathscr{G}$P is the circle through P centered at O.

40. Let $\mathscr{G}$ be the group of all translations along a line t. If P lies on t, prove that $\mathscr{G}$P $= t$. Suppose P doesn't lie on t. Show that $\mathscr{G}$P is the unique parallel to t through P in case the plane is Euclidean, and show that $\mathscr{G}$P is the *equidistant curve* to t through P if the plane is hyperbolic.

41. Let Ω be an ideal point, and A be an ordinary point in the hyperbolic plane. Define the *horocycle* through A about Ω to consist of A and all points A′ such that the perpendicular bisector of AA′ passes through Ω. Prove that this horocycle is the orbit of A under the group $\mathscr{G}$ of parallel displacements about Ω. (Hint: Use Proposition 9.17.)

42. Let $l \| m$, "meeting" at ideal point Ω, let A ı l, and let B be the foot of the perpendicular from A to m. By the method of Major Exercise 11, Chapter 6, there is a unique point P on m such that PQ $\cong$ AB, where Q is the foot of the perpendicular from P to l. Let t be the perpendicular bisector of AP. Prove that t is the *symmetry axis* of l and m — i.e., that $R_t(l) = m$ — and hence, if we fix A and Ω and let m vary through all lines through Ω, the locus of such *corresponding points* P (as Gauss called them) fills out the horocycle through A centered at Ω. (Hint: Show that AB meets PQ

in a point C, and use Major Exercise 5, Chapter 6, and AAA to show that $\triangle CPB \cong \triangle CAQ$ and hence C I t and $\sphericalangle PAQ \cong \sphericalangle APB$. Deduce that t lies between l and m and so passes through Ω. See Figure 9.32.)

43. Prove that the symmetry axes t, u, v of the pairs of sides of a trebly asymptotic triangle are concurrent in a point G that has the properties described in Exercise K-13 of Chapter 7. See Figure 7.48. (Hint: Show first that t meets u at some point G, then that $R_t(u) = v$, so that $G = R_t(G)$ lies on v also.) Show that $R_v R_u R_t = R_u$. (Hint: Use the theorem on three reflections.) We have succeeded in proving the result of Exercise K-13 without resorting to Euclidean geometry.

44. Prove that all horocycles are congruent to one another. (Hint: Use the fact that all rays are congruent to one another — Lemma 9.3.)

45. In the Poincaré upper half-plane model, let t be the upper half of the y axis. Show that the equidistant curves of t are the nonvertical rays in the upper half-plane emanating from 0. (Hint: See Example 4, p. 356.)

46. In the Poincaré upper half-plane model, show that the horocycles about ∞ are the horizontal lines in the upper half-plane. (Hint: See Example 3, p. 356.) Show that the horocycles about x_0 are the circles in the upper half-plane tangent to the x axis at x_0. (Hint: Use the inversion $z \rightarrow \bar{z}^{-1}$ in the unit circle to map ∞ to 0 and the horocycles about ∞ to the horocycles about 0; then use the parallel displacement $z \rightarrow z + x_0$.)

47. In the Poincaré upper half-plane model, prove that the equidistant curves are either (1) nonvertical rays emanating from a point on the x axis; or (2) intersections with the upper half-plane of circles cutting the x axis in two points with the centers of the circles not lying on the x axis. (Hint: Use a real linear fractional transformation to map the upper-half of the y axis onto any other Poincaré line, and apply Exercises 45 and 37.)

48. In the Poincaré upper half-plane model, show that a hyperbolic circle is represented by a Euclidean circle. (Hint: Use the isomorphism E with

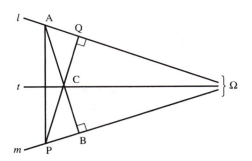

FIGURE 9.32

the Poincaré disk model and Exercise P-5, Chapter 7.) Find the Euclidean radius and Euclidean center of the hyperbolic circle of hyperbolic center i passing through $2i$. (Hint: See Example 5, p. 357)

49. Use the Poincaré upper half-plane model to demonstrate that in the hyperbolic plane, three points lie either on a line, a circle, an equidistant curve, or a horocycle.

50. In this exercise, verify the following translation of geometric statements into algebraic equations in the group of motions (which leads to new proofs of geometric theorems — see Bachmann, 1973):
    (1) P lies on $l \iff (H_P R_l)^2 = I$.
    (2) $l \perp m \iff (R_l R_m)^2 = I$.
    (3) l, m, n belong to a pencil $\iff (R_l R_m R_n)^2 = I$.
    (4) l is the perpendicular bisector of AB $\iff R_l H_B R_l H_A = I$.
    (5) Let m and n be intersecting lines. Then l is a bisector of one of the angles formed by m and $n \iff R_l R_m R_l R_n = I$. (Describe l geometrically in case this equation holds and $m \parallel n$.)
    (6) M is the midpoint of AB $\iff H_M H_B H_M H_A = I$.
    (7) $l \perp \overleftrightarrow{AB} \iff (R_l H_A H_B)^2 = I$.
    (8) $(H_A R_l R_m)^2 = I \iff$ A lies on a common perpendicular to l and m.
    (9) Given A $\ne$ B. In Euclidean geometry, $H_A R_l H_A = H_B R_l H_B \iff l = \overleftrightarrow{AB}$ or $l \parallel \overleftrightarrow{AB}$. In hyperbolic geometry, the equation holds $\iff l = \overleftrightarrow{AB}$.
    (10) In hyperbolic geometry, A, B, C are collinear $\iff (H_A H_B H_C)^2 = I$.
    (11) Assume A, B, C not collinear and that G is the centroid of $\triangle ABC$ (the point of intersection of the medians). Then in Euclidean geometry, $H_P H_C H_P H_B H_P H_A = I \iff P = G$. (Hint: Recall that $H_P H_Q$ is the translation by a vector of length $2\overline{PQ}$ in the direction of ray $\overrightarrow{QP}$.)
    (12) Given $\square ABCD$ in the Euclidean plane. Then $H_A H_B H_C H_D = I \iff \square ABCD$ is a parallelogram.

51. A set $\mathscr{S}$ of transformations is called *invariant* under a group $\mathscr{G}$ of transformations if for every $S \in \mathscr{S}$ and $T \in \mathscr{G}$, TST^{-1} belongs to $\mathscr{S}$. (In case $\mathscr{S}$ is a subgroup of $\mathscr{G}$ and is invariant under $\mathscr{G}$, then $\mathscr{S}$ is called a *normal subgroup*.) For instance, you showed in Exercise 20 that the sets of half-turns and reflections are each invariant under the group of motions. Determine whether each of the following sets is invariant under the indicated groups:
    (i) $\mathscr{S}$ = all rotations about one given point, $\mathscr{G}$ = all motions.
    (ii) $\mathscr{S}$ = all rotations about all points, $\mathscr{G}$ = all motions.
    (iii) $\mathscr{S}$ = all translations along one given line, $\mathscr{G}$ = all motions.
    (iv) $\mathscr{S}$ = all translations along all lines, $\mathscr{G}$ = all motions.
    (v) (Euclidean geometry) $\mathscr{S}$ = all motions, $\mathscr{G}$ = all similarities.

   (vi) (Euclidean geometry) $\mathscr{S}$ = all dilations, $\mathscr{G}$ = all similarities.

  (vii) (Hyperbolic geometry) $\mathscr{S}$ = all parallel displacements, $\mathscr{G}$ = all motions.

 (viii) $\mathscr{S}$ = all glides, $\mathscr{G}$ = all motions.

  (ix) (Euclidean geometry) $\mathscr{S}$ = all direct motions, $\mathscr{G}$ = all similarities.

   (x) (Euclidean geometry) $\mathscr{S}$ = all rotations about O, $\mathscr{G}$ = all similarities having O as a fixed point.

  (xi) (Euclidean geometry) $\mathscr{S}$ = all translations along t, $\mathscr{G}$ = all similarities having t as an invariant line.

52. In 1809, B. F. Thibaut attempted to prove Euclid's parallel postulate with the following argument using rotation; find the flaw. Given any triangle $\triangle ABC$. Let $D * A * B$, $A * C * E$, $C * B * F$ (see Figure 9.33). Rotate $\overleftrightarrow{AB}$ about A through $\sphericalangle DAC$ to $\overleftrightarrow{AC}$; then at C, rotate $\overleftrightarrow{AC}$ to $\overleftrightarrow{BC}$ through $\sphericalangle ECB$; finally, at B, rotate $\overleftrightarrow{BC}$ to $\overleftrightarrow{AB}$ through $\sphericalangle FBA$. After these three rotations, $\overleftrightarrow{AB}$ has returned to itself, and has thus been rotated through $360°$. Adding up the individual angles of rotation gives

$$[180° - (\sphericalangle A)°] + [180° - (\sphericalangle C)°] + [180° - (\sphericalangle B)°] = 360°$$

so that the angle sum of $\triangle ABC$ is $180°$, and Euclid's parallel postulate follows.

53. Given $\triangle ABC$ in the hyperbolic plane, let T_1 be the translation along $\overleftrightarrow{AB}$ taking A to B, T_2 the translation along $\overleftrightarrow{BC}$ taking B to C, and T_3 the translation along $\overleftrightarrow{AC}$ taking C to A. Show that $T_3 T_2 T_1$ is a rotation about A through $d°$, where $d°$ is the defect of $\triangle ABC$.

54. Given $\triangle ABC$ that is not a right triangle. Let $a = \overleftrightarrow{BC}$, $b = \overleftrightarrow{AC}$, $c = \overleftrightarrow{AB}$, and let D, E, F be the feet of the perpendiculars from A, B, C to a, b, c. We know from Proposition 9.19 that the product $G = R_a R_b R_c$ of the reflections in the sides of $\triangle ABC$ is a glide. Assume the geometry is Euclidean. Prove that the axis of G is the line $t = \overleftrightarrow{DF}$. (Hint: Show that right triangles $\triangle ABE$ and $\triangle ACF$ are similar, then that $\triangle AEF \sim$

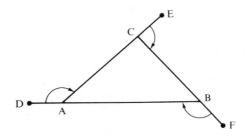

FIGURE 9.33

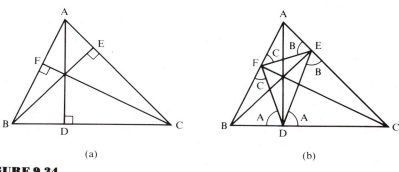

(a)   (b)

FIGURE 9.34

△ABC; hence, $\angle AEF \cong \angle B$, $\angle AFE \cong \angle C$. Applying the same argument to △BFD and △CED, deduce $\angle AFE \cong \angle BFD$, $\angle AEF \cong \angle CED$, $\angle BDF \cong \angle CDE$. From these congruences deduce that line t is invariant under G. See Figure 9.34.)

55. With the same notation as in the previous exercise, assume that △ABC is acute-angled, and let $G = TR_t$, where T is a translation along t. Show that this translation is in the direction $\overrightarrow{FD}$ through a distance equal to the perimeter of △DEF. (Hint: Determine the image of F under G.)

56. With the same notation as above, determine the axis of G in case △ABC is a right triangle. (Hint: It depends on whether or not $\angle B$ is right; use Hjelmslev's Lemma, p. 339, applied to line $\overleftrightarrow{AB}$.)

57. Given three lines a, b, c in the Euclidean plane such that $a \parallel c$, b meets c at A, and b meets a at C. (See Figure 9.35.)

   (i) If b is perpendicular to both a and c, prove that b is the axis of the glide $R_a R_b R_c$ and $R_a R_b R_c = R_a R_c R_b = R_b R_a R_c$.

   (ii) If b is not perpendicular to a and c, let D, F be the feet of the

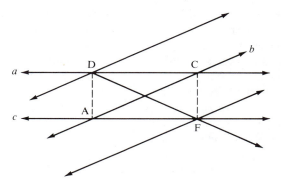

FIGURE 9.35

perpendiculars from A, C to a, c. Prove that $\overleftrightarrow{DF}$ is the axis of glide $R_a R_b R_c$. (Hint: Use Hjelmslev's lemma to show that F and D lie on the axis.) Prove that the axis of glide $R_a R_c R_b$ (respectively, $R_c R_a R_b$) is the line through D (respectively, F) parallel to b. (Hint: Find the midpoint of C and its image, and the midpoint of A and its image.)

58. Given ideal point Ω in the hyperbolic plane and distance d, define a transformation T that sends point P onto the unique point P' on ray PΩ such that $\overline{PP'} = d$. (The analogous transformation in a Euclidean plane would be a translation through distance d.) Describe T explicitly in the Poincaré upper half-plane model when $\Omega = \infty$ (Hint: The Poincaré distance from $x + iy$ to $x + iy'$ is $|\log y/y'|$.) Show that T is not a collineation by showing that the image under T of a line not through ∞ is not a line.

59. Instead of generating the dihedral group D_n by one reflection and one rotation, show that it can be generated by two reflections, and give the three basic relations for these generators $(n > 1)$.

60. Find the symmetry group of each of the following infinite patterns (describe the group by generators and relations):
    (i) ... L L L L L ...
    (ii) ... L Γ L Γ L ...
    (iii) ... V V V V V ...
    (iv) ... N N N N N ...
    (v) ... V Λ V Λ V ...
    (vi) ... D D D D D ...
    (vii) ... H H H H H ...

61. Which pairs of groups in the previous exercise are isomorphic? (Hint: See Coxeter, 1969, p. 48.)

62. Refer to the model in Exercise 35, Chapter 3, in which length has been changed along the x axis, causing the SAS criterion to fail. Show that the only similarities that are automorphisms of this model are the ones that leave the x axis invariant.

63. Suppose the model in the previous exercise is modified so that length along three nonconcurrent lines is converted to three different units of measurement from the unit on all other lines. Show that the identity is the only automorphism of this model (hence "nothing can be moved" in this model, everything is invariant, and Klein's *Programme* gives no insight into the "geometry").

64. Define an *incidence plane* to be a model of incidence geometry which is *two-dimensional* in the following sense: if l, m, n are three lines forming a triangle and P any point, then there exists a line t through P such that t meets $l \cup m \cup n$ in at least two points. (See Figure 9.36.) Show that a model of both the incidence and betweenness axioms is automatically

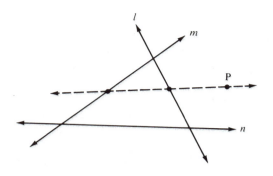

FIGURE 9.36

two-dimensional. Let T be a one-to-one mapping of the set of points of an incidence plane onto itself such that if O, P, Q are collinear, then their images O', P', Q' are collinear. Prove that, conversely, if O', P', Q' are collinear, then so are O, P, Q. Is there a model of incidence geometry that is not two-dimensional in which this converse fails? (I don't know the answer.)

65. Prove the following analogue of Proposition 9.16(c) in one-dimensional projective geometry over an arbitrary field K: for any two triples of points on the projective line $\mathscr{P}^1(K)$, there is a unique projective transformation mapping one triple onto the other. (Hint: This is an exercise in two-dimensional linear algebra, based on the fact that there is a unique nonsingular 2×2 matrix mapping one pair of linearly independent vectors onto another — see Ewald, 1971, p. 215.)

66. Given four points P_1, P_2, P_3, P_4 on the projective line $\mathscr{P}^1(K)$, let the homogeneous coordinates of P_i be $[s_i, t_i]$ for $i = 1, 2, 3, 4$. Define the *cross-ratio* (P_1P_2, P_3P_4) by

$$\frac{\begin{vmatrix} s_1 & t_1 \\ s_3 & t_3 \end{vmatrix} \begin{vmatrix} s_2 & t_2 \\ s_4 & t_4 \end{vmatrix}}{\begin{vmatrix} x_2 & t_2 \\ s_3 & t_3 \end{vmatrix} \begin{vmatrix} s_1 & t_1 \\ s_4 & t_4 \end{vmatrix}} = (P_1P_2, P_3P_4),$$

where the four terms in this ratio are 2×2 determinants (which are not zero because the points are distinct). If a projective transformation maps point P_i onto P_i', $i = 1, 2, 3, 4$, prove that the cross-ratio is preserved: $(P_1P_2, P_3P_4) = (P_1'P_2', P_3'P_4')$. (Hint: If M is a matrix of the projective transformation, each determinant occurring in the formula for $(P_1'P_2', P_3'P_4')$ is the product of $\det M$ with the corresponding determinant in the formula for (P_1P_2, P_3P_4).)

67. Since the determinants in the formula for cross-ratio of Exercise 66 may be negative (when K is a subfield of the field $\mathbb{R}$ of real numbers), this cross-ratio is not the same as the positive cross-ratio defined on p. 248. Show that for points $\neq \infty$, this cross-ratio is the same as the *signed* cross-ratio defined in Exercises H-4 and H-7, Chapter 7. (Hint: Use inhomogeneous coordinates $[s_i, 1]$.)

68. Deduce from Exercises 65 ad 66 that if (P_1, P_2, P_3) and (P'_1, P'_2, P'_3) are any two triplets of points on $\mathscr{P}^1(K)$, if T is the unique projective transformation carrying the first triple onto the second, and if P_4 is any fourth point, then the image P'_4 of P_4 under T is uniquely determined by the equation

$$(P_1P_2, P_3P_4) = (P'_1P'_2, P'_3P'_4).$$

This shows that the cross-ratio is the fundamental invariant of the one-dimensional projective group.

Note: The remaining exercises are in Euclidean geometry.

69. Prove that the medians of a triangle concur at a point G that is $\frac{2}{3}$ the distance from each vertex to the opposite midpoint. (Hint: Use analytic geometry.) For which triangles does the Euler line (see Problem 8, p. 319) pass through a vertex?

70. Show that the center U of the 9-point circle is the harmonic conjugate of the circumcenter O with respect to the orthocenter H and the centroid G (see Problems 9 and 8, pp. 320 and 319).

71. Show that the dilation T' with the centroid G as center and ratio -2 also maps the 9-point circle onto the circumcircle (hence, in the terminology of Exercise 29, G and H are the two *centers of similitude* of these circles).

72. Show that the distance from the circumcenter O of a triangle to a side is half the distance from the orthocenter H to the opposite vertex. (Hint: See Problem 8, p. 319.)

73. Justify all the assertions in the solutions to Problems 1 – 9, pp. 315 – 321 that have not been justified there.

74. Finish the solution to Problem 4, p. 317, by showing that if X, Y, Z are the feet of the altitudes, then Y and Z lie on X_1X_2. (Hint: Use the congruences given in the hint to Exercise 54.)

75. Given line l, two points A and B on the same side of l, and a positive number d. Find a segment XY on l of length d such that the polygonal path AXYB is as short as possible. (Hint: Use the same method as in the solution to Problem 1, p. 315, with a glide reflection along l through distance d instead of the reflection in l.)

76. Report on Feuerbach's theorem, which states that the 9-point circle is tangent to the inscribed circle and the three escribed circles (see H. Eves, 1972; or D. C. Kay, 1969).

77. Let $\triangle ABC$ be any triangle whose largest angle is $< 120°$. For any point P, let $d(P) = \overline{PA} + \overline{PB} + \overline{PC}$. Prove that there is a unique point P_0 at which the function $d(P)$ achieves its minimum value, that P_0 lies in the interior of $\triangle ABC$, that it can be constructed with straightedge and compass, that

$$120° = (\measuredangle AP_0B)° = (\measuredangle BP_0C)° = (\measuredangle CP_0A)°,$$

and that

$$d(P_0)^2 = a^2 + b^2 - 2ab \cos\left(C + \frac{\pi}{3}\right),$$

where $a = \overline{BC}$, $b = \overline{AC}$. (Hint: Rotate $\triangle ABC$ about each of its vertices through $60°$. See Kay (1969), p. 271, where P_0 is called *the Fermat point*, or H. Rademacher and O. Toeplitz, *The Enjoyment of Mathematics*, Princeton University Press, 1957, p. 33.)

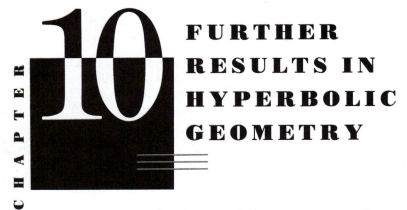

FURTHER RESULTS IN HYPERBOLIC GEOMETRY

> The theorems of this geometry appear to be paradoxical and, to the uninitiated, absurd; but calm, steady reflection reveals that they contain nothing at all impossible.
>
> C. F. GAUSS

In this chapter we will penetrate more deeply into the "strange new universe" of hyperbolic geometry. To facilitate this adventure, we will sometimes glance quickly at certain unusual forms, not pausing for rigorous proofs of existence; at other times we will be more rigorous but allow ourselves to operate within the Klein or Poincaré models where our familiar Euclidean tools are available, instead of laboriously reasoning directly from the hyperbolic axioms. Hopefully, by the end of this excursion, the terrain will become sufficiently familiar to you that you will be comfortable enough to make further explorations with other guides.

AREA AND DEFECT

In 1799, in answer to a letter from Farkas Bolyai in which Bolyai claimed to have proved Euclid's fifth postulate, Gauss wrote:

> . . . the way in which I have proceeded does not lead to the desired goal, the goal that you declare you have reached, but instead to a doubt of the validity of [Euclidean] geometry. I have certainly achieved

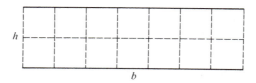

FIGURE 10.1 Area = bh.

results which most people would look upon as proof, but which in my eyes prove almost nothing; if, for example, one can prove that there exists a right triangle whose area is greater than any given number, then I am able to establish the entire system of [Euclidean] geometry with complete rigor. Most people would certainly set forth this theorem as an axiom; I do not do so, though certainly it may be possible that, no matter how far apart one chooses the vertices of a triangle, the triangle's area still stays within a finite bound. I am in possession of several theorems of this sort, but none of them satisfy me.[1]

One of the most surprising facts in hyperbolic geometry is that there is an upper limit to the possible area a triangle can have, even though there is *not* an upper limit to the lengths of the sides of the triangle.

To see how this can be, we have to review the way in which area is calculated in Euclidean geometry. The simplest figure is a rectangle, whose area we calculate as the length of the base times the length of the side (Figure 10.1). This formula is arrived at by noticing that exactly bh unit squares fill up the interior of the rectangle, where a unit square is a square whose side has length one. Keep in mind that the unit of length is arbitrary, so that if we measure area in square inches, we get a different number than if we measure in square feet (but the latter number is always proportional to the former, the proportionality factor being $144 = 12^2$).

From the area of a rectangle we can calculate the area of a right triangle. A diagonal of a rectangle divides it into two congruent right triangles, and since we want congruent triangles to have the same area, the area of the right triangle must be half the area of the rectangle. (See Figure 10.2.)

We can decompose the interior of an arbitrary triangle into the union of the interiors of two right triangles by dropping an appropriate

[1] R. Bonola (1955).

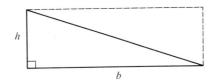

FIGURE 10.2 Area $= \frac{1}{2}bh$.

altitude. Since we want the area of the whole to be the sum of the area of its parts, we find the area of the triangle to be $\frac{1}{2}b_1h + \frac{1}{2}b_2h$, and since $b = b_1 + b_2$, we again get half the base times the height for the area of a general triangle (Figure 10.3).

You can see that by taking b and h to be sufficiently large, the area $\frac{1}{2}bh$ can be made as large as you like.

So why doesn't this work equally well in hyperbolic geometry? Because the whole system of measuring area is based on square units, and as we have seen (Lemma 6.1) rectangles (in particular, squares) do not exist in hyperbolic geometry.

What then does "area" mean in hyperbolic geometry? We can certainly say intuitively that it is a way of assigning to every triangle a certain positive number called its *area*, and we want this area function to have the following properties:

1. *Invariance under congruence*. Congruent triangles have the same area.
2. *Additivity*. If a triangle T is split into two triangles T_1 and T_2 by a segment joining a vertex to a point of the opposite side, then the area of T is the sum of the areas of T_1 and T_2 (Figure 10.4).

Having defined area, we then ask how it is calculated. It can be proved rigorously that in hyperbolic geometry the area of a triangle cannot be calculated as half the base times the height (see Moise, 1990, p. 411). So how do you calculate it? Here we find one of the

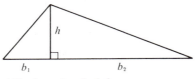

FIGURE 10.3 Area $= \frac{1}{2}bh$ where $b = b_1 + b_2$.

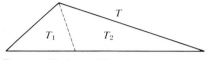

FIGURE 10.4 Area T = area T_1 + area T_2.

most beautiful aspects of mathematics, a direct relationship between two concepts that at first seem totally unrelated. You may have recognized this relationship in reading area properties 1 and 2, for in Theorem 4.7 and Exercise 1, Chapter 4, we proved that the *defect* also has these properties. Recall that in hyperbolic geometry the angle sum of any triangle is always less than 180° (Theorem 6.1), so that if we define the defect to be 180° minus the angle sum, we get a positive number. When a mathematician sees two functions with the same properties, he suspects they are related. Gauss discovered this relationship as early as 1794 (he was only 17 years old), and called it the first theorem in the subject.[2]

THEOREM 10.1. In hyperbolic geometry there is a positive constant k such that for any $\triangle ABC$

$$\text{area}(\triangle ABC) = \frac{\pi}{180} \, k^2 \times \text{defect}(\triangle ABC).$$

For the proof, which is not difficult although it is somewhat lengthy, see Moise (1990, p. 413). The theorem says that the area of any triangle is proportional to its defect, with proportionality constant $(\pi/180)k^2$. This constant depends on the unit of measurement, i.e., on whichever triangle is taken to have area equal to 1.

We can now see why there is an upper limit to the area of all triangles. Namely, the defect measures how much the angle sum is

---

[2] This theorem seems to imply that Euclidean and hyperbolic area theories have little in common. Yet J. Bolyai discovered a wonderful theorem on area that is valid in *both* geometries. Define the area of a polygon to be the sum of the areas of the triangles used to triangulate the polygon. (It is not difficult to show that this definition does not depend on the choice of triangulation.) Bolyai's theorem states that two polygons S and S' have the same area if and only if for some n, polygon S (respectively, S') has a triangulation $\{T_1, \ldots, T_n\}$ (respectively, $\{T'_1, \ldots, T'_n\}$) such that $T_j \cong T'_j$ for all $j = 1, \ldots, n$. (For a proof, see E. E. Moise (1990, pp. 394–410). To appreciate the depth of this theorem, try to prove it for two rectangles in the Euclidean plane.)

less than 180. Since the angle sum can never get below 0, the defect can never get above 180. Thus, we have the following corollary.

COROLLARY. In hyperbolic geometry the area of any triangle is at most πk^2.

Of course, there is no finite triangle whose area equals the maximal value πk^2, although we can approach this area as closely as we wish (and achieve it with an infinite trebly asymptotic triangle). However, J. Bolyai showed how to construct a *circle* of area πk^2 and a regular 4-sided polygon with a 45° angle that also has this area (see Exercise 29 or Bonola, 1955, pp. 106–110).

If we interpret hyperbolic geometry in our physical world, it is clear that since defects of terrestrial triangles are immeasurably small, while their areas are measurable, the proportionality constant k^2 must be extremely large. According to Kulczycki (1961, pp. 153–155) the measurements of the parallaxes of fixed stars "elicit that the constant k is not less than about six hundred trillion miles." Thus, $k^2 > 36 \times 10^{28}$. These measurements are based on the fact that the right triangle whose one side is half the major axis of the earth's orbit around the sun and whose opposite vertex is at a fixed star has defect less than the parallax of the star (Figure 10.5). Of course, since Einstein, we do not use hyperbolic geometry to model the geometry of the universe.

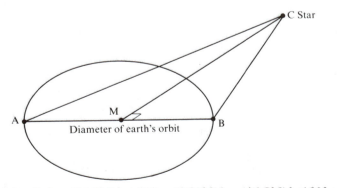

FIGURE 10.5 Defect $(\triangle ACM)° = 90° - (\angle CAM)° - (\angle ACM)° < 90° - (\angle CAM)° =$ parallax.

THE ANGLE OF PARALLELISM

Recall that given any line *l* and any point P not on *l*, there exist limiting parallel rays $\overrightarrow{PX}$ and $\overrightarrow{PY}$ to *l* that are situated symmetrically about the perpendicular PQ from P to *l* (Figure 10.6; see Chapter 6). We proved that ⊀XPQ ≅ ⊀YPQ (Theorem 6.6), so either of these angles can be called the *angle of parallelism* for P with respect to *l*.

It is not difficult to show that the number α of radians in the angle of parallelism depends only on the distance *d* from P to Q, not on the particular line *l* or the particular point P (see Major Exercise 5, Chapter 6). In the notation used after the proof of Theorem 6.6, $\alpha = (\pi/180)\Pi(PQ)°$.

The following formula relating α and *d* was discovered by J. Bolyai and Lobachevsky.

THEOREM 10.2. Formula of Bolyai-Lobachevsky:

$$\tan\frac{\alpha}{2} = e^{-d/k}$$

where *k* is the constant whose square occurs in the proportionality factor of area to defect in Theorem 10.1.

This is certainly one of the most remarkable formulas in all of mathematics, and it is astonishing how few mathematicians know it. In this formula the number *e* is the base for the natural logarithms (*e* is approximately 2.718 . . .), and tan $\alpha/2$ is the trigonometric tangent of half of α. Proofs of this formula will be found in Kulczycki (1961, §20) and in Borsuk and Szmielew (1960, Chapter 6, §26). We have checked this formula for the Poincaré disk model (where *k* = 1) — see Theorem 7.2, p. 256.

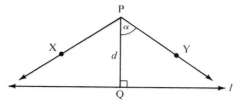

FIGURE 10.6

CYCLES

The various proofs of the Bolyai-Lobachevsky formula (Theorem 10.2) all make use of a curve that is peculiar to hyperbolic geometry, called either a *limiting curve* or a *horocycle* in the literature. It is obtained as follows (see also Exercises 41–44, Chapter 9).

Start with a line l, point Q on l, and erect perpendicular $\overleftrightarrow{PQ}$ to l through Q. Then consider the circle δ with center P and radius PQ, which is tangent to l at Q (see Figure 10.7). Now let P recede from l along the perpendicular. The circle δ will increase in size, remaining tangent to l, and will approach a limiting position as P recedes arbitrarily far from Q. In Euclidean geometry the limiting position of δ would just be the line l, but in hyperbolic geometry the limiting position of δ is a new curve h called a *limiting curve* or *horocycle*.

We can visualize this in the Poincaré model as follows. Let l be a diameter of the Euclidean circle γ whose interior represents the hyperbolic plane, and let Q be the center of γ. It can be proved that the hyperbolic circle with hyperbolic center P is represented by a Euclidean circle whose Euclidean center R lies between P and Q (Exercise P-5, Chapter 7); see Figure 10.8.

As P recedes from Q toward the ideal point represented by S, R is pulled up to the Euclidean midpoint of SQ, so that the horocycle h is a Euclidean circle tangent to γ at S and tangent to l at Q. It can be shown in general that all horocycles are represented in the Poincaré model by Euclidean circles inside γ and tangent to γ. Moreover, all the Poincaré lines passing through the ideal point S are orthogonal to h; a hyperbolic ray from a point of h out to the ideal point S is called a *diameter* of h (see Chapter 9, Exercises 41, 44, and 46).

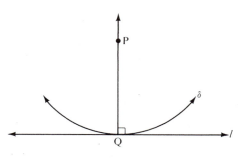

FIGURE 10.7

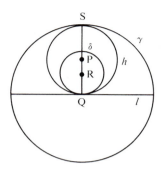

FIGURE 10.8

In the Poincaré model two horocycles tangent to γ at S are said to be concentric (see Figure 10.9). It is by studying the ratio of corresponding arcs on concentric horocycles that the Bolyai-Lobachevsky formula can be derived (see Kulczycki, 1961, §18).

There is an analogous construction in hyperbolic space called the *horosphere.* Instead of taking the limit of circles to get the horocycle, one takes the limit of spheres to get the horosphere or *limiting surface.*

Another strange curve in hyperbolic geometry that has no Euclidean counterpart is the *equidistant curve* (Figure 10.10). Start with a line *l* and a point P not on *l*. Consider the locus of all points on the same side of *l* as P and at the same perpendicular distance from *l* as P. In Euclidean geometry this locus would just be the unique line through P parallel to *l*, but in hyperbolic geometry it is not a line, it is the *hypercycle,* or *equidistant curve,* through P.

In the Poincaré model let A and B be the ideal endpoints of *l*. It turns out that the equidistant curve to *l* through P is represented by

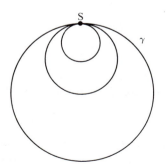

FIGURE 10.9 Concentric horocycles.

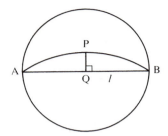

FIGURE 10.10   Equidistant curve.

the arc of the Euclidean circle passing through A, B, and P (Chapter 9, Exercises 40, 45, and 47). This curve is orthogonal to all Poincaré lines perpendicular to the line *l*.

In the Euclidean plane three points either lie on a uniquely determined line or on a uniquely determined circle. Not so in the hyperbolic plane — look at the Poincaré model. A Euclidean circle represents:

1. A *hyperbolic circle* if it is entirely inside γ;
2. A *horocycle* if it is inside γ except for one point where it is tangent to γ;
3. An *equidistant curve* if it cuts γ nonorthogonally in two points;
4. A *hyperbolic line* if it cuts γ orthogonally.

It follows that in the hyperbolic plane three noncollinear points lie either on a circle, a horocycle, or a hypercycle accordingly as the perpendicular bisectors of the triangle are "concurrent" in an ordinary, ideal, or ultra-ideal point — see the last section of this chapter.

THE PSEUDOSPHERE

One of the difficulties with the Poincaré model is that, although it faithfully represents angles of the hyperbolic plane (i.e., it is a *conformal model*), it distorts distances. So it is natural to ask whether another model exists that also represents hyperbolic lengths faithfully by Euclidean lengths. If there is such a model, it would be called *isometric*. An equally natural idea is to seek as a model some surface in Euclidean three-dimensional space. The lines of the hyperbolic plane would

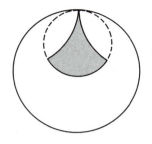

FIGURE 10.11

then be represented by *geodesics* on the surface, and we would expect the surface to be curved so as to mirror our expectation that hyperbolic lines are "really curved." (See Appendix A for the definition of "geodesic." If a shortest path exists on a surface between two given points, it must be an arc of a geodesic; but conversely, an arc of a geodesic need not be the shortest path, for on a sphere, there are two arcs of one great circle joining two points, and if the points are not antipodal, one arc is shorter.)

A difficult theorem of Hilbert (see Do Carmo, 1976, p. 446) states that it is impossible to embed the entire hyperbolic plane isometrically as a surface in Euclidean three-space. On the contrary, it *is* possible to embed the Euclidean plane isometrically in hyperbolic space, as the surface of the horosphere (see Kulczycki, 1961, §17). This result, proved by both J. Bolyai and Lobachevsky, was already recognized by Wachter in 1816.[3]

But all is not lost. It turns out to be possible to embed a portion of the hyperbolic plane isometrically in Euclidean space, the portion called a *horocyclic sector,* bounded by an arc of a horocycle and the two diameters cutting off this arc. Such a sector in the Poincaré model is shown in Figure 10.11.

The surface that represents this region isometrically is called a *pseudosphere*. It is obtained by rotating a curve called a *tractrix* around its asymptote. It looks like an infinitely long horn. (In this representation the two diameters of the horocyclic arc have been identified —

---

[3] For those who understand this language, strangely enough there does exist a continuously differentiable embedding of the hyperbolic plane into Euclidean three-space. This was proved in 1955 by N. Kuiper, using analytic methods (see *Indagationes Mathematicae*, **17**: 683). It is known that no "nice," e.g., C², embeddings exist.

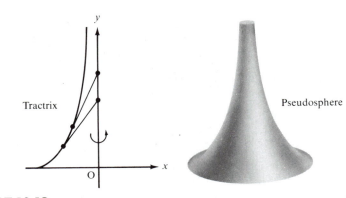

Tractrix

Pseudosphere

FIGURE 10.12

pasted together — so that the mapping of the sector into Euclidean space is actually only an embedding of the region between the diameters.) The tractrix is characterized by the fact that the tangent from any point on the curve to the vertical asymptote has constant length a.[4] See Figure 10.12.

F. Minding was the first to publish an article about the pseudosphere, in 1839, but Gauss had written an unpublished note about it around 1827, calling it "the opposite of a sphere." Curiously, neither of them recognized that it could be used to prove the consistency of hyperbolic geometry, as Beltrami did in 1868. The pseudosphere is related to the hyperbolic plane as a cylinder is to the Euclidean plane.

The representation on the pseudosphere enables us to give some geometric meaning to the fundamental constant k that appears in Theorems 10.1 and 10.2. The point is that there is a way (discovered by Gauss) of measuring the *curvature* of any surface. We cannot give the precise definition, since it involves a knowledge of differential geometry (see Coxeter, 1969; Hilbert and Cohn-Vossen, 1952; or Appendix A of this book). In general, the curvature K varies from point to point, being close to zero at points where the surface is rather flat, large at points where the surface bends sharply. For some surfaces the curvature is the same at all points, so naturally these are called *surfaces of constant curvature K*. An important property of such surfaces is that figures can be moved around on them without changing size or shape.

---

[4] The Dutch physicist Huygens called the tractrix the "dog curve" because it resembles the curve described by the nose of a dog being dragged reluctantly on a leash.

In 1827, Gauss proved (for constant K) a fundamental formula relating the curvature, area, and angular measure. He took a geodesic triangle $\triangle ABC$ with vertices A, B, and C and sides geodesic segments. By integration, he calculated the area of the triangle. He determined that, if $(\sphericalangle A)^r$ denotes the radian measure of angle A, then

$$K \times \text{area } \triangle ABC = (\sphericalangle A)^r + (\sphericalangle B)^r + (\sphericalangle C)^r - \pi.$$

He then showed what this meant by considering the three possible cases:

Case 1. K is positive, hence, both sides of the equation are positive. In this case Gauss' formula shows that the angle sum in radians of a geodesic triangle is greater than π (or in degrees, greater than $180°$), and that the area is proportional to the *excess* (the number on the right side of the equation), with a proportionality factor of $1/K$. An example could be the surface of a sphere of radius r, whose curvature is $K = 1/r^2$. The larger the radius, the smaller the curvature, and the more the surface resembles a plane. (Gauss' formula in the special case of a sphere was already discovered by Girard in the seventeenth century.) According to a theorem of H. Liebmann, H. Hopf, and W. Rinow, spheres are the only complete surfaces of constant positive curvature in Euclidean three-space, so that the elliptic plane cannot be embedded in Euclidean three-space either.

Case 2. $K = 0$. In this case Gauss' formula shows that the angle sum in radians is equal to π. An example would be the Euclidean plane; another example is an infinitely long cylinder.

Case 3. K is negative. In this case Gauss' formula shows that the angle sum in radians is less than π, and the area is proportional to the *defect*. An example of such a surface is the *pseudosphere*. Since the pseudosphere represents a portion of the hyperbolic plane isometrically, we can compare Gauss' formula with the formula in Theorem 10.1 relating area to defect. The comparison gives $K = -1/k^2$. Thus, $-1/k^2$ is the *curvature of the hyperbolic plane*.

Here we can recapture the analogy with case 1 by setting $r = ik$, where $i = \sqrt{-1}$. Then $K = 1/r^2$, so that the hyperbolic plane can be described as a "sphere of imaginary radius $r = ik$," as Lambert noticed (see Chapter 5).

Finally, notice that as k gets very large, the curvature K approaches zero, and the geometry of the surface resembles more and more the geometry of the Euclidean plane. It is in this sense that Euclidean geometry is a "limiting case" of hyperbolic geometry.

We will also see in the next sections that the geometry of an "infinitesimal region" in the hyperbolic plane is Euclidean.

HYPERBOLIC TRIGONOMETRY

Trigonometry is the study of the relationships among the sides and angles of a triangle. We reviewed a few formulas of Euclidean trigonometry in Exercises 22 and 23 of Chapter 5. There, the theory of similar triangles, which is only valid in Euclidean geometry, was used to establish definitions. It defined, for example, the *sine* of an acute angle to be the ratio of the opposite side to the hypotenuse in any right triangle having that acute angle as one of its angles (similarly for the *cosine, tangent,* and other trigonometric functions). We will need to use these functions for hyperbolic trigonometry. What can this mean, when these functions have been defined in Euclidean geometry?

An evasive answer is to do hyperbolic trigonometry only in a conformal Euclidean model such as a Poincaré model; since angles are measured in the Euclidean way, the sine, cosine, and tangent of an angle have their usual Euclidean meaning. This answer will have to suffice for those readers who have not yet studied infinite series.

The rigorous answer is to redefine these trigonometric functions purely analytically, without reference to geometry, and then apply them to the different geometries. The definition is in terms of the Taylor series expansions:

$$\sin x = \sum_{n=0}^{\infty} (-1)^n \frac{x^{2n+1}}{(2n+1)!} \qquad \cos x = \sum_{n=0}^{\infty} (-1)^n \frac{x^{2n}}{(2n)!}$$

(and tan $x =$ sin x/cos x, etc.). In good calculus texts, one can find proof that these series converge for all x (by the ratio test). Good calculus texts will also show how to develop all the familiar formulas for these *circular functions* (so called because $x =$ cos θ and $y =$ sin θ are parametric equations for the Cartesian unit circle $x^2 + y^2 = 1$; in particular, this equation and the Pythagorean theorem imply that a right triangle in the Euclidean plane with legs of length sin θ and cos θ has hypotenuse of length 1, so that sin θ is indeed the ratio of opposite side to hypotenuse for the appropriate acute angle in that triangle).

Hyperbolic trigonometry involves, in addition to the circular functions, the *hyperbolic functions*, defined analytically by

(1) $$\sinh x = \frac{e^x - e^{-x}}{2} \qquad \cosh x = \frac{e^x + e^{-x}}{2}$$

(and tanh $x =$ sinh x/cosh x, etc.). These functions were introduced by Lambert. Their graphs are shown in Figure 10.13. The hyperbolic

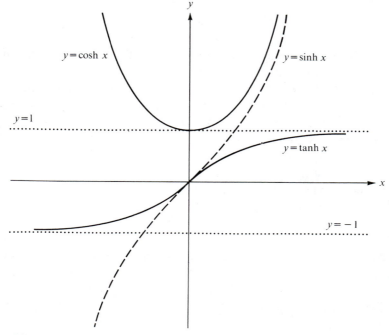

FIGURE 10.13

sine and cosine have the Taylor series expansions

$$(2) \qquad \sinh x = \sum_{n=0}^{\infty} \frac{x^{2n+1}}{(2n+1)!} \qquad \cosh x = \sum_{n=0}^{\infty} \frac{x^{2n}}{(2n)!},$$

which are obtained from the expansion of the circular sine and cosine by omitting the coefficients $(-1)^n$; this can be seen by recalling that the Taylor series for the exponential function is

$$e^x = \sum_{n=0}^{\infty} \frac{x^n}{n!}.$$

In fact, by introducing the imaginary number $i = \sqrt{-1}$ we see that

$$\sinh x = i \sin \frac{x}{i} \qquad \cosh x = \cos \frac{x}{i}.$$

The name "hyperbolic functions" stems from the identity

$$(3) \quad \cosh^2 x - \sinh^2 x = \frac{e^{2x} + 2 + e^{-2x}}{4} - \frac{e^{2x} - 2 + e^{-2x}}{4} = 1$$

from which the parametric equations $x = \cosh \theta$ and $y = \sinh \theta$ give one branch of the hyperbola $x^2 - y^2 = 1$ in the Cartesian plane. (Here the number θ has the geometric interpretation of twice the area bounded by the hyperbola, the x axis, and the line joining the origin to the point $(\cosh \theta, \sinh \theta)$. There is an analogous interpretation for θ for the circular functions when we replace the hyperbola with the circle.)

(You may well wonder what this hyperbola in the Cartesian model for Euclidean geometry has to do with hyperbolic geometry! Nothing, so far as I know. Felix Klein coined the names "hyperbolic" and "elliptic" geometries because lines in these geometries have two and zero ideal points at infinity, respectively; this is analogous to affine hyperbolas and ellipses, which have two and zero points at infinity, respectively. A Euclidean line has only one ideal point, and this is analogous to an affine parabola, which has one point at infinity.)

Here is a list of identities for hyperbolic and circular functions that will be used in the sequel. (Define $\tanh x = \sinh x/\cosh x$.)

Hyperbolic	Circular
$\cosh^2 x - \sinh^2 x = 1$	$\cos^2 x + \sin^2 x = 1$
$1 - \tanh^2 x = \operatorname{sech}^2 x$	$1 + \tan^2 x = \sec^2 x$
$\sinh (x \pm y) = \sinh x \cosh y$ $\pm \cosh x \sinh y$	$\sin (x \pm y) = \sin x \cos y$ $\pm \cos x \sin y$
$\cosh (x \pm y) = \cosh x \cosh y$ $\pm \sinh x \sinh y$	$\cos (x \pm y) = \cos x \cos y$ $\mp \sin x \sin y$
$\tanh (x + y) = \dfrac{\tanh x + \tanh y}{1 + \tanh x \tanh y}$	$\tan (x + y) = \dfrac{\tan x + \tan y}{1 - \tan x \tan y}$
$\sinh^2 \dfrac{x}{2} = \dfrac{\cosh x - 1}{2}$	$\sin^2 \dfrac{x}{2} = \dfrac{1 - \cos x}{2}$
$\cosh^2 \dfrac{x}{2} = \dfrac{\cosh x + 1}{2}$	$\cos^2 \dfrac{x}{2} = \dfrac{1 + \cos x}{2}$
$\tanh^2 \dfrac{x}{2} = \dfrac{\cosh x - 1}{\cosh x + 1}$	$\tan^2 \dfrac{x}{2} = \dfrac{1 - \cos x}{1 + \cos x}$
$\tanh \dfrac{x}{2} = \dfrac{\sinh x}{\cosh x + 1}$	$\tan \dfrac{x}{2} = \dfrac{\sin x}{1 + \cos x}$
$= \dfrac{\cosh x - 1}{\sinh x}$	$= \dfrac{1 - \cos x}{\sin x}$
$\sinh x = 2 \sinh \dfrac{x}{2} \cosh \dfrac{x}{2}$	$\sin x = 2 \sin \dfrac{x}{2} \cos \dfrac{x}{2}$
$\sinh x \pm \sinh y$ $= 2 \sinh \tfrac{1}{2} (x \pm y) \cosh \tfrac{1}{2} (x \mp y)$	$\sin x \pm \sin y$ $= 2 \sin \tfrac{1}{2} (x \pm y) \cos \tfrac{1}{2} (x \mp y)$
$\cosh x + \cosh y$ $= 2 \cosh \tfrac{1}{2} (x + y) \cosh \tfrac{1}{2} (x - y)$	$\cos x + \cos y$ $= 2 \sin \tfrac{1}{2} (x + y) \cos \tfrac{1}{2} (x - y)$
$\cosh x - \cosh y$ $= 2 \sinh \tfrac{1}{2} (x + y) \sinh \tfrac{1}{2} (x - y)$	$\cos x - \cos y$ $= 2 \sin \tfrac{1}{2} (x + y) \sin \tfrac{1}{2} (x - y)$

We are going to state the formulas of hyperbolic trigonometry under the simplifying assumption that $k = 1$ (where k is the constant in Theorems 10.1 and 10.2 and in case 3, p. 397); this can be shown to mean that we have chosen our unit of length so that the ratio of the length of corresponding arcs on concentric horocycles is equal to e when the distance between the horocycles is 1. This choice is entirely analogous to the choice of unit of angle measure such that a right angle has (radian) measure $\pi/2$ — it makes the formulas come out nicely. For example, the area of a triangle is *equal* to its defect under this convention (Theorem 10.1), provided we now (and henceforth) measure defect in *radians* instead of degrees. Furthermore, the fundamental formula of Bolyai-Lobachevsky for the radian measure of the angle of parallelism (Theorem 10.2) becomes

$$(4) \qquad\qquad \Pi(x) = 2 \arctan e^{-x}.$$

Straightforward calculation using double angle formulas for the circular functions then yields the following formulas:

$$(5) \qquad\qquad \sin \Pi(x) = \operatorname{sech} x = 1/\cosh x$$
$$(6) \qquad\qquad \cos \Pi(x) = \tanh x$$
$$(7) \qquad\qquad \tan \Pi(x) = \operatorname{csch} x = 1/\sinh x.$$

Thus, the function Π provides a link between the hyperbolic and circular functions.

Given $\triangle ABC$, we will use the standard notation $a = \overline{BC}$, $b = \overline{AC}$, $c = \overline{AB}$ for the lengths of the sides. We will write expressions such as $\cos A$ to abbreviate "cosine of the number of radians in $\sphericalangle A$," and we repeat that this does *not* mean the ratio of adjacent side to hypotenuse in a hyperbolic triangle. We will develop the first formulas (10) of hyperbolic trigonometry from the Poincaré model, i.e., using Euclidean geometry; after that we can deduce the remaining formulas without referring to the Poincaré model. For segments that are part of diameters of the absolute circle κ, there is ambiguity in the notation for length; as in Chapter 7, we will write $\overline{AB}$ for the Euclidean length and $d(AB)$ for the Poincaré length. We will take κ to have radius 1. For a segment OB with one endpoint at the center O of κ, the proof of Lemma 7.4, p. 255, showed that

$$e^{d(OB)} = \frac{1 + \overline{OB}}{1 - \overline{OB}}.$$

Writing, for brevity, $x = d(OB)$ and $t = \overline{OB}$ in this formula, a little algebra gives the basic relations

$$(8) \qquad\qquad \sinh x = \frac{2t}{1 - t^2} \qquad \cosh x = \frac{1 + t^2}{1 - t^2}$$

so that

(9)
$$\tanh x = \frac{2t}{1 + t^2} = F(t),$$

where F is the isomorphism of the Poincaré model onto the Klein model defined in Chapter 7, p. 258.

THEOREM 10.3. Given any *right* triangle $\triangle ABC$, with $\sphericalangle C$ right, in the hyperbolic plane (with $k = 1$). Then

(10)
$$\sin A = \frac{\sinh a}{\sinh c} \qquad \cos A = \frac{\tanh b}{\tanh c}$$

(11)
$$\cosh c = \cosh a \cosh b = \cot A \cot B$$

(12)
$$\cosh a = \frac{\cos A}{\sin B}$$

Proof:
Before indicating a proof of this theorem, let us compare these formulas to the formulas for a Euclidean right triangle. The first equality in formula (11) is the hyperbolic analogue of the Pythagorean theorem; for if we expand both sides in Taylor series using (2), the formula becomes

$$1 + \tfrac{1}{2}c^2 + \cdots = 1 + \tfrac{1}{2}(a^2 + b^2) + \cdots$$

And if we neglect the higher-order terms (when $\triangle ABC$ is sufficiently small), this reduces to

$$c^2 \approx a^2 + b^2.$$

Similarly (when $\triangle ABC$ is sufficiently small), formula (10) becomes approximately

$$\sin A \approx \frac{a}{c} \qquad \cos A \approx \frac{b}{c}.$$

Let us be more precise: Consider right triangles with fixed $\sphericalangle A$ and with $c \to 0$. Then by Proposition 4.5, $a \to 0$ (since $a < c$). By formula (2) and the geometric series formula,

$$\frac{1}{\sinh c} = \frac{1}{c(1 + u)} = \frac{1}{c}(1 - u + u^2 - u^3 + \cdots),$$

where $\lim_{c \to 0} u = 0$. Thus,

$$\frac{\sinh a}{\sinh c} = \frac{a}{c}\left(1 + \frac{a^2}{3!} + \frac{a^4}{5!} + \cdots\right)(1 - u + u^2 - u^3 + \cdots)$$

and we see that

$$\lim_{c \to 0} \frac{a}{c} = \lim_{c \to 0} \frac{\sinh a}{\sinh c} = \sin A.$$

A similar argument applies to cos A. So *it is appropriate to say that the hyperbolic trigonometry of "infinitesimal" triangles is Euclidean.*

Formula (12) and the second equality in formula (11) have no counterparts in Euclidean geometry because in Euclidean geometry the angles do not determine the lengths of the sides.

All the geometry of a right triangle is incorporated in formula (10), for all the other formulas follow from (10) by pure algebra and identities. Namely, the identity $\sin^2 A + \cos^2 A = 1$ and (10) give

$$1 = \frac{\tanh^2 b}{\tanh^2 c} + \frac{\sinh^2 a}{\sinh^2 c}$$

$$\sinh^2 c = \cosh^2 c \tanh^2 b + \sinh^2 a$$

$$1 + \sinh^2 c = \cosh^2 c \left(\frac{\sinh^2 b}{\cosh^2 b}\right) + 1 + \sinh^2 a$$

$$\cosh^2 c \cosh^2 b = \cosh^2 c \sinh^2 b + \cosh^2 a \cosh^2 b$$

$$\cosh^2 c (\cosh^2 b - \sinh^2 b) = \cosh^2 a \cosh^2 b$$

$$\cosh^2 c = \cosh^2 a \cosh^2 b,$$

which gives the first equality in (11). Applying (10) to B instead of A gives

$$\sin B = \frac{\sinh b}{\sinh c}$$

so that we get (12):

$$\frac{\cos A}{\sin B} = \frac{\tanh b}{\tanh c}\frac{\sinh c}{\sinh b} = \frac{\cosh c}{\cosh b} = \cosh a.$$

Multiplying this by the analogous formula for $\cosh b$ yields the second equality in (11).

Finally, to prove formula (10), we do the geometry under the assumption that vertex A of the right triangle coincides with the center O of the absolute (this can always be achieved by a suitable inversion as in the proof of SAS, p. 254. The points B′, C′ in Figure 10.14 are the images of B, C under the isomorphism F (see also Figure 7.33). From Euclidean right triangle $\triangle B'C'O$ and formula (9) we get

$$\cos A = \frac{\overline{OC'}}{\overline{OB'}} = \frac{\tanh b}{\tanh c}.$$

Let B″ be the other intersection of $\overrightarrow{OB}$ with the orthogonal circle κ_1 containing Poincaré line $\overleftrightarrow{BC}$. By Proposition 7.5, p. 246, B″ is the inverse of B in κ, so that

$$\overline{BB''} = \overline{OB''} - \overline{OB} = \frac{1}{t} - t = \frac{1 - t^2}{t} = \frac{2}{\sinh x}$$

in the notation of formula (8). In the standard notation,

$$\overline{BB''} = \frac{2}{\sinh c} \quad \text{and} \quad \overline{CC''} = \frac{2}{\sinh b}.$$

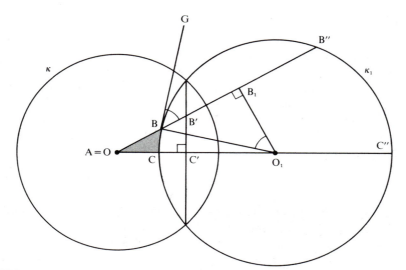

FIGURE 10.14

Let B_1 be the foot of the perpendicular from the center O_1 of κ_1 to BB'', so that B_1 is the midpoint of BB'' (base of isoceles triangle). Then $\sphericalangle BO_1B_1 \cong \sphericalangle B$ $(= \sphericalangle GBB_1$ in Figure 10.14, where $\overset{\leftrightarrow}{GB}$ is the tangent to κ_1 at B), because both these angles are complements of $\sphericalangle B_1BO_1$. Hence,

$$\sin B = \frac{\overline{BB_1}}{\overline{O_1B}} = \frac{\overline{BB''}}{2\overline{O_1C}} = \frac{\overline{BB''}}{\overline{CC''}} = \frac{\sinh b}{\sinh c}.$$

Since $\sphericalangle B$ is an arbitrary acute angle in a right triangle, we can relabel, interchanging A and B, to get the second formula in (10). ■

THEOREM 10.4. For any $\triangle ABC$ in the hyperbolic plane (with $k = 1$ and standard notation for the sides)

(13) $\qquad \cosh c = \cosh a \cosh b - \sinh a \sinh b \cos C$

(14) $\qquad \dfrac{\sin A}{\sinh a} = \dfrac{\sin B}{\sinh b} = \dfrac{\sin C}{\sinh c}$

(15) $\qquad \cosh c = \dfrac{\cos A \cos B + \cos C}{\sin A \sin B}.$

Formula (13) is the *hyperbolic law of cosines* and formula (14) is the *hyperbolic law of sines;* they are analogous to the Euclidean laws and reduce to the latter for "infinitesimal" triangles as before. Formula (15) has no Euclidean analogue.

This theorem can be proved by dropping an altitude to create two right triangles and by applying the preceding theorem, some algebra, and identities such as

$$\cosh(x \pm y) = \cosh x \cosh y \pm \sinh x \sinh y.$$

We leave the details as an exercise.

Note on Elliptic Geometry. Analogously, elliptic geometry with $k = 1$ can be developed from its model on a sphere of radius 1 with antipodal points identified (see Kay, 1969, Chapter 10). The elliptic law of cosines is

$$\cos c = \cos a \cos b + \sin a \sin b \cos C$$

and the elliptic law of sines is

$$\frac{\sin A}{\sin a} = \frac{\sin B}{\sin b} = \frac{\sin C}{\sin c}.$$

In a triangle with right angle at C, the elliptic analogue of formula (10) is

$$\sin A = \frac{\sin a}{\sin c} \qquad \cos A = \frac{\tan b}{\tan c}.$$

In each of these formulas, if we replace the ordinary trigonometric functions by the corresponding hyperbolic trigonometric functions whenever the argument is the length of a side, we get the formulas above for hyperbolic geometry with $k = 1$.

CIRCUMFERENCE AND AREA OF A CIRCLE

THEOREM 10.5. The circumference C of a circle of radius r is given by $C = 2\pi \sinh r$.

Proof:
Of course C is defined as the limit $\lim\limits_{n \to \infty} p_n$ of the perimeter p_n of the regular n-gon inscribed in the circle (Figure 10.15). Recall first

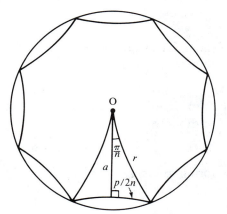

FIGURE 10.15

how the formula $C = 2\pi r$ is derived in Euclidean geometry. From Figure 10.15 and Euclidean trigonometry we see that

$$p_n = r2n \sin \frac{\pi}{n} = r2n \left[\frac{\pi}{n} - \frac{1}{3!} \left(\frac{\pi}{n} \right)^3 + \frac{1}{5!} \left(\frac{\pi}{n} \right)^5 - \cdots \right]$$

$$= 2\pi r - \frac{2r\pi^2}{n^2} \left[\frac{\pi}{3!} - \frac{1}{5!} \left(\frac{\pi}{n} \right)^3 + \cdots \right]$$

$$\lim_{n \to \infty} p_n = 2\pi r.$$

In the hyperbolic case, we use instead formula (10) of Theorem 10.3 to get

$$\sinh (p/2n) = \sinh r \sin (\pi/n),$$

which expanded in series becomes

$$\frac{p}{2n} \left[1 + \frac{1}{3!} \left(\frac{p}{2n} \right)^2 + \frac{1}{5!} \left(\frac{p}{2n} \right)^4 + \cdots \right]$$

$$= \frac{\pi}{n} \sin r \left[1 - \frac{1}{3!} \left(\frac{\pi}{n} \right)^2 + \frac{1}{5!} \left(\frac{\pi}{n} \right)^4 - \cdots \right]$$

(where $p = p_n$ for typographical simplicity). Multiplying both sides by $2n$ and taking $\lim_{n \to \infty}$ gives the formula we seek. (Note once more that for a circle of "infinitesimal radius," the hyperbolic formula reduces to the Euclidean formula.) ∎

This theorem enables us to rewrite the *law of sines* (14) *in a form that is valid in neutral geometry.*

COROLLARY (J. Bolyai). The sines of the angles of a triangle are to one another as the circumference of the circles whose radii are equal to the opposite sides.

Bolyai denoted the circumference of a circle of radius r by Or and wrote this result in the form

$$Oa : Ob : Oc = \sin A : \sin B : \sin C.$$

Consider next formulas for *area*. By Theorem 10.1 and our convention $k = 1$, the area K of a triangle is equal to its defect in radians, i.e.,

$$K = \pi - A - B - C.$$

Let us calculate this defect for a right triangle with right angle at C, *so that*
$K = \pi/2 - (A + B)$.

THEOREM 10.6. $\tan K/2 = \tanh a/2 \cdot \tanh b/2$.

(For Euclidean geometry, the formula for area K is $K/2 = a/2 \cdot b/2$.)

Proof:
Here are the main steps in the proof:

$$\tanh^2 \frac{a}{2} \tanh^2 \frac{b}{2} = \frac{(\cosh a) - 1}{(\cosh a) + 1} \frac{(\cosh b) - 1}{(\cosh b) + 1}$$

$$= \frac{1 - \sin(A + B)}{1 + \sin(A + B)} \frac{\cos (A - B)}{\cos (A - B)}$$

$$= \frac{1 - \cos K}{1 + \cos K}$$

$$= \tan^2 \frac{K}{2}.$$

Steps 1 and 4 are just identities for $\tanh^2 (x/2)$ and $\tan^2 (x/2)$, respectively. Step 2 follows from substituting formula (12) for $\cosh a$ and $\cosh b$ and doing a considerable amount of algebra using trigonometric identities.[5] And step 3 just uses the identity $\cos (\pi/2 - x) = \sin x$. ■

THEOREM 10.7.   The area of a circle of radius r is $4\pi \sinh^2 (r/2)$.

Proof:
Here again we define the area A of a circle to be the limit $\lim\limits_{n \to \infty} K_n$ of the area K_n of the inscribed regular n-gon. Referring to Figure 10.15 again, using the previous theorem, and writing a, K, p, for a_n, K_n, p_n, we have

$$\tan \frac{K}{4n} = \tanh \frac{p}{4n} \tanh \frac{a}{2}.$$

[5] See Exercise 5. From now on, you will be offered the opportunity to exercise your algebraic technique to fill in such gaps.

If we multiply both sides by $4n$ and pass to the limit as $n \to \infty$, we obtain

$$(16) \qquad\qquad A = C \tanh \frac{r}{2},$$

using $\lim_{n \to \infty} p_n = C$, $\lim_{n \to \infty} a_n = r$, continuity of tan and tanh, and the series

$$4n \tan \frac{K}{4n} = K + \frac{K}{3}\left(\frac{K}{4n}\right)^2 + \cdots$$

$$4n \tanh \frac{p}{4n} = p - \frac{p}{3}\left(\frac{p}{4n}\right)^2 + \cdots.$$

Then we substitute in (16) the formula for C from Theorem 10.5 and use the identities

$$\tanh \frac{r}{2} = \frac{\sinh r}{\cosh r + 1}$$

$$\sinh^2 r = \cosh^2 r - 1$$

$$2 \sinh^2 \frac{r}{2} = \cosh r - 1$$

to obtain Theorem 10.7. ∎

By expanding this formula in a series we can see how much larger is the area of a hyperbolic circle than a Euclidean circle with the same radius:

$$A = \pi \left(r^2 + \frac{r^4}{12} + \cdots \right).$$

Note on Elliptic Geometry. The formulas for circumference and area of a circle of radius r are

$$C = 2\pi \sin r$$
$$A = 4\pi \sin^2 (r/2).$$

Bolyai's formula is valid in elliptic geometry (so it is indeed a theorem in *absolute geometry*).

SACCHERI AND LAMBERT QUADRILATERALS

We next consider a Saccheri quadrilateral with base b, legs of length a, and summit of length c. You showed in Exercise 1, Chapter 6, that $c > b$. We now make this more precise.

THEOREM 10.8. For a Saccheri quadrilateral,

$$\sinh \frac{c}{2} = \cosh a \sinh \frac{b}{2}.$$

(Since $\cosh^2 a = 1 + \sinh^2 a > 1$, this tells us that $\sinh (c/2) > \sinh (b/2)$, hence $c > b$.)

Proof
Theorem 10.8 is proved by letting $d = \overline{AB'}$ and $\theta = (\sphericalangle A'AB')^r$ in Figure 10.16, applying formula (13) from Theorem 10.4 to get

$$\cosh c = \cosh a \cosh d - \sinh a \sinh d \cos \theta.$$

using formulas (10) and (11) from Theorem 10.3 to get

$$\cos \theta = \sin \left(\frac{\pi}{2} - \theta \right) = \frac{\sinh a}{\sinh d}$$
$$\cosh d = \cosh a \cosh b,$$

and eliminating d to obtain

$$\cosh c = \cosh^2 a \cosh b - \sinh^2 a$$
$$= \cosh^2 a(\cosh b - 1) + 1.$$

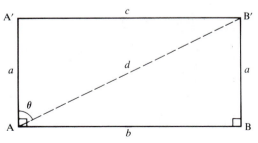

Finally, the identity

$$2 \sinh^2 \frac{x}{2} = \cosh x - 1$$

gives the result. ∎

COROLLARY. Given a Lambert quadrilateral, if c is the length of a side adjacent to the acute angle and b is the length of the opposite side, then

$$\sinh c = \cosh a \sinh b,$$

where a is the length of the other side adjacent to the acute angle (in particular, $c > b$).

The corollary follows from representing the Lambert quadrilateral as half of a Saccheri quadrilateral (see Figure 10.17). ∎

There are additional remarkable formulas for Lambert quadrilaterals that we will derive next. They are based on the concept of *complementary segments:* these are segments whose lengths x, x^* are related by

(17) $$\Pi(x) + \Pi(x^*) = \frac{\pi}{2}.$$

The geometric meaning of this equation is shown in Figure 10.18, where the "fourth vertex" of the Lambert quadrilateral is the ideal point Ω.

If we apply our earlier formulas (4) through (7) for the angle of parallelism, we obtain

(18) $$\sinh x^* = \operatorname{csch} x$$

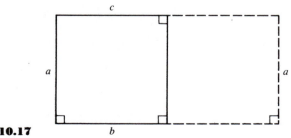

FIGURE 10.17

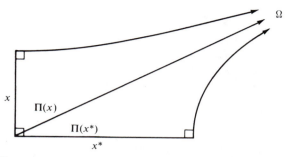

FIGURE 10.18

(19) $$\cosh x^* = \coth x$$
(20) $$\tanh x^* = \operatorname{sech} x$$

(21) $$\tanh \frac{x^*}{2} = e^{-x}.$$

For example: $\sinh x^* = \cot \Pi(x^*) = \tan \Pi(x) = \operatorname{csch} x$ by (7); formula (21) follows from (18), (19), and the identity

$$\tanh (t/2) = (\sinh t)/(1 + \cosh t).$$

THEOREM 10.9 (Engel's Theorem). There exists a right triangle with the parameters shown in Figure 10.19 if and only if there exists a Lambert quadrilateral with the parameters shown in Figure 10.20. Note that PQ is a complementary segment to the segment (not shown) whose angle of parallelism is ⦡A.

The geometric meaning of Engel's theorem is shown in Figure 10.21. It includes J. Bolyai's parallel construction (Figure 6.16), for if

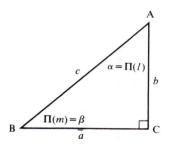

FIGURE 10.19

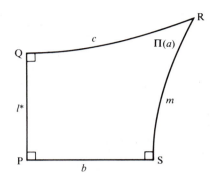

FIGURE 10.20

$B = X$ is the point between R and S such that $PX \cong QR$, Engel's theorem says $(\sphericalangle BAC)^r = \Pi(\overline{PQ}^*)$, and since $(\sphericalangle QPX)^r = \pi/2 - (\sphericalangle BAC)^r$, $(\sphericalangle QPX)^r = \Pi(\overline{PQ})$, i.e., $\overrightarrow{PX}$ is limiting parallel to $\overrightarrow{QR}$.

Engel's theorem also says that the ray emanating from R limiting parallel to $\overrightarrow{SP}$ makes an angle with $\overrightarrow{RS}$ that is congruent to $\sphericalangle ABC$; and that the ray emanating from X limiting parallel to $\overrightarrow{SP}$ makes an angle with $\overrightarrow{XS}$ that is congruent to the acute $\sphericalangle R$ of the Lambert quadrilateral.

Proof:
For the proof, start with a Lambert quadrilateral labeled as in

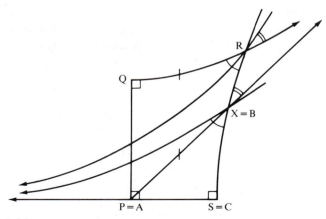

FIGURE 10.21

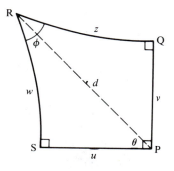

FIGURE 10.22

Figure 10.22. We've already shown that

(i) $\sinh w = \cosh z \sinh v$
(i′) $\sinh z = \cosh w \sinh u.$

Let $\theta = (\sphericalangle SPR)^r$, $d = \overline{PR}$. By Theorem 10.3, $\sinh w = \sin \theta$ $\sinh d = \cos (\pi/2 - \theta)$ $\sinh d = \tanh v \cosh d = \tanh v$ (cosh u cosh w), so that

(ii) $\tanh w = \tanh v \cosh u$

and by symmetry

(ii′) $\tanh z = \tanh u \cosh v.$

Let $\phi = (\sphericalangle R)^r$. By the law of sines and Theorem 10.3,

$$\frac{\sin \phi}{\sin \overline{QS}} = \frac{\sin (\sphericalangle QSR)^r}{\sinh z} = \frac{\cos (\sphericalangle PSQ)^r}{\sinh z} = \frac{\tanh u}{\tanh \overline{QS} \sinh z}$$

so by (i′) and Theorem 10.3 we have

(iii) $\sin \phi = \dfrac{\tanh u \cosh \overline{QS}}{\sinh z} = \dfrac{\tanh u (\cosh u \cosh v)}{\sinh u \cosh w} = \dfrac{\cosh v}{\cosh w}$

and by symmetry

(iii′) $\sin \phi = \dfrac{\cosh u}{\cosh z}.$

Now let X be the point between R and S such that $\overline{PX} = z$, and consider right triangle $\triangle PSX$ (Figure 10.21). By (i′), (ii′), and (iii′), respectively, we get (using Theorem 10.3)

$$\sin\ (\sphericalangle\text{PXS})^r = \frac{\sinh u}{\sinh z} = \text{sech}\ w$$

$$\cos\ (\sphericalangle\text{XPS})^r = \frac{\tanh u}{\tanh z} = \text{sech}\ v = \tanh v^*$$

$$\cosh \overline{\text{XS}} = \frac{\cosh z}{\cosh u} = \csc \phi$$

so that

$$(\sphericalangle\text{PXS})^r = \Pi(w)$$
$$(\sphericalangle\text{XPS})^r = \Pi(v^*)$$
$$\Pi(\overline{\text{XS}}) = \phi$$

by formulas (5), (6), and (5), respectively. Thus, if we relabeled P as A, X as B, and S as C, we would obtain right triangle $\triangle$ABC corresponding to our given Lambert quadrilateral as asserted.

Conversely, given right triangle $\triangle$PSX, we can recover $\square$PQRS by setting R equal to the unique point on $\overrightarrow{\text{SX}}$ such that $\Pi(\overline{\text{RS}}) = (\sphericalangle\text{PXS})^r$ and setting Q equal to the foot of the perpendicular from R to the line through P perpendicular to $\overleftrightarrow{\text{PS}}$. ■

The correspondence in Theorem 10.9 provides for a whole series of existence theorems. For example, it says that from the existence of a right triangle with parameters $(a, \Pi(m), c, \Pi(l), b)$ we can deduce the existence of a Lambert quadrilateral with parameters $(l^*, c, \Pi(a), m, b)$, as in Figure 10.20, ordering the parameters by a clockwise progression in the figure. Now read the parameters backwards! This gives

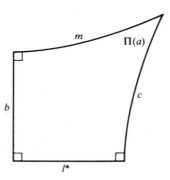

FIGURE 10.23

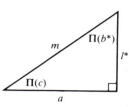

FIGURE 10.24

Figure 10.23, from which we deduce the existence of a second right triangle in Figure 10.24, having parameters $(a, \Pi(c), m, \Pi(b^*), l^*)$.

We can continue the process of reading these parameters backwards, obtaining a second Lambert quadrilateral, etc. We then end up with the existence of five Lambert quadrilaterals and four other right triangles which are implied by the existence of the first right triangle. The results are summarized in the following tabulation.

| $\triangle ABC$, $\angle C$ right | | | | | Lambert $\square SPQR$, $\angle R$ acute | | | | |
BC	$\angle B$	AB	$\angle A$	AC	PQ	QR	$\angle R$	RS	SP
a	$\Pi(m)$	c	$\Pi(l)$	b	l^*	c	$\Pi(a)$	m	b
a	$\Pi(c)$	m	$\Pi(b^*)$	l^*	c^*	m	$\Pi(l^*)$	b^*	a
l^*	$\Pi(m)$	b^*	$\Pi(a^*)$	c^*	m^*	b^*	$\Pi(c^*)$	a^*	l^*
c^*	$\Pi(b^*)$	a^*	$\Pi(l)$	m^*	b	a^*	$\Pi(m^*)$	l	c^*
m^*	$\Pi(a^*)$	l	$\Pi(c)$	b	a	l	$\Pi(b)$	c	m^*

Note also that since Theorem 10.3 gave us formulas showing how a right triangle is uniquely determined by any two of its five parameters, Theorem 10.9 gives us the same result for a Lambert quadrilateral (e.g., starting with u and v, w is given by (ii), z by (ii'), and ϕ by (iii) in the proof of Theorem 10.9).

COORDINATES IN THE HYPERBOLIC PLANE

Choose perpendicular lines through an origin O and fix coordinate systems on each of them so that they can be called the u axis and v axis. For any point P, let U and V be the perpendicular projections of P on

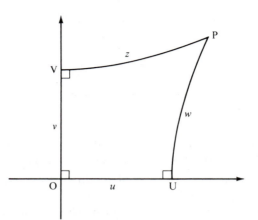

FIGURE 10.25

these axes, and let u and v be the respective coordinates of U and V. We then have a Lambert quadrilateral □UOVP. We label the remaining sides with coordinates w, z such that

$$(22) \qquad \tanh w = \tanh v \cosh u$$
$$\tanh z = \tanh u \cosh v$$

(see Figure 10.25). Formulas (ii) and (ii′) in the proof of Theorem 10.9 showed that if P is in the first quadrant (i.e., $u > 0$ and $v > 0$), then $w = \overline{PU}$ and $z = \overline{PV}$. We also set

$$(23) \qquad x = \tanh u, \qquad y = \tanh v$$
$$(24) \qquad T = \cosh u \cosh w, \qquad X = xT, \qquad Y = yT.$$

Then we call (u, v) the *axial coordinates*, (u, w) the *Lobachevsky coordinates*, (x, y) the *Beltrami coordinates*, and (T, X, Y) the *Weierstrass coordinates* of point P. The latter two are the most important coordinate systems, as is shown by the next long theorem.

THEOREM 10.10 (still assuming $k = 1$). Assigning to each point P its pair (x, y) of Beltrami coordinates gives an isomorphism of the hyperbolic plane onto the Beltrami-Klein model. In particular, $Ax + By + C = 0$ is an equation of a line in Beltrami coordinates if and only if $A^2 + B^2 > C^2$, and every line has such an equation. The distance $\overline{P_1P_2}$ between two points is given in terms of Beltrami coordinates by

(25)
$$\cosh \overline{P_1 P_2} = \frac{p_1 \cdot p_2}{\|p_1\| \|p_2\|},$$

where $p_i = (1, x_i, y_i)$, the inner product $p_1 \cdot p_2$ is defined by

$$p_1 \cdot p_2 = 1 - x_1 x_2 - y_1 y_2,$$

and $\|p_i\| = \sqrt{p_i \cdot p_i}$. Similarly, if $A_i x + B_i y + C_i = 0$ are the equations of two lines l_i, $i = 1, 2$, intersecting in a nonobtuse angle of radian measure θ, then

(26)
$$\cos \theta = \frac{l_1 \cdot l_2}{\|l_1\| \|l_2\|},$$

where now the inner product is defined by

$$l_1 \cdot l_2 = A_1 A_2 + B_1 B_2 - C_1 C_2$$

and $\|l_i\| = \sqrt{l_i \cdot l_i}$ (in particular, $0 = l_1 \cdot l_2$ is the necessary and sufficient condition for the lines to be perpendicular).

Assigning to each point P its triple (T, X, Y) of Weierstrass coordinates maps the hyperbolic plane onto the locus

$$T^2 - X^2 - Y^2 = 1, \qquad T \geq 1,$$

which is one of the two sheets of a hyperboloid in Cartesian three-space. The equation of a line in Weierstrass coordinates is linear homogeneous (i.e., of the form $AX + BY + CT = 0$).

Before giving the proof, note that the Weierstrass representation gives one interpretation of the hyperbolic plane as a "sphere of imaginary radius i." Namely, if we replace the usual positive definite quadratic form $X^2 + Y^2 + T^2$ (that measures distance squared from the origin) with the indefinite quadratic form $X^2 + Y^2 - T^2$, then the sphere of radius i with respect to this "distance" has equation

$$X^2 + Y^2 - T^2 = i^2 = -1,$$

which is the equation of a hyperboloid. This indefinite metric is the three-dimensional analogue of the metric determined by the form $X^2 + Y^2 + Z^2 - T^2$ in four-dimensional space-time, which is used for special relativity (see Taylor and Wheeler, 1992). Note that the "lines" in the Weierstrass model are intersections with the sheet of the hyperboloid of Euclidean planes through the origin. To picture this model, just imagine one branch of the hyperbola $T^2 - X^2 = 1$ in the (T, X) plane rotated around the T axis. See Figure 7.19.

Proof:

The proof of Theorem 10.10 is based on the trigonometry of Lambert quadrilaterals obtained in the preceding theorem.

As the graph in Figure 10.13 showed, $u \to \tanh u$ is a one-to-one mapping of the entire real line onto the open interval $(-1, 1)$. That the pairs (x, y) of Beltrami coordinates satisfy $x + y^2 < 1$ follows from the fact that the perpendiculars to the axes at U and V intersect if and only if $|u| < |v|^*$ (see Figure 10.18), i.e.,

$$\tanh^2 u < \tanh^2 |v|^* = \operatorname{sech}^2 v = 1 - \tanh^2 v$$

(using formula (20)).

To derive the distance formula, introduce *polar coordinates* (r, θ) for point P in Figure 10.25 defined by

$$r = \overline{OP}$$
$$\theta = \begin{cases} \measuredangle(\text{XOP})^r & \text{if } v \geq 0 \\ -\measuredangle(\text{XOP})^r & \text{if } v \leq 0. \end{cases}$$

The relations with axial coordinates are then

(27)
$$\begin{aligned} \tanh r \cos \theta &= \tanh u = x \\ \tanh r \sin \theta &= \tanh v = y \end{aligned}$$

by formula (10) for the cosine of an angle in a right triangle and the identity $\sin \theta = \cos (\pi/2 - \theta)$. Hence,

$$\tanh^2 r = \tanh^2 u + \tanh^2 v = x^2 + y^2.$$

From the identity $\operatorname{sech}^2 r = 1 - \tanh^2 r$, we get

$$\cosh r = (1 - x^2 - y^2)^{-1/2} = \|p\|^{-1}$$

if $p = (1, x, y)$, which is the distance formula when $P_1 = P$ and $P_2 = O$. For general P_1 and P_2, (27) gives

$$\cos (\theta_2 - \theta_1) = \cos \theta_1 \cos \theta_2 + \sin \theta_1 \sin \theta_2$$
$$= \frac{x_1 x_2 + y_1 y_1}{\tanh r_1 \tanh r_2}.$$

Suppose first O, P_1, P_2 are collinear, so that $\cosh \overline{P_1 P_2} = \cosh (r_1 \pm r_2)$. Since $\cos (\theta_2 - \theta_1) = \pm 1$,

$$\begin{aligned} \cosh \overline{P_1 P_2} &= \cosh r_1 \cosh r_2 - \sinh r_1 \sinh r_2 \cos (\theta_2 - \theta_1) \\ &= \cosh r_1 \cosh r_2 \left[1 - \tanh r_1 \tanh r_2 \cos (\theta_2 - \theta_1)\right]. \end{aligned}$$

But this formula also holds when O, P_1, P_2 are not collinear by the law of cosines (13). Substituting the two preceding formulas gives the desired formula (25).

To show that the mapping $P \rightarrow (x, y)$ sends hyperbolic lengths onto Klein lengths means reconciling formula (25) with the formula in Exercise K-14, Chapter 7. This follows from a calculation based on the formula

$$(28) \quad \tanh \overline{P_1 P_2} = \frac{[(x_1 - x_2)^2 + (y_1 - y_2)^2 - (x_1 y_2 - x_2 y_1)^2]^{1/2}}{p_1 \cdot p_2}$$

and the identity

$$(29) \qquad \operatorname{arctanh} t = \frac{1}{2} \ln \frac{1+t}{1-t}.$$

Formula (28) is obtained from (25) by means of the identity $\tanh^2 t = 1 - \cosh^{-2} t$. (The term in brackets on the right side of (28) could be written as $(p_1 \cdot p_2)^2 - \|p_1\|^2 \|p_2\|^2$. Incidentally, the $\frac{1}{2}$ occurring in formula (29) explains why the factor $\frac{1}{2}$ appeared in the formula for Klein length in Theorem 7.4, p. 268.)

Because $P \rightarrow (x, y)$ is an isometry, it is a collineation, so lines in the hyperbolic plane are mapped onto chords of the absolute in the Klein model, which have linear equations as described in the theorem.

The formula (26) for $\cos \theta$ is an assertion about angle measure in the Klein model, once we pass to that model by means of the isomorphism $P \rightarrow (x, y)$. Suppose the two lines meet at point P_0 with coordinates (x_0, y_0) and suppose we write the ith line as $\overleftrightarrow{P_0 P_i}$, where P_i has coordinates (x_i, y_i), $i = 1, 2$. Then the coefficients in the equation for the ith line are given by $A_i = y_1 - y_0$, $B_i = x_0 - x_i$, $C_i = x_i y_0 - y_i x_0$. Suppose $P_0 = O$, the center of the absolute. Then formula (26) reduces to

$$\cos \theta = \frac{x_1 x_2 + y_1 y_2}{(x_1^2 + y_1^2)^{1/2}(x_2^2 + y_2^2)^{1/2}},$$

which is the Euclidean formula for the cosine of the angle $\sphericalangle P_1 O P_2$. But the Klein model is conformal at the special point O, so we have verified (26) in this case.

If $P_0 \neq O$, let us find a hyperbolic motion T such that $T(O) = P_0$, and let $Q_i = T^{-1}(P_i)$. Since T preserves angle measure, all we then have to do is show that formula (26) is equal to the cosine of $\sphericalangle Q_1 O Q_2$. The natural candidate for T is the reflection across the perpendicular bisector of OP_0. We need two lemmas (which are generalized in Exercise 9).

LEMMA 10.1. The coordinates of the Klein midpoint M of O and P are

$$\left(\frac{x}{1+\|p\|}, \frac{y}{1+\|p\|}\right)$$

where $\|p\| = \sqrt{1 - x^2 - y^2}$ and P has coordinates (x, y).

Proof

Let $r = \overline{OP}$; we've seen that $\cosh r = \|p\|^{-1}$, $x = \tanh r \cos \theta$, $y = \tanh r \sin \theta$. The coordinates (x', y') of M are then given by $x' = \tanh (r/2) \cos \theta$, $y' = \tanh (r/2) \sin \theta$; i.e., $x' = x \tanh (r/2)/\tanh r$, $y' = y \tanh (r/2)/\tanh r$. But

$$\frac{\tanh (r/2)}{\tanh r} = \frac{\sinh r}{\cosh r + 1} \cdot \frac{\cosh r}{\sinh r}$$

$$= \left(1 + \frac{1}{\cosh r}\right)^{-1} = (1 + \|p\|)^{-1}. \blacksquare$$

LEMMA 10.2. The perpendicular bisector of OP_0 has equation $x_0 x + y_0 y + \|p_0\| - 1 = 0$, where $\|p_0\| = \sqrt{1 - (x_0^2 + y_0^2)}$ and P_0 has coordinates (x_0, y_0).

Proof:

The perpendicular bisector of OP_0 passes through the midpoint and has slope $-x_0/y_0$ (since Klein perpendicularity is the same as Euclidean perpendicularity when one chord is a diameter of the absolute). $\blacksquare$

If we now apply the general formula for reflection in the Klein model that you checked in Exercise K-16, Chapter 7, then Lemma 10.2 implies that reflection across the perpendicular bisector of OP_0 is given by

$$x' = \frac{x[\|p_0\|^2 - \|p_0\|] - x_0(x_0 x + y_0 y + \|p_0\| - 1)}{\|p_0\|^2 - \|p_0\| + [\|p_0\| - 1](x_0 x + y_0 y + \|p_0\| - 1)}$$

$$y' = \frac{y[\|p_0\|^2 - \|p_0\|] - y_0(x_0 x + y_0 y + \|p_0\| - 1)}{\|p_0\|^2 - \|p_0\| + [\|p_0\| - 1](x_0 x + y_0 y + \|p_0\| - 1)}.$$

Using these formulas, another long calculation shows that formula (26) is equal to the cosine of $\sphericalangle Q_1 O Q_2$.

As a check on the formula, note that $\cos\theta = 0$ if and only if $A_1 A_2 + B_1 B_2 + C_1(-C_2) = 0$, which equation says line l_1 passes through the pole $(A_2, B_2, -C_2)$ of line l_2.

We leave the assertions about the Weierstrass coordinates as an exercise. ∎

THE CIRCUMSCRIBED CYCLE OF A TRIANGLE

You learned in Exercise 9, Chapter 5, that the existence of a circumscribed circle for every triangle is equivalent to the Euclidean parallel postulate. The circumscribed circle exists if and only if the perpendicular bisectors of the sides are concurrent in an ordinary point (Exercise 12, Chapter 6). In Exercise 13, Chapter 6, and Major Exercise 7, Chapter 6, you showed that the perpendicular bisectors are always "concurrent" in an ideal or ultra-ideal point if the circumscribed circle does not exist.

In the ultra-ideal case, you showed (see Figure 6.26) that the vertices A, B, C of the given triangle are all equidistant from the common perpendicular t to the perpendicular bisectors. This implies that they lie on an equidistant curve having t as an axis. According to our definition of "equidistant curve," it is required that A, B, C all lie on the same side of t.

Some authors (e.g., Coxeter, Sommerville) define "equidistant curve" differently; i.e., they define it to be the locus of all points at the same distance from an axis t, no matter which side of t. These authors would designate our "equidistant curve" one of the two "branches" of theirs. Let us call the equidistant curve of Coxeter and Sommerville a *doubly equidistant curve*, indicating the union of two equidistant curves having the same axis, each being the reflection of the other across the axis. In Exercise 11(a), Chapter 6, you showed that every triangle is circumscribed by three doubly equidistant curves whose axes are the lines that join pairs of midpoints of the sides (Figure 6.24).

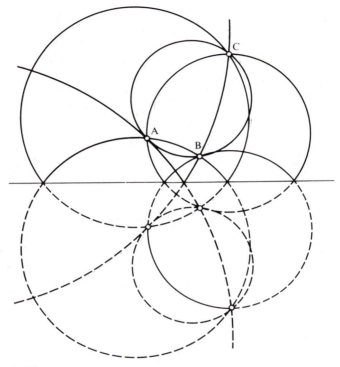

FIGURE 10.26

Refer to the Poincaré upper half-plane model: the Euclidean circle through A, B, and C is a hyperbolic circle if it lies entirely in the upper half-plane (compare Exercise P-5, Chapter 7, and Exercise 48, Chapter 9); it is a horocycle with ideal center Ω if it is tangent to the x axis at Ω (Exercise 46, Chapter 9), and its arc in the upper half-plane is an equidistant curve otherwise (Exercise 47, Chapter 9).

Figure 10.26 shows the three doubly equidistant curves and a hyperbolic circle circumscribing $\triangle ABC$ in this model.

The next theorem gives trigonometric criteria to decide which type of cycle circumscribes $\triangle ABC$.

THEOREM 10.11. With standard notation for $\triangle ABC$, let a be the length of a longest side, so that $\sphericalangle A$ is a largest angle. The cycle

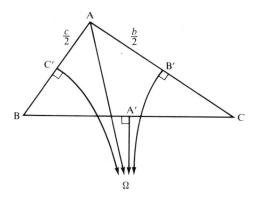

FIGURE 10.27

circumscribing $\triangle ABC$ is a

$$
\left.\begin{array}{l}
\text{Circle} \\
\text{Horocycle} \\
\text{Equidistant curve}
\end{array}\right\} \Leftrightarrow \sinh\frac{a}{2}\left\{\begin{array}{c}< \\ = \\ >\end{array}\right\}\sinh\frac{b}{2}+\sinh\frac{c}{2}
$$

$$
\Leftrightarrow (\sphericalangle A)^r\left\{\begin{array}{c}< \\ = \\ >\end{array}\right\}\Pi\left(\frac{b}{2}\right)+\Pi\left(\frac{c}{2}\right).
$$

Proof:
Consider first the case where the perpendicular bisectors are asymptotically parallel through an ideal point Ω. According to Lemma 6.3, p. 215, Figure 10.27 holds, where A', B', C' are the midpoints. This shows that $(\sphericalangle A)^r = (\sphericalangle C'A\Omega)^r + (\sphericalangle B'A\Omega)^r = \Pi(c/2) + \Pi(b/2)$.

In case the perpendicular bisectors have a common perpendicular t, Figure 10.28 holds.

Since $\sphericalangle C'A\Omega > \sphericalangle C'A\Lambda$ and $\sphericalangle B'A\Omega > \sphericalangle B'A\Sigma$, we see that

$$(\sphericalangle A)^r > (\sphericalangle C'A\Lambda)^r + (\sphericalangle B'A\Sigma)^r = \Pi(c/2) + \Pi(b/2).$$

In case the perpendicular bisectors meet, we must have

$$(\sphericalangle A)^r < \Pi(c/2) + \Pi(b/2),$$

since this is the only other possibility. Thus, the second criterion is established.

The derivation of the first criterion in terms of hyperbolic sines from the second criterion involves a calculation using identities and our earlier formulas. First, by the hyperbolic law of cosines (13),

$$\cos A = \frac{\cosh b \cosh c - \cosh a}{\sinh b \sinh c}$$

$$= \frac{\left(2 \sinh^2 \dfrac{b}{2} + 1\right)\left(2 \sinh^2 \dfrac{c}{2} + 1\right) - \left(2 \sinh^2 \dfrac{a}{2} + 1\right)}{4 \sinh \dfrac{b}{2} \cosh \dfrac{b}{2} \sinh \dfrac{c}{2} \cosh \dfrac{c}{2}}$$

$$= \frac{2 \sinh^2 \dfrac{b}{2} \sinh^2 \dfrac{c}{2} + \sinh^2 \dfrac{b}{2} + \sinh^2 \dfrac{c}{2} - \sinh^2 \dfrac{a}{2}}{2 \sinh \dfrac{b}{2} \sinh \dfrac{c}{2} \cosh \dfrac{b}{2} \cosh \dfrac{c}{2}}.$$

Second, by the identity for $\cos (x + y)$ and formulas (5) and (6),

$$\cos \left[\Pi\left(\frac{b}{2}\right) + \Pi\left(\frac{c}{2}\right)\right]$$

$$= \cos \Pi\left(\frac{b}{2}\right) \cos \Pi\left(\frac{c}{2}\right) - \sin \Pi\left(\frac{b}{2}\right) \sin \Pi\left(\frac{c}{2}\right)$$

$$= \tanh \frac{b}{2} \tanh \frac{c}{2} - \frac{1}{\cosh \dfrac{b}{2} \cosh \dfrac{c}{2}}$$

$$= \frac{\sinh \dfrac{b}{2} \sinh \dfrac{c}{2} - 1}{\cosh \dfrac{b}{2} \cosh \dfrac{c}{2}}.$$

The first criterion then follows from these equations after some algebra. ∎

COROLLARY. An isosceles triangle whose base is not longer than its sides (in particular, an equilateral triangle) has a circumscribed circle.

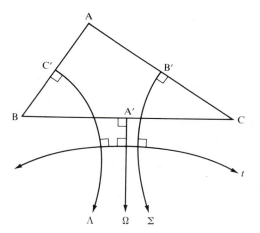

FIGURE 10.28

If the base is longer than the sides, then the circumscribed cycle is a

$$
\left.\begin{array}{l}
\text{Circle} \\
\text{Horocycle} \\
\text{Equidistant curve}
\end{array}\right\} \Leftrightarrow \cosh a \left\{\begin{array}{l} < \\ = \\ > \end{array}\right\} 4 \cosh b - 3
$$

where a is the length of the base, and b the length of a side.

We leave the proof for Exercise 10.

Our final theorem gives a lovely formula relating the radius of the circumscribed circle to the area (which equals the defect) of a triangle.

THEOREM 10.12. If $\triangle ABC$ has a circumscribed circle of radius R, then with standard notation, the area K of $\triangle ABC$ is given by

(30)
$$
\sin \frac{K}{2} = \frac{\tanh \dfrac{a}{2} \tanh \dfrac{b}{2} \tanh \dfrac{c}{2}}{\tanh R}
$$

Note. If we only look at the leading terms in the power series expansion of sin and tanh (i.e., we only look at an "infinitesimal"

hyperbolic triangle), this formula reduces to the Euclidean formula

$$K = \frac{abc}{4R}.$$

In Euclidean geometry we could replace K by $\frac{1}{2}bc \sin A$ and solve for R; in hyperbolic geometry, Exercise 28 provides a formula for R purely in terms of the sides of the triangle.

Here is a proof of the Euclidean formula. Choose B to be a vertex such that the diameter BD of the circumscribed circle κ intersects side AC. Then $\angle D$ of $\triangle BDC$ and $\angle A$ subtend the same arc BC of κ, so $\sin A = \sin D = a/2R$ (since $\angle BCD$ is right, being inscribed in a semicircle). Substitute for $\sin A$ in $K = \frac{1}{2}bc \sin A$ to get the formula.

The proof of Theorem 10.12 will be indicated in Exercises 20 – 28.

REVIEW EXERCISE

All statements in these exercises refer to hyperbolic geometry (unless explicit mention of other geometries is made.) Which of the following statements are correct?

(1) The area of a triangle is proportional to its defect.
(2) The angle of parallelism $\Pi(x)$ in radians relates the circular and hyperbolic functions by means of an equation such as $\tanh x = \cos \Pi(x)$.
(3) In all right triangles having a fixed number of radians for $\angle A$ (and standard notation, right angle at C), the ratio a/c is the same, and is called the sine of $\angle A$.
(4) J. Bolyai discovered a formulation of the law of sines that is valid in neutral geometry.
(5) The segment length x^* complementary to x is uniquely determined by the formula $\Pi(x^*) = \pi/2 - \Pi(x)$.
(6) The equations relating Beltrami coordinates to Lobachevsky coordinates are $x = \tanh u$ and $y = \tanh v$.
(7) With standard notation, if a is a largest side of $\triangle ABC$, then the cycle circumscribing $\triangle ABC$ is a circle if and only if $\sinh (a/2) > \sinh (b/2) + \sinh (c/2)$.

(8) The representation by Weierstrass coordinates helps make sense of Lambert's description of the hyperbolic plane as a "sphere of imaginary radius i."

(9) The curvature of the hyperbolic plane is $1/k^2$, where k^2 times the defect in radians equals the area of a triangle.

(10) The analogue of the Pythagorean theorem is the formula $\cosh c = \cosh a \cosh b$ (for a right triangle with right angle at C, standard notation, $k = 1$).

EXERCISES

1. Verify all the identities for hyperbolic functions listed in the table on p. 401.

2. Verify formulas (5), (6), (7) in which Π provides a link between hyperbolic and circular functions. Graph the function $\Pi(x)$.

3. The proof of Theorem 10.3 required a complicated argument using the Poincaré model. Give a different proof using the Klein model. (Hint: According to the note at the end of Chapter 7, you can assume A = O, the center of the absolute. Show that $\cos A = \overline{AC}/\overline{AB}$ and $\sin A = \overline{BC}/\overline{AB}$ (Euclidean lengths), $\overline{AB} = \tanh c$ (where $c = d'(AB)$ is the Klein length), $\overline{AC} = \tanh b$, and $\overline{BC} = (1 - \overline{AC}^2)^{1/2} \tanh a = \tanh a/\cosh b$ (use Theorem 7.4). Conclude by deducing the formula $\cosh c = \cosh a \cosh b$ from the Pythagorean theorem. See Figure 10.29.)

4. Prove Theorem 10.4.

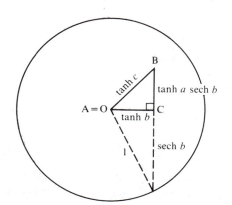

FIGURE 10.29

5. Verify step 2 in the proof of Theorem 10.6.
6. Verify formulas (18) through (21) for complementary lengths. Graph the function

$$f(x) = x^* = \ln \frac{e^x + 1}{e^x - 1}.$$

7. Prove the assertions about Weierstrass coordinates in Theorem 10.10. (Hint: Derive the equation of a line in Weierstrass coordinates from the equation of a line in Beltrami coordinates.)
8. Verify formulas (28) and (29) in the proof of Theorem 10.10 and use them to reconcile with the distance formula in Exercise K-14, Chapter 7.
9. Generalize Lemmas 10.1 and 10.2 by showing that if (x_1, y_1) and (x_2, y_2) are distinct points in the Klein model, then the midpoint and perpendicular bisector of the segment they determine are given respectively by

$$\left(\frac{x_1 s_2 + x_2 s_1}{s_1 + s_2}, \frac{y_1 s_2 + y_2 s_1}{s_1 + s_2} \right)$$

$$(x_1 s_2 - x_2 s_1)x + (y_1 s_2 - y_2 s_1)y + (s_1 - s_2) = 0$$

where $s_i = \sqrt{1 - x_i^2 - y_i^2}$, $i = 1, 2$. (Hint: Use Lemma 10.2 to find the point Q in the Cartesian plane where the perpendicular bisectors of OP_1 and OP_2 meet. Then joining Q to the pole of $\overleftrightarrow{P_1 P_2}$ gives the perpendicular bisector of $P_1 P_2$, and intersecting it with $P_1 P_2$ gives the midpoint.)
10. Prove the corollary to Theorem 10.11.
11. In a right triangle with right angle at C, prove that the circumscribed cycle is a

$$\left.\begin{array}{l} \text{Circle} \\ \text{Horocycle} \\ \text{Equidistant curve} \end{array}\right\} \Leftrightarrow \frac{a}{2} \left\{\begin{array}{l} < \\ = \\ > \end{array}\right\} \left(\frac{b}{2}\right)^*.$$

(Hint: Apply the second criterion of Theorem 10.11 with ∢C the largest angle, using the fact that Π is a decreasing function. Or else argue directly from the definition of complementary lengths.)
12. Verify A. P. Kotelnikov's rule for remembering the relations among the parts of a right triangle with the right angle at C, standard notation (x^* denoting the complementary length to x): in Figure 10.30, *the sine of each angle is equal to the product of the tangents of the two adjacent angles and is equal to the product of the cosines of the two opposite angles.* For example,

$$\sin A = \tan \bar{\Pi}(a^*) \tan \Pi(c) = \cos \Pi(b^*) \cos B.$$

(This rule is the hyperbolic analogue of the rule John Napier published in 1614 for the trigonometry of a right triangle on a unit sphere in

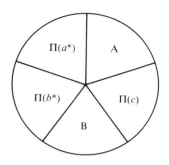

FIGURE 10.30

Euclidean space. For Napier's rule, use a, b, A', c', B' in cyclic order, where A' denotes the complementary angle to ∢A and $c' = \pi/2 - c$. J. Bolyai and Lobachevsky discovered that spherical trigonometry in hyperbolic space is the same as in Euclidean space.)

13. In Euclidean geometry, every circle can be inscribed in a triangle (the tangents at three appropriate points on the circle meet to form a triangle). Show that in hyperbolic geometry (with $k = 1$), an inscribed circle of a triangle must have diameter less than ln 3. (Hint: In Figure 10.32, show that $(∢AIB')^r + (∢BIC')^r + (∢CIA')^r = \pi$, and that each of these three angles is less than $\Pi(r)$; then apply the Bolyai-Lobachevsky formula to find x such that $\Pi(x) = \pi/3$.)

14. Show that (with $k = 1$) a trebly asymptotic triangle has an inscribed circle of diameter ln 3 (see Exercise K-13, Chapter 7). Show that the area of this circle is

$$2\pi \left(\frac{2}{\sqrt{3}} - 1 \right)$$

and its circumference is $2\pi/\sqrt{3}$ (use Theorems 10.5 and 10.7).

15. Show that for any three positive numbers α, β, γ such that $\pi < \alpha + \beta + \gamma$, there exists a triangle having these numbers as the radian measures of its angles. (Hint: Use Theorem 10.4.)

16. In a singly asymptotic triangle ABΩ, if $c = \overline{AB}$, then

$$\cosh c = \frac{\cos A \cos B + 1}{\sin A \sin B}.$$

(Hint: Let C approach Ω in formula (15), Theorem 10.4. For a proof without using continuity, note that when ∢A and ∢B are acute,

$$c = \Pi^{-1}(\alpha) + \Pi^{-1}(\beta),$$

where $\alpha = (∢A)^r$, $\beta = (∢B)^r$—see Figure 10.31.) Show that the gen-

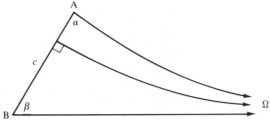

FIGURE 10.31

eralization of the Bolyai-Lobachevsky formula to the case when $\beta < \pi/2$ is

$$\tan \frac{\alpha}{2} = e^{-c} \cot \frac{\beta}{2}.$$

17. Write down equations which show how the side and the angle of an equilateral triangle determine each other.

18. (a) In a right triangle with standard notation and right angle at C, show that $\tan A = \tanh a / \sinh b$.

   (b) Deduce that in an isosceles triangle with base b and side a, summit at B, and one base angle at A,

$$\tanh a \cos \frac{B}{2} = \tan A \sinh \frac{b}{2}$$

$$\sin A \cosh \frac{b}{2} = \cos \frac{B}{2}.$$

(Hint: Drop the altitude to the base.)

19. In a right triangle $\triangle ABC$ with right angle at C (and standard notation), show that

$$\sin K = \frac{\sinh a \sinh b}{1 + \cosh a \cosh b},$$

where $K =$ the area $=$ the defect of $\triangle ABC$. (Hint: Use Theorem 10.3 and trigonometric identities.)

20. Given $\triangle ABC$, if h is the length of the altitude from vertex B, show that (in standard notation) the product $\sinh b \sinh h$ is independent of the choice of which vertex is labeled B; this is the hyperbolic analogue of the Euclidean theorem that bh is constant. (Hint: Show that $\sinh b \sinh h = S \sinh a \sinh b \sinh c$, where S is the constant ratio occurring in the law of sines.) The next exercises will shed light on the

geometric significance of the constant $\frac{1}{2}$ sinh b sinh h, which we will denote by H (for Heron); by the hint, $2H = $ sin C sinh a sinh $b = $ sin B sinh c sinh $a = $ sin A sinh b sinh c.

In Exercises 21 – 28, s will denote the *semiperimeter* $\frac{1}{2}(a + b + c)$ of $\triangle ABC$.

21. Show that (in standard notation)

$$\sin \frac{A}{2} = \sqrt{\frac{\sinh\,(s - b)\,\sinh\,(s - c)}{\sinh b\,\sinh c}}$$

$$\cos \frac{A}{2} = \sqrt{\frac{\sinh s\,\sinh\,(s - a)}{\sinh b\,\sinh c}}.$$

(Hint: Square both sides; use identities and Theorem 10.4.)

22. Show that $H = \sqrt{\sinh s\,\sinh\,(s - a)\,\sinh\,(s - b)\,\sinh\,(s - c)}$. (Hint: Use the identity sin A $= 2$ sin (A/2) cos (A/2).

23. "Infinitesimally," the Heron is equal to $\sqrt{s(s - a)(s - b)(s - c)}$. Show that in Euclidean geometry, this quantity is equal to the area of $\triangle ABC$. (Hint: See Coxeter, 1969, p. 12.)

24. Suppose the inscribed circle of $\triangle ABC$ has radius r and touches BC at A', CA at B', and AB at C'. Show that in neutral geometry, $\overline{AB'} = s - a = \overline{AC'}$, $\overline{BC'} = s - b = \overline{BA'}$, $\overline{CA'} = s - c = \overline{CB'}$ (see Figure 10.32). (Hint: Review the construction of the inscribed circle in Exercise 16, Chapter 6.)

25. Deduce from Exercise 24 that in hyperbolic geometry

$$\tanh r \sinh s = H$$

whereas in Euclidean geometry $rs = $ the area of $\triangle ABC$. (Hint: In hyperbolic geometry, use Exercises 18, 21, 22, and 24 to compute tan (A/2) sinh $(s - a)$; in Euclidean geometry, add up the areas of triangles IAB, IBC, and ICA in Figure 10.32.)

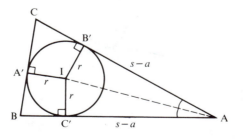

FIGURE 10.32

26. Prove Gauss' equations:

$$\cosh \tfrac{1}{2} c \sin \tfrac{1}{2} (A + B) = \cosh \tfrac{1}{2} (a - b) \cos \tfrac{1}{2} C$$
$$\cosh \tfrac{1}{2} c \cos \tfrac{1}{2} (A + B) = \cosh \tfrac{1}{2} (a + b) \sin \tfrac{1}{2} C$$
$$\sinh \tfrac{1}{2} c \sin \tfrac{1}{2} (A - B) = \sinh \tfrac{1}{2} (a - b) \cos \tfrac{1}{2} C$$
$$\sinh \tfrac{1}{2} c \cos \tfrac{1}{2} (A - B) = \sinh \tfrac{1}{2} (a + b) \sin \tfrac{1}{2} C.$$

(Hint: Use identities such as

$$\sinh x + \sinh y = 2 \sinh \tfrac{1}{2} (x + y) \cosh \tfrac{1}{2} (x - y)$$

and analogous identities for the circular functions, then apply the half-angle formulas of Exercise 21.)

27. Show that a hyperbolic analogue of Heron's Euclidean area formula in Exercise 23 is the formula

$$\sin \frac{K}{2} = \frac{H}{2 \cosh \dfrac{a}{2} \cosh \dfrac{b}{2} \cosh \dfrac{c}{2}},$$

where K = area = defect of $\triangle ABC$. (Hint: Use Gauss' equations, the identity $\sin K/2 = \cos \tfrac{1}{2}(A + B + C)$, other trigonometric identities, and the formula $H = \tfrac{1}{2} \sin C \sinh a \sinh b$.)

28. If $\triangle ABC$ has a circumscribed circle of radius R, show that

$$\tanh R = \frac{2 \sinh \dfrac{a}{2} \sinh \dfrac{b}{2} \sinh \dfrac{c}{2}}{H},$$

which by Exercise 27 is equivalent to formula (30) of Theorem 10.12. (Hint: In Figure 10.33 $\sin A = \sin (\beta' + \gamma')$, $H = \tfrac{1}{2} \sin A \sinh b \sinh c$;

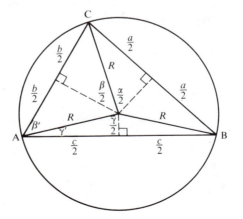

FIGURE 10.33

use Theorem 10.3 to determine $\cos \gamma'$ and $\cos \beta'$, and use Exercise 18 to determine $\sin \gamma'$ and $\sin \beta'$, obtaining with the help of identities the formula

$$H \tanh R = \left[\cos \frac{\gamma}{2} \sinh \frac{b}{2} + \cos \frac{\beta}{2} \sinh \frac{c}{2} \right] 2 \sinh \frac{b}{2} \sinh \frac{c}{2}.$$

Show finally that the term in brackets equals $\sinh a/2$ by using Theorem 10.3 to derive expressions for $\sinh a/2$, $\sinh b/2$, and $\sinh c/2$ and plugging in $\sin (\gamma + \beta)/2 = \sin \alpha/2$.)

29. Let $\Omega\Sigma\Lambda$ be a trebly asymptotic triangle and $\triangle ABC$ its *pedal triangle* formed by the feet of the perpendiculars from each ideal vertex to the opposite side. Since all trebly asymptotic triangles are congruent to one another by Proposition 9.16(c), $\triangle ABC$ is equilateral. Show that in the Poincaré upper half-plane model, the radian measure θ of an angle of $\triangle ABC$ is given by $\tan \frac{1}{2}\theta = \frac{1}{2}$ or $\tan \theta = \frac{4}{3}$, $\sin \theta = \frac{4}{5}$, $\cos \theta = \frac{3}{5}$, and the length c of a side is given by $\cosh c = \frac{3}{2}$. Deduce that *a circle whose radius is a side of $\triangle ABC$ has area equal to π.* Show further that the Heron H, the circumradius R, and the inradius r of $\triangle ABC$ are given by

$$H = \tfrac{1}{2}$$
$$\tanh R = \tfrac{1}{2}$$
$$\tanh r = \tfrac{1}{4}.$$

(Hint: There are many ways to obtain these results using the previous exercises and the models. In the Poincaré upper half-plane model, taking $\Omega = -1$, $\Sigma = \infty$, $\Lambda = 1$ gives $A = i$, $B = 1 + 2i$, $C = -1 + 2i$. Show that $\overleftrightarrow{BC}$ is the upper semicircle of $x^2 + y^2 = 5$, $\overleftrightarrow{AB}$ the upper semicircle of $(x - 2)^2 + y^2 = 5$, and that the tangents to these circles at B have slopes $-\frac{1}{2}$, $\frac{1}{2}$, respectively.

This and the double angle formulas yield the assertions about θ. Exercises 18(a), 20, 25, and 28 can then be applied. Or use the Klein model, choosing $\Omega\Sigma\Lambda$ so that the origin is the incenter and circumcenter of $\triangle ABC$.)

Exercises 30–33 develop some more trigonometric formulas that William Thurston uses in his application of hyperbolic geometry to topology; see Casson and Bleiler (1988).

30. Given a quadrilateral with two adjacent right angles, labeled as in Figure 10.34. Prove that

$$\cosh d = \cosh a \cosh b \cosh c - \sinh a \sinh c.$$

(Hint: Imitate the proof of Theorem 10.8.)

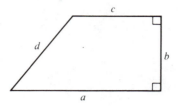

FIGURE 10.34

31. Given a pentagon with at least four right angles, labeled as in Figure 10.35. Prove that

$$\sinh e \cosh a = \sinh c \cosh b.$$

(Hint: Drop altitude h from the fifth vertex to the opposite side and apply the corollary to Theorem 10.8 to the two Lambert quadrilaterals obtained.)

32. Given a pentagon with five right angles (in Exercise K-5, Chapter 7, you showed that such pentagons exist), labeled as in Figure 10.36. Prove that

$$\cosh d = \sinh a \sinh b = \coth c \coth e.$$

(Hint: Here is one method; perhaps you can find one simpler: first apply Exercise 31 five times to all the combinations of four consecutive sides; then by considering the broken lines in Figure 10.36 and Exercise 30 and Theorem 10.3, deduce the equations

$$\cosh a \cosh b \left(\frac{\cosh c}{\cosh e}\right) - \sinh a \left(\frac{\sinh c}{\cosh e}\right)$$
$$= \cosh d = \cosh a \cosh b \left(\frac{\cosh e}{\cosh c}\right) - \sinh b \left(\frac{\sinh e}{\cosh c}\right).$$

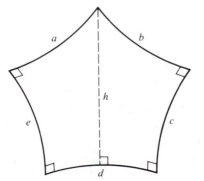

FIGURE 10.35

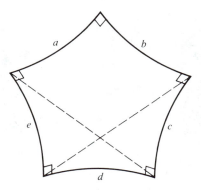

FIGURE 10.36

Substitute for the terms in parentheses from four of the earlier equations and do some algebra using the identity $\cosh^2 x = 1 + \sinh^2 x$ to get the result.)

33. In an all right-angled hexagon (again review Exercise K-5, Chapter 7) labeled as in Figure 10.37, prove the following remarkable analogues of the laws of sines and cosines given in Theorem 10.4 for a triangle:

$$\frac{\sinh a}{\sinh \alpha} = \frac{\sinh b}{\sinh \beta} = \frac{\sinh c}{\sinh \gamma}$$

$$\cosh a = \frac{\cosh \beta \cosh \gamma + \cosh \alpha}{\sinh \beta \sinh \gamma}.$$

(Hint: First show that the common perpendicular segment of length h falls inside the hexagon as shown; then apply Exercise 32 to the two all right-angled pentagons to get two expressions for $\cosh h$ and obtain the "law of sines"; for the "law of cosines," apply Exercise 32 to get expressions for $\cosh a$, $\cosh \beta$, $\cosh \alpha$, and $\cosh h$, use Exercise 31 to eliminate $\sinh c_1$, and use the identity for $\sinh \gamma_2 = \sinh (\gamma - \gamma_1)$.)

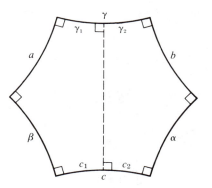

FIGURE 10.37

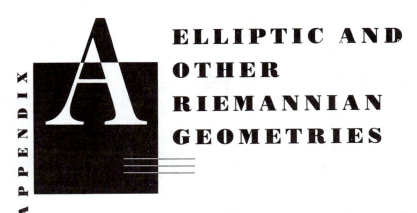

The dissertation submitted by Herr Riemann offers convincing evidence . . . of a creative, active, truly mathematical mind, and of a gloriously fertile imagination.

C. F. GAUSS

ELLIPTIC GEOMETRY

In Euclidean geometry there is exactly one parallel to a line *l* through a point P not on *l*; in hyperbolic geometry there is more than one parallel. A third geometry could be studied, one in which there is *no* parallel to *l* through P, i.e., a geometry in which parallel lines do not exist.

However, if we simply add the latter as a new parallel axiom to replace the other parallel axioms, the system we get is inconsistent. In Corollary 2 to Theorem 4.1 we proved that parallel lines do exist in neutral geometry, so that we would get a contradiction by adding such a parallel axiom.

To avoid this, we have to modify some of our other axioms. We can see what modifications need to be made by thinking of the surface of a sphere and interpreting "line" as "great circle." Then, indeed, there are no parallel lines. But other things change as well. It is impossible to talk about one point B being "between" two other points A and C on a circle. So all the axioms of betweenness have to be scrapped. They are replaced instead by seven axioms of *separation*. In Figure A.1, A and C

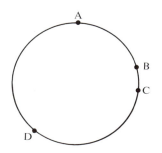

FIGURE A.1

separate B and D on the circle, since you can't get from B to D without crossing either A or C.

Let us designate the undefined relation "A and C separate B and D" by the symbol (A, C | B, D). The separation axioms are then:

SEPARATION AXIOM 1. If (A, B | C, D), then points A, B, C, and D are collinear and distinct.

SEPARATION AXIOM 2. If (A, B | C, D), then (C, D | A, B) and (B, A | C, D).

SEPARATION AXIOM 3. If (A, B | C, D), then not (A, C | B, D).

SEPARATION AXIOM 4. If points A, B, C, and D are collinear and distinct, then (A, B | C, D) or (A, C | B, D) or (A, D | B, C).

SEPARATION AXIOM 5. If points A, B, and C are collinear and distinct, then there exists a point D such that (A, B | C, D).

SEPARATION AXIOM 6. For any five distinct collinear points A, B, C, D, and E, if (A, B | D, E), then either (A, B | C, D) or (A, B | C, E).

To state the last axiom, we recall the notion of a *perspectivity* from one line onto another (from Chapter 7). Let *l* and *m* be any two lines and O a point not on either of them. For each point A on *l* the line $\overleftrightarrow{OA}$ intersects *m* in a unique point A' (Figure A.2; remember the elliptic parallel property); the one-to-one correspondence that assigns A' to A for each A on *l* is called the *perspectivity* from *l* to *m* with center O.

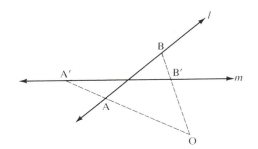

FIGURE A.2

SEPARATION AXIOM 7. Perspectivities preserve separation; i.e., if $(A, B | C, D)$, with *l* the line through A, B, C, and D, and if A′, B′, C′, and D′ are the corresponding points on line *m* under a perspectivity, then $(A', B' | C', D')$.

Without the notion of betweenness we have to carefully reformulate all the geometry using that relation. For example, the *segment* AB consists of the points A and B and all points between them. Yet this doesn't make sense on a circle. We can only talk about the segment ABC determined by *three* collinear points: it consists of the points A, B, and C and all the points not separated from B by A and C.

Similarly, we have to redefine the notion of a triangle, since its sides are no longer determined by the three vertices (see Figure A.3).

Once these notions have been redefined, the axioms of congruence and continuity all make sense when rephrased, and can be left intact.

There is still a difficulty with Incidence Axiom 1, which asserts that two points do not lie on more than one line. This is false for great circles on the sphere, since antipodal points (such as the poles) lie on infinitely many lines.

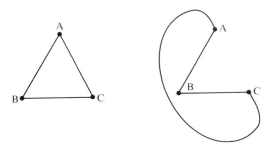

FIGURE A.3 Two different "triangles" with the same vertices.

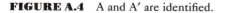

FIGURE A.4 A and A′ are identified.

Klein saw that the way to remedy this is to *identify* antipodal points; i.e., just as we interpret "line" to mean "great circle" in this model, we interpret "point" to mean "pair of antipodal points" (Figure A.4). This means that *in our imagination* we have pasted together two antipodal points so that they coalesce into a single point. It can be proved, as you might guess, that such pasting cannot actually be carried out in Euclidean three-dimensional space. But we can still identify antipodal points in our minds — every time we move from one to the other we think of ourselves as being back at the original point.

In making these identifications we discover another surprising property: a line no longer divides the plane into two sides, for you can "jump across" a great circle by passing from a given point to its now equal antipodal point that used to be on the other side. If we cut out a strip from this plane, it will look like a Möbius strip, which has only one side (see Figure A.5). The technical name for this property of "onesidedness" in *nonorientability*.

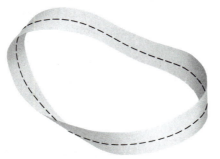

FIGURE A.5 Möbius strip.

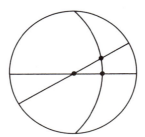

FIGURE A.6

To sum up, the axioms of plane elliptic geometry consist of the same incidence, congruence, and continuity axioms as neutral geometry (with the new definitions of segment, triangle, etc.). The betweenness axioms are replaced by separation axioms, and the parallel postulate is replaced by an axiom stating that no two lines are parallel. The model, which shows that elliptic geometry is just as consistent as Euclidean geometry, consists of the great circles on the sphere with antipodal points identified.[1]

As you might expect from this model, it is a theorem in elliptic geometry that *lines have finite length*. Moreover, all the lines perpendicular to a line *l* are not parallel to each other but are concurrent, i.e., all the perpendiculars to *l* have a point in common called *the pole of l*. In the model, for instance, the pole of the equator is the north (or, what is the same, the south) pole.

Another model for plane elliptic geometry (due to Klein) is *conformal* — like the Poincaré model for hyperbolic geometry, angles are accurately represented by Euclidean angles. In this model "points" are the Euclidean points inside the unit circle in the Euclidean plane as well as pairs of antipodal points on the circle; "lines" are either diameters of the unit circle or arcs of Euclidean circles that meet the unit circle at the ends of a diameter (see Coxeter, 1968, Section 14.6). This representation shows clearly that the angle sum of a triangle is greater than 180° in elliptic geometry (see Figure A.6).

Elliptic geometry becomes even more interesting when you pass from two to three dimensions. In three dimensions orientability is restored and a new kind of parallelism occurs. Two lines are called

---

[1] The geometry of the sphere itself is sometimes misleadingly called "double elliptic geometry."

Clifford-parallel if they are equidistant from each other; the lines are joined to each other by a continuous family of common perpendicular segments of the same length. Such lines cannot lie in a plane (in an elliptic plane two lines must intersect), so they are skew lines. Moreover, in general, in elliptic space for any point P not on a line *l* there exist exactly two lines through P that are Clifford-parallel to *l*, called the *right and left Clifford parallels* to *l* through P. We say "in general" because there is a special line *l**, called the *absolute polar* of *l*: if P lies on *l**, there is only one Clifford parallel to *l* through P, which is *l**. (Naturally, this is difficult to visualize! See Coxeter, 1968, Chapter 7.)

Elliptic space is finite but unbounded — finite because all lines have finite length and look like circles, and unbounded because there is no boundary, just as on the surface of a sphere there is no boundary. In a universe having this geometry, with light rays traveling along elliptic lines, you could conceivably look through a very powerful telescope and see the back of your own head! (Although you might have to wait a few billion years for the light to travel all the way around.)

RIEMANNIAN GEOMETRY

It is impossible to rigorously explain the ideas of Riemannian geometry without using the language of the differential and integral calculus, so in this appendix we can only attempt to understand very roughly the intuitive idea. The basic notion we shall attempt to grasp is *curvature*.

The two simplest smooth one-dimensional figures in the Euclidean plane are a line and a circle. We think of a line as being "straight" i.e., not curved, so that if we were to assign any numerical curvature k to a line, we would assign the value $k = 0$. A circle γ, on the other hand, is certainly "curved," and how much it is curved depends on its radius r. The larger the radius of γ, the more γ approaches a line, i.e., the less it curves; it is therefore natural to define the curvature k of a circle of radius r by $k = 1/r$ (the curvature is inversely proportional to the radius); see Figure A.7.

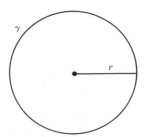

FIGURE A.7 Curvature $k = 1/r$.

Consider next an arbitrary smooth curve γ in the Euclidean plane. By "smooth curve" we mean one that has a continuously turning tangent line at each point. The tangent to a point on a circle is the perpendicular to the radius at that point. We can define the tangent to a point P on general curve γ as follows. P and another point Q on γ determine a line l. We fix point P and let Q approach P. The limiting position achieved by line l as Q approaches P is by definition the *tangent line t* to γ at P (see Figure A.8).

Besides the fixed point P, we can also consider two other points P_1 and P_2 on γ. These three points determine a circle δ. Fix P and let P_1 and P_2 both approach P along γ. The limiting position of circle δ as P_1 and P_2 approach P is the circle that "best fits" the curve γ at P, and is called the *osculating circle* to γ at P (from the Latin *osculari*, "to kiss"); see Figure A.9. It is reasonable to define the *curvature* of γ at P as the curvature of its osculating circle at P, i.e., the reciprocal $k = 1/r$ of the radius r of the osculating circle (r is also called the *radius of curvature* of γ at P).

FIGURE A.8

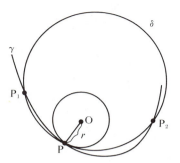

It is clear from Figure A.10 that the osculating circle will vary in size as we move along the curve γ, so that the curvature k varies from point to point along γ. Notice also that the tangent to γ at a point P is also the tangent to the osculating circle at P.

Notice also in Figure A.10 that the osculating circle may be on different sides of the curve γ. It is convenient to *redefine curvature* so that it is positive on one side of γ and negative on the other. Once this is done, it becomes clear that in Figure A.10 there must be a point I between A and B on γ at which the curvature is zero, since we assume γ is smooth enough for the curvature to vary continuously. This point I is called a *point of inflection,* and at such a point the osculating "circle" degenerates into a line, the tangent line at I (see Figure A.11).

What we have said about plane curves applies equally well to curves in Euclidean space, with the following modifications. The osculating circle lies in a unique plane through P called the *osculating plane* Π of γ at P (except in the degenerate case of a point P at which γ has curvature zero). Since Π varies with P if γ is not a plane curve, we no longer have a smooth way to assign positive and negative values to the

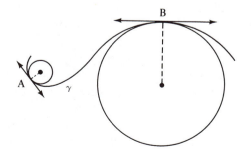

FIGURE A.10

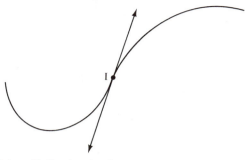

FIGURE A.11   Point of inflection $k = 0$.

curvature. We assign a *curvature vector* k lying in the osculating plane, emanating from P and perpendicular to the tangent line to γ at P (which also lies in the osculating plane), of length $1/r$ and pointing toward the center of the osculating circle (called the *center of curvature*). If γ has curvature zero at P, we take k to be the zero vector.

We next pass to smooth surfaces embedded in Euclidean three-dimensional space. "Smooth" now means that there is a continuously turning *tangent plane* for each point P of the surface. The tangent plane T at P separates Euclidean three-space into two half-spaces, one of which we arbitrarily consider positive and the other negative. Then we can define a *signed curvature k* to any curve γ through P lying on the given surface Σ which is positive or negative according as the center of curvature of γ at P lies in the positive or negative half-space. (For these signs to vary smoothly with P, we have to *orient* the surface, which can always be done locally — i.e., in a neighborhood of P.)

Consider the line through P that is perpendicular to T, called the *normal line* at P. A plane that contains the normal line intersects the surface in a plane curve. We can imagine this plane rotating around the normal line, and as it does so, it will cut out different curves on the surface passing through P. We have already explained how to define the curvature at P for each of these normal sections. In general, these curvatures will vary as we rotate around the normal line. (In the special case of a sphere these curvatures are constant and equal to the reciprocal of the radius of the sphere, since the curves cut out are all great circles on the sphere.) It can be proved by methods of differential geometry that these curvatures achieve a maximum value k_1 and a minimum value k_2 as we rotate, and the corresponding normal sections (called *principal curves*) are perpendicular to each other. The product

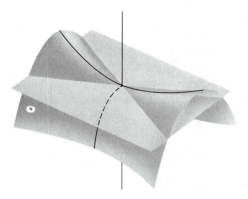

FIGURE A.12

$K = k_1 k_2$ of these maximum and minimum curvatures is now called the *Gaussian curvature* (after Gauss, who first studied it), or simply "the curvature," of the surface at the point P. Once again, K will in general change as P varies over the surface; if K happens to stay constant, we obtain the three geometries discussed in Chapter 10, according as K is negative (pseudosphere), zero (plane), or positive (sphere).

In Figure A.12, the tangent plane, normal line, and principal curves for a saddle-shaped surface are shown. For point P on this surface the Gaussian curvature will be a negative number, according to our convention, since the osculating circles for the two principal curves lie on different sides of the tangent plane. On the other hand, for the egg-shaped surface in Figure A.13, the two principal curves lie on the same side of the tangent plane, so the Gaussian curvature is positive.

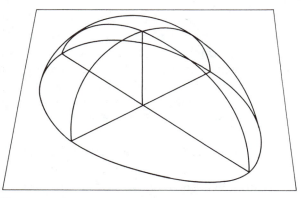

FIGURE A.13

Refer back to the picture of a pseudosphere in Figure 10.12, which is obtained by revolving a tractrix around its asymptote. At any point on this surface, it can be shown that the two principal curves are the (horizontal) circle of revolution through that point and the (vertical) tractrix through that point. Since these curves lie on opposite sides of the tangent plane, we see that the surface curvature is negative. As we move up the surface, the circle shrinks and its curvature increases indefinitely, while the tractrix flattens and its curvature decreases to zero (the curvature of its asymptote); this makes it plausible that the product K of the principal curvatures could stay constant (but of course the proof of that requires a calculation).

Similarly, in the case of a cylinder obtained, say, by rotating a vertical line around a parallel vertical line, the principal curves at any point are the horizontal circle and the vertical line, which has curvature $k_2 = 0$; hence, the Gaussian curvature $K = k_1 k_2$ at any point on a cylinder will also be zero (see Figure A.14). We can better grasp this surprising result if we think of a cylinder as a "rolled-up plane." Surely, in any sensible definition of surface curvature a flat plane should be assigned zero curvature. In the process of "rolling-up" a rectangular plane strip the arc lengths and angles between curves on the strip are not changed, and in this sense the "intrinsic geometry" is not changed. Gauss was looking for a definition of surface curvature that depended only on the intrinsic geometry of the surface, not on the particular way the surface was embedded in Euclidean three-space. He was able to prove that his curvature K did not change if the surface was subjected to a "bending" in which arc lengths and angles of all curves on the surface are left invariant. Thus, K describes the intrinsic curvature of the surface independent of the way it is bent to fit into Euclidean three-space. This is all the more remarkable because the principal curvatures k_1 and k_2 may change under such a "bending"; nevertheless, their product $K = k_1 k_2$ stays the same. Gauss was so excited about this result that he named it the *theorema egregrium,* "the extraordinary theorem." In a letter to the astronomer Hansen he wrote: "These investigations deeply affect many other things; I would go so far as to say they are involved in the metaphysics of the geometry of space."

Gauss also solved the problem of determining this intrinsic curvature K without reference to the ambient three-space. Imagine a two-dimensional creature living on a surface and having no conception of a

FIGURE A.14   Gaussian curvature of a cylinder is zero.

third dimension, being unable to conceive of the normal lines we used to define the curvature K. How could this creature calculate K? We will have to use the language of the differential calculus to give Gauss' answer.

In the Euclidean plane a point is determined by its x and y coordinates. If these coordinates are subjected to infinitesimal changes denoted dx and dy, then the point moves an infinitesimal distance ds whose square is given by the Pythagorean formula $ds^2 = dx^2 + dy^2$. Now on a smooth surface a point will also be determined locally by two coordinates x and y. If these coordinates are subjected to infinitesimal changes dx and dy, then the point moves a distance ds on the surface whose square is given by the more complicated expression

$$ds^2 = E\,dx^2 + 2F\,dx\,dy + G\,dy^2,$$

where E, F, and G may vary as the point varies. The functions E, F and G could in principle be determined by the two-dimensional creature making measurements on his surface. Gauss then showed that his curvature K is given in terms of E, F, and G by a not-too-complicated formula (see Lanczos, 1970, p. 183). Thus, the creature could also calculate K from this formula and discover that his world was curved, although he would have difficulty visualizing what that might mean.

Although this talk about a two-dimensional creature may seem bizarre, it is not, as Riemann demonstrated. Riemann reasoned that we are in an entirely analogous situation, living in a three-dimensional universe in which an infinitesimal change in distance ds is given by an analogous formula involving the three infinitesimals dx, dy, and dz:

$$ds^2 = g_{11}\, dx^2 + g_{22}dy^2 + g_{33}dz^2 + 2g_{23}dy\, dz + 2g_{31}dz\, dx + 2g_{12}dx\, dy.$$

From this formula Riemann was able to define a "curvature tensor" analogous to the Gaussian curvature for a surface, only more complicated: Gauss' curvature involved only a single number K, whereas Riemann's depended on six different numbers. Riemann discovered this curvature almost accidentally in his research on heat transfer. In fact, he developed such a curvature tensor for abstract geometries of any dimension n, and Einstein was able to apply Riemann's ideas to his four-dimensional space-time continuum.

So we are in the same position as that poor two-dimensional creature. We can make measurements to determine the Riemannian curvature of our universe. Astronomers have been performing such measurements. If we find that the Riemannian curvature is not zero, we know that the geometry is not Euclidean. However, this does not mean that our space is embedded in some higher-dimensional physical space in which it is somehow "curved." When we say, loosely, that "space is curved," we mean only that its geometric properties differ from the properties of Euclidean space in a very specific way given by Riemann's formulas.

It was in his 1854 inaugural lecture *Ueber die Hypothesen welche der Geometrie zugrunde liegen* ("On the hypotheses that form the foundation for geometry") that Riemann introduced the idea of an *n*-dimensional space whose intrinsic geometry is determined by a quadratic formula for the infinitesimal change in distance *ds*. Such a structure is now called a *Riemannian manifold* (see Spivak, 1970, or Do Carmo,

1976, for the definition). Different manifolds yield different geometries, so that Riemann discovered an infinite number of new geometries.

Let us return to the two-dimensional case of a surface Σ. For each point P on Σ and each curve γ lying on Σ and passing through P, the curvature vector $\mathbf{k}$ decomposes naturally into a vector sum

$$\mathbf{k} = \mathbf{k}_n + \mathbf{k}_g$$

of its projection $\mathbf{k}_n$ on the normal line and its projection $\mathbf{k}_g$ on the tangent plane T to Σ at P; these projections are called, respectively, the *normal* and *tangential curvature vectors*. The length of $\mathbf{k}_g$ is called the *geodesic curvature k_g* of the curve γ at P relative to Σ. We call γ a *geodesic* if $k_g = 0$. In that case, γ has *zero curvature relative to the surface* Σ; it may have nonzero curvature relative to Euclidean three-space, but then its curvature vector points along the normal line to the surface at P. Another way to describe a geodesic γ that has nonzero $\mathbf{k} = \mathbf{k}_n$ is that its osculating plane Π at P is perpendicular to the tangent plane T (since Π contains the normal line at P). From this description we see immediately that the geodesics on a sphere are its great circles ($\Pi \perp T$ iff Π passes through the center of the sphere); Figure A.15.

This description of geodesic curvature refers to the ambient Euclidean three-space. But F. Minding in 1830 showed that it too is an intrinsic quantity for Σ: it depends only on the functions E, F, and G and the curve γ. Hence the notion of "geodesic" can be defined on a Riemannian manifold. And it gives us the correct interpretation of the heretofore confusing term *straight line* on such a manifold. Riemann recognized that the geodesics would be of finite length if the curvature of the surface was greater than some positive constant.

A proposed alternative interpretation of the term "straight line

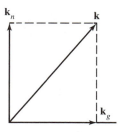

FIGURE A.15

segment" is "the shortest path on the surface joining two points on the surface." It can be proved that *if such a shortest path exists, it must be an arc of a geodesic.* But a shortest path may not exist: let Σ be a punctured plane or a punctured sphere; two points on opposite sides of the puncture cannot be joined by a shortest path on Σ. On *complete* manifolds (meaning that every Cauchy sequence of points converges), it can be proved that shortest paths always exist (theorem of Hopf-Rinow). However, an arc of a geodesic need not be the shortest path: consider the longer great-circular arc joining two nonantipodal points on a sphere, or consider the helical arc joining two points on a vertical line on the cylinder in Figure A.16. So the definition of "geodesic" (found in several books by nonmathematicians) as "the shortest path" is inadequate because it excludes such arcs.

It is important to note that, although we have emphasized Riemann's generalization of Gauss' ideas from two dimensions to higher dimensions, Riemann's formulation gives new information about *surfaces* that cannot be embedded in Euclidean three-space. For example, the hyperbolic plane can be described as a complete two-dimensional Riemannian manifold of constant negative curvature, and the elliptic plane can be described as a complete two-dimensional Riemannian manifold of constant positive curvature on which any two points lie on a unique geodesic. Neither of these manifolds can be analytically embedded in Euclidean three-space.

Some idea of Riemann's influence on modern mathematics can be gleaned from the following list of concepts, methods, and theorems

FIGURE A.16   A helix is a geodesic on a cylinder.

that have been named after him: Riemannian curvature of Riemannian manifolds, Riemann integral, Riemann-Lebesgue lemma, Riemann surfaces, Riemann-Roch theorem, Riemann matrices, Riemann hypothesis about the Riemann zeta function, Riemann's method in the theory of trigonometrical series, Riemann's method for hyperbolic partial differential equations, Riemann mapping theorem, and Cauchy-Riemann equations.[2]

[2] For the story of Riemann's difficult life, see Bell (1961); for a mathematically precise treatment of Riemannian geometry, including an explanation of what Riemann said in his famous 1854 inaugural dissertation lecture, see Spivak (1970), Vol. 2; for applications to the theory of relativity, see Lanczos (1970); for an intuitive discussion of how gravity can be explained in terms of the curvature of space, see the last part of Taylor and Wheeler (1992).

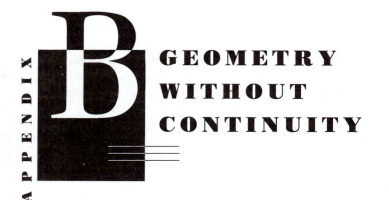

B GEOMETRY WITHOUT CONTINUITY

Any serious student should become familiar with the
great discovery, made at the end of the last century,
that large parts of geometry do not depend on continuity.

H. BUSEMANN AND P. J. KELLY

If Euclid could have returned to life at the beginning of the twentieth
century and heard the criticisms made of the gaps in his work, he
undoubtedly would have accepted the incidence, betweenness, and
congruence axioms that replaced his first four postulates. A model of
those axioms is called an H-*plane* ("H" for Hilbert, Hessenberg, and
Hjelmslev). However, it is easy to imagine Euclid bristling at Dedekind's continuity axiom, although he could be expected to admit
Archimedes' axiom and the elementary continuity principle.

Much work has been done in the last century to develop geometries
without continuity assumptions. The most comprehensive such treatise is F. Bachmann's *Aufbau der Geometrie aus dem Spiegelungsbegriff*
(Construction of Geometry based on the Concept of Reflection,
1973), in which plane geometries based only on axioms of incidence,
perpendicularity, and reflections are studied.[1] Bachmann is justified
in calling this study *plane absolute geometry,* since it includes elliptic,
hyperbolic, and Euclidean geometries as special cases (as well as other
unusual geometries). Bachmann has succeeded in dispensing with
betweenness axioms as well as continuity axioms. It is unfortunate

---

[1] For a presentation of Bachmann's axioms in English, see Ewald (1971). I. M. Yaglom, in
the foreword to the Russian translation of Bachmann's book, calls it "indisputably the most
significant development in the foundations of geometry in decades." You were introduced to
Bachmann's methods on pp. 340–342 and in Exercise 50 of Chapter 9.

that he calls models of his axioms "metric" planes — since "metric" spaces" have a quite different meaning in which a "metric" measures distance by real numbers; I propose calling them *absolute planes*.

The key ideas for this study are:

1. Embed the absolute plane in a projective plane.
2. Show that the points and lines in this projective plane have homogeneous coordinates from some field K (as in Major Exercise 10, Chapter 2).
3. Show that two lines are perpendicular if and only if their coordinates satisfy a particular homogeneous quadratic equation.
4. Use algebra to get further information about the absolute plane.

This program has been completely successful in classifying H-planes (the problem of determining all absolute planes remains unsolved). In particular, it is shown that (up to isomorphism) the only H-planes in which Archimedes' axiom and the elementary continuity principle hold are either the Cartesian model or the Klein model over an Archimedean Euclidean field K (Archimedean ordered field in which every positive element has a square root). This result is particularly remarkable because it gives the existence of limiting parallel rays in the non-Euclidean case without appealing to Dedekind's continuity axiom (compare the proof of Theorem 6.6, p. 196).

Let us describe the embedding briefly. The intuitive idea comes from our discussion of the Klein model, in which new "ideal" and "ultra-ideal" points are introduced so that lines that were previously parallel now meet when extended through these new points (see Major Exercise 13, Chapter 6, and p. 240). Abstractly, the new points are simply pencils of lines. The old points are in one-to-one correspondence with the pencils of the first kind, i.e., A corresponds to the pencil $p(A)$ of all lines through A.

An ultra-ideal point is a pencil of the second kind; it is a pole $P(t)$, which for a fixed line t consists of all the perpendiculars to t. But the pencils of the third kind must be described carefully to avoid circular reasoning. We described such a pencil in Chapter 9 (p. 340) as consisting of all lines through a fixed ideal point, the ideal point being defined as an equivalence class of limiting parallel rays. Since we don't know limiting parallel rays exist now, we instead use the theorem on three reflections (Proposition 9.19) as our definition: the pencil $p(lm)$ determined by parallel lines l and m which do not have a common perpendicular consists of all lines n such that the product

$$R_l R_m R_n$$

is a reflection. Certain properties that were previously obvious now require a considerable amount of ingenuity to prove — for example, if $h, k \in p(lm)$, then $p(hk) = p(lm)$. Hjelmslev discovered how to prove that a pencil of the first kind $p(D)$ and a pencil of the third kind $p(a'c')$ have a unique line b in common (see Figure B.1). He dropped perpendiculars a, c from D to a', c' at points A, C, then dropped perpendicular d from D to $\overleftrightarrow{AC}$. Then the line b is uniquely determined by the equation

$$R_a R_b R_c = R_d.$$

We now know the new points of our projective plane. Our old lines a can be extended to new lines $l(a)$ so that any pencil containing a is by definition incident with $l(a)$. But we need more new lines to fill out our projective plane. For example, for any old point A, the polar $l(A)$, consisting of all poles of lines through A, should be a new line; in the Klein model, this line lies entirely outside the absolute circle (in the Cartesian model, it's the line at infinity); see Figure B.2.

But how are we to describe the lines in the Klein model that are tangent to the absolute circle, and how are we to verify the axioms for a projective plane? To accomplish this, Hjelmslev discovered a remarkable device: he fixed a point O and fixed a pair u, v of nonperpendicu-

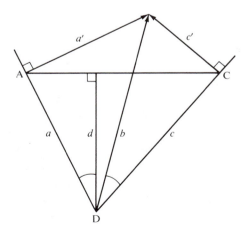

FIGURE B.1

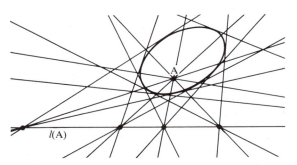

FIGURE B.2 The polar of a point.

lar lines through O—think of these lines as determining an acute angle θ. He then defined a transformation that fixed O and sent any A ≠ O to the midpoint A* of the segment joining A to its image under rotation $R_u R_v$ about O through angle 2θ (Figure B.3); this transformation is called the *half-rotation* (or *snail map*) about O corresponding to u, v.

Hjelmslev observed that in the Klein model, half-rotations extend to collineations of the projective plane, and that any projective line except $l(O)$ could be mapped onto an extended Klein line $l(a)$ by a suitable half-rotation about O. So he proposed to call a set of pencils a "line" if it is mapped onto some $l(a)$ by some half-rotation about O, or if it is $l(O)$. With this definition, he was then able to verify the axioms for a projective plane.

The execution of idea 2—construction of the field K of coordinates—requires even more technique. The key tool is a complicated theorem of Hessenberg, which generalizes the Euclidean theorem that tells when a quadrilateral can be circumscribed by a

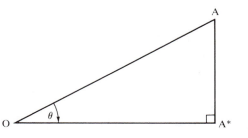

FIGURE B.3

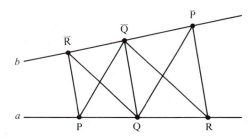

FIGURE B.4 Pappus' theorem.

circle. The method of constructing K is the standard method for any affine plane in which Pappus' theorem holds (Ewald, 1971, Chapter 3); here the affine plane is obtained by removing the polar $l(O)$ of O, and Pappus' theorem states: Let P, Q, R lie on a, $\overline{P}, \overline{Q}, \overline{R}$ lie on b, such that a does not meet b in any of these six points (Figure B.4). If $\overline{PQ} \parallel \overline{PQ}$ and $\overline{QR} \parallel \overline{QR}$, then $\overline{PR} \parallel \overline{PR}$. Bachmann's approach is to verify Pappus' theorem by brute force; there is another approach due to Lingenberg that I believe gives more insight (see Lenz, 1967, p. 206 ff., on the Euclidean "pseudoplane"). Pappus' theorem implies that the definition given for "lines" in our projective plane does not depend on the choice of O.

If we started with an H-plane, it is then possible to define an order on K in terms of the betweenness relation in the H-plane. We need only specify the set P of positive scalars (since $x < y \Leftrightarrow y - x > 0$).

Choose any two points A, B from the H-plane. For any third point X collinear with A, B, the ratio

$$AX : BX$$

is defined to be that unique scalar in K which, multiplying the vector from B to X, gives the vector from A to X. We then define P to consist of 1 and all ratios AX : BX as X runs over all third points on the affine line $\overleftrightarrow{AB}$ that do *not* lie between A and B (this includes all ideal and ultra-ideal points on $\overleftrightarrow{AB}$). It can be shown from invariance of the ratio under parallel projection that this definition of P does not depend on the choice of A, B.

Moreover, a theorem of Menelaus (compare Exercise H-5, Chapter 7) can be proved for these ratios, and can be used to demonstrate that P has all the properties required for a set of positive numbers that makes K an *ordered field* (see Lenz, 1967, p. 223 ff.). In turn, this enables us to extend the betweenness relation to all triples of collinear points in the affine plane, and to show that the H-plane is embedded as a convex open subset of the affine plane, containing the origin O. The embedding is locally Euclidean at O in that perpendicularity for lines through O has the familiar meaning from Cartesian analytic geometry.

It can be proved (from free mobility — Lemma 9.3, p. 358) that the field K is Pythagorean (the sum of two squares is a square), but K is not Euclidean unless the elementary continuity principle holds (a neat algebraic criterion for this geometric property).

As for idea 3 on our list, further argument shows that there is a constant $k \in K$ such that lines having homogeneous coordinates $[a_1, a_2, a_3]$, $[b_1, b_2, b_3]$ are perpendicular if and only if

$$a_1 b_1 + a_2 b_2 + k a_3 b_3 = 0.$$

Here "perpendicularity" has been extended to all pairs of lines in the projective plane, and there may exist certain lines (called *isotropic*) that are perpendicular to themselves (e.g., when $k = -1$, the line $[0, 1, -1]$, i.e., $y = 1$). To each line $[a_1, a_2, a_3]$ is associated its *pole* $[a_1, a_2, k a_3]$ (except for line at infinity $[0, 0, 1]$ when $k = 0$), and we see that the perpendiculars to this line are precisely the lines passing through its pole; isotropic lines pass through their own poles, and for $k \neq 0$, the locus of poles of isotropic lines is given by the affine equation

$$x^2 + y^2 = -k^{-1}$$

and may be called *the absolute*. The points (x, y) in the H-plane satisfy the inequality $|k| (x^2 + y^2) < 1$.

The fourth angle of a Lambert quadrilateral is acute, right, or obtuse accordingly as $k < 0$, $k = 0$, or $k > 0$. If $k = 0$ and the H-plane fills up the affine plane, we obtain the *generalized Euclidean plane* over K. However, if $k = 0$, if K is non-Archimedean, and if we restrict our H-plane points to those (x, y) that are infinitesimal, we obtain Dehn's *semi-Euclidean plane,* in which the Euclidean parallel postulate fails but rectangles exist.

If $k = -1$, the absolute is the unit circle, and if every point in its interior belongs to the H-plane, we obtain the *generalized hyperbolic plane* over K, provided K is Euclidean. In case K is non-Archimedean, we can again restrict (x, y) to infinitesimal values, obtaining Schur's *semihyperbolic plane,* in which limiting parallel rays do not exist, even though the elementary continuity principle holds (the ideal points fill up an entire region inside and on the absolute).

If neither k nor $-k$ is a square (so K is not Euclidean), yet every sum $1 + kx^2$ is either a square or k times a square, the points (x, y) satisfying $x^2 + y^2 < |k^{-1}|$ form a *semielliptic* H-plane, in which parallel lines have a unique common perpendicular (the absolute is empty).

Finally, if K is non-Archimedean, $k = 1$, and (x, y) are again restricted to infinitesimal values, we obtain Dehn's *non-Legendrian plane,* in which the fourth angle of a Lambert quadrilateral is obtuse.

You can see that there are some unusual H-planes, most of them living on non-Archimedean fields; on Archimedean fields, only Euclidean, hyperbolic, and semielliptic planes survive. For the complete algebraic classification of H-planes (discovered by W. Pejas, 1960), see Hessenberg and Diller (1967). For an introduction to absolute plane geometry in English, see Chapter 5 of Bachmann, Behnke, and Fladt (1974). Finally, for an expanded version of this appendix, see my article "Euclidean and Non-Euclidean Geometries without Continuity," *American Mathematical Monthly,* November 1979.

SUGGESTED
FURTHER
READING

1. If you want to fill in the few gaps in this book (such as the proof of Theorem 4.3), see Borsuk and Szmielew (1960), where it is proved that their axioms for Euclidean and hyperbolic geometries (equivalent to ours) are categorical. For hyperbolic area theory, see Moise (1990). For more on hyperbolic constructions and trigonometry, see Martin (1982). For hyperbolic space, see Fenchel (1989). Lively presentations of the non-Euclidean revolution and its philosophical implications can be found in Gray (1989) and in Trudeau (1987). A delightful spoof on Lewis Carroll and an introduction to the Poincaré model are in Marta Sved (1991). More advanced Euclidean geometry will be found in Kay (1969), Coxeter and Greitzer (1967), and Pedoe (1970).

2. For more on the history of non-Euclidean and other geometries, see Rosenfeld (1988) and the survey article by Milnor (1982). The classic by Bonola (1955) includes translations of the original articles by Bolyai and Lobachevsky. For a biting account of the confused reaction to non-Euclidean geometry by nineteenth-century scholars, see Freudenthal (1962).

3. You will want to know more about projective geometry, the study of which will illuminate the mysterious cross-ratio used to define length in the Klein and Poincaré models (Chapter 7). Projective

geometry, as a science independent of Euclidean geometry, blossomed in the first half of the nineteenth century; in the latter half, Klein and Cayley showed that projective geometry plays a dominant role in the classification of the other geometries (by means of projective metrics). For introductory expositions, see Coxeter (1974 and 1960), Ewald (1971), or Hughes and Piper (1973). For advanced treatments, see Coxeter (1968), Klein (1968), or Cartan (1950). For superb illustrations, see Whicher (1971).

4. Once you have mastered the calculus, you can begin differential geometry using Do Carmo (1976) or O'Neill (1966) and then advance to Helgason (1962) and Spivak (1970).

5. If you are curious about the application of non-Euclidean geometry to general relativity and cosmology, see Lanczos (1970) or Taylor and Wheeler (1992) or the fine introductory article by Roger Penrose "The Geometry of the Universe," in *Mathematics Today: Twelve Informal Essays*, L. A. Steen, ed., Springer-Verlag, New York, 1978. Applications to classical mechanics will be found in Arnold and Avez (1968), as well as Bottema and Roth (1979).

6. Learn some group theory and then read about its connections to geometry in Artin (1957), Bachmann, Behnke, and Fladt (1974), Benson and Grove (1970), and Schwerdtfeger (1962).

7. Few mathematicians know about the interaction between geometry and art; see Bouleau (1963), Ghyka (1977), Hambidge (1967), and Rhodos (1967) for this subject.

8. For use of hyperbolic geometry in advanced research, see Casson and Bleiler (1988), Gallo and Porter (1987), Milnor (1982), and Fenchel (1989).

BIBLIOGRAPHY

INTRODUCTORY AND GENERAL

Aleksandrov, A. 1969. "Abstract Spaces," in vol. 3 of *Mathematics: Its Content, Methods and Meaning.* A. Aleksandrov *et al.*, eds., 2nd ed., Cambridge, Mass.: MIT Press.

Bouleau, C. 1963. *The Painter's Secret Geometry,* New York: Harcourt.

Courant, R., and H. Robbins. 1941. *What Is Mathematics?* Oxford University Press.

Gamow, G. 1956. "The Evolutionary Universe," *Scientific American,* **195** (September): 136–154 (Offprint no. 211).

Ghyka, M. 1977. *The Geometry of Art and Life,* New York: Dover.

Hambidge, J. 1967. *The Elements of Dynamic Symmetry,* New York: Dover.

Hilbert, D., and S. Cohn-Vossen. 1952. *Geometry and the Imagination,* New York: Chelsea.

Kline M., 1979. *Mathematics: An Introduction to Its Spirit and Use: Readings from Scientific American,* New York, W. H. Freeman and Company. (See especially the articles on geometry by Kline.)

Rhodos. 1967. *The Secrets of Ancient Geometry and Its Use,* 2 vols., Copenhagen.

HISTORY AND BIOGRAPHY

Bell, E. T. 1934. *The Search for Truth,* New York: Reynal and Hitchcock.

———. 1961. *Men of Mathematics,* New York: Simon and Schuster.

———. 1969. "Father and Son, Wolfgang and Johann Bolyai," in *Memorable Personalities in Mathematics: Nineteenth Century,* Stanford, Calif.: School Mathematics Study Group.

Bonola, R. 1955. *Non-Euclidean Geometry,* New York: Dover.

Boyer, C. B. 1968. *A History of Mathematics,* New York: Wiley.

Dodgson, C. L. (Lewis Carroll). 1890. *Curiosa Mathematica: A New Theory of Parallels,* London: Macmillan.

Dunnington, G. W. 1955. *Carl Friedrich Gauss: Titan of Science,* New York: Hafner.

Engel, F., and P. Stäckel. 1895. *Theorie der Parallellinien von Euklid bis auf Gauss,* Leipzig: Teubner.

Freudenthal, H. 1962. "The Main Trends in the Foundations of Geometry in the Nineteenth Century," in *Logic, Methodology and Philosophy of Science,* E. Nagel, P. Suppes, and A. Tarski, eds., Stanford, Calif.: Stanford University Press, pp. 613–621.

Gray, J. 1989. *Ideas of Space: Euclidean, Non-Euclidean, and Relativistic,* 2nd ed., New York: Oxford University Press.

Hall, T. 1970. *C. F. Gauss: A Biography,* Cambridge, Mass: MIT Press.

Heath, T. L. 1956. *Euclid's Elements,* New York: Dover.

Lanczos, C. 1970. *Space through the Ages,* New York: Academic Press.

Nagel, E. 1939. "Formation of Modern Conceptions of Formal Logic in the Development of Geometry," *Osiris* (no. 7): 142–224.

Pont, J.-C. 1986. *L'Aventure des parallèles,* Bern: Lang.

Reid, C. 1970. *Hilbert,* New York: Springer-Verlag.

Rosenfeld, B. A. 1988. *A History of Non-Euclidean Geometry,* tr. A. Shenitzer, New York: Springer-Verlag.

Saccheri, G. 1970. *Euclides ab omni naevo vindicatus,* tr. G. B. Halsted, New York: Chelsea.

Schmidt, F., and P. Stäckel. 1972. *Briefwechsel zwischen C. F. Gauss und W. Bolyai,* New York: Johnson Reprint Corp.

Stäckel, P. 1913. *Wolfgang und Johann Bolyai, Geometrische Untersuchungen,* 2 vols., Leipzig: Teubner.

Van der Waerden, B. L. 1961. *Science Awakening,* New York: Oxford University Press.

PHILOSOPHICAL

DeLong, H. 1970. *A Profile of Mathematical Logic*, Reading, Mass.: Addison-Wesley.

Fang, J. 1976. *The Illusory Infinite — A Theology of Mathematics*, Memphis: Paideia.

Grünbaum, A. 1968. *Geometry and Chronometry in Philosophical Perspective*, Bloomington: University of Minnesota Press.

Hadamard, J. 1945. *The Psychology of Invention in the Mathematical Field*, Princeton, N.J.: Princeton University Press.

Hardy, G. H. 1940. *A Mathematician's Apology*, New York: Cambridge University Press.

Hempel, C. G. 1945. "Geometry and Empirical Science," *American Mathematical Monthly*, **52**:7 – 17; also in vol. 3, *World of Mathematics*, J. R. Newman, ed., pp. 1619 – 1646.

Meschkowski, H. 1965. *Evolution of Mathematical Thought*, San Francisco: Holden-Day.

Poincaré, H. 1952. *Science and Hypothesis*, New York: Dover. Originally published in French, 1902.

Polanyi, M. 1964. *Personal Knowledge*, New York: Harper and Row.

Putnam, H. 1977. *Mathematics, Matter and Method*, New York: Cambridge University Press.

Renyi, A. 1967. *Dialogues on Mathematics*, San Francisco: Holden-Day.

Stein, H. 1970. "On the Paradoxical Time-Structures of Gödel," *Journal of the Philosophy of Science* **37**: 589 ff.

Torretti, R. 1978. *Philosophy of Geometry from Riemann to Poincaré*, Hingham, Mass.: D. Reidel.

Zeeman, E. C. 1974. "Research, Ancient and Modern," *Bulletin of the Institute of Mathematics and Its Applications*, Warwick University, England **10**:272 – 281.

MATHEMATICALLY ELEMENTARY TO MODERATE

Adler, I. 1966. *A New Look at Geometry*, New York: John Day Co.

Artzy, R. 1965. *Linear Geometry*, Reading, Mass.: Addison-Wesley.

Beck, A., M. Bleicher, and D. Crowe. 1969. *Excursions into Mathematics*, New York: Worth.

Coxeter, H. S. M. 1969. *Introduction to Geometry*, 2nd ed., New York: Wiley.

——. 1987. *Projective Geometry*, 2nd ed., New York: Cambridge University Press.

——. 1960. *The Real Projective Plane*, 2nd ed., New York: Cambridge University Press.

Coxeter, H. S. M., and S. L. Greitzer. 1967. *Geometry Revisited*, New York: Random House.

Eves, H. 1972. *A Survey of Geometry*, revised ed., Boston: Allyn and Bacon.

Golos, E. B. 1968. *Foundations of Euclidean and Non-Euclidean Geometry*, New York: Holt, Rinehart and Winston.

Hartshorne, R. 1967. *Foundations of Projective Geometry*, Menlo Park, Calif.: Benjamin Cummings.

Hessenberg, G., and J. Diller. 1967. *Grundlagen der Geometrie*, Berlin: de Gruyter.

Jones, A., A. Morris, and K. Pearson. 1991. *Abstract Algebra and Famous Impossibilities*, New York: Springer-Verlag.

Kay, D. C. 1969. *College Geometry*, New York: Holt, Rinehart and Winston.

Klein, F. 1948. *Geometry*, part 2 of *Elementary Mathematics from an Advanced Standpoint*, New York: Dover.

Klein, F. 1956. *Famous Problems of Elementary Geometry*, New York: Dover.

Kleven, D. J. 1978. "Morley's Theorem and a Converse," *American Mathematical Monthly* **85**:100–105.

Kulczycki, S. 1961. *Non-Euclidean Geometry*, Elmsford, N.Y.: Pergamon.

Kutuzov, B. V. 1960. *Geometry*, Stanford, Calif.: School Mathematics Study Group.

Luneburg, R. K. 1947. *Mathematical Analysis of Binocular Vision*, Princeton, N.J.: Princeton University Press.

Martin, G. E. 1982. *The Foundations of Geometry and the Non-Euclidean Plane*, New York: Springer-Verlag.

Meschkowski, H. 1964. *Noneuclidean Geometry*, New York: Academic Press.

Millman, R. S., and G. D. Parker. 1991. *Geometry: A Metric Approach with Models*, 2nd ed., New York: Springer-Verlag.

Pedoe, D. 1970. *A Course of Geometry*, New York: Cambridge University Press.

Peressini, A. L. and D. R. Sherbert. 1971. *Topics in Modern Mathematics for Teachers*, New York: Holt.

Prenowitz, W., and M. Jordan. 1965. *Basic Concepts of Geometry*, New York: Blaisdell.

Sved, M. 1991. *Journey into Geometries*, Washington, D.C.: Mathematical Association of America.

Taylor, E. F., and J. A. Wheeler. 1992. *Spacetime Physics*, 2nd ed., New York: W. H. Freeman and Company.

Trudeau, R. J. 1987. *The Non-Euclidean Revolution*, Boston: Birkhäuser.

Weyl, H. 1952. *Symmetry*, Princeton, N.J.: Princeton University Press.

Whicher, O. 1971. *Projective Geometry: Creative Polarities in Space and Time*, London: Rudloph Steiner Press.

Yaglom, I. M. 1979. *A Simple Non-Euclidean Geometry and Its Physical Basis*, tr. A. Shenitzer, New York: Springer-Verlag.

MATHEMATICALLY ADVANCED

Arnold, V. I., and A. Avez. 1968. *Ergodic Problems of Classical Mechanics*, Reading, Mass.: W. A. Benjamin.

Artin, E. 1957. *Geometric Algebra*, New York: Wiley.

Bachmann, F. 1973. *Aufbau der Geometrie aus dem Spiegelungsbegriff*, New York: Springer-Verlag.

Bachmann, F., H. Behnke, and K. Fladt, eds. 1974. *Geometry*, vol. 2 of *Fundamentals of Mathematics*, Cambridge, Mass.: MIT Press.

Benson, C. T., and L. C. Grove. 1970. *Finite Reflection Groups*, Belmont, Calif.: Bogdon and Quigley.

Blumenthal, L. M., and K. Menger. 1970. *Studies in Geometry*, New York: W. H. Freeman and Company.

Borsuk, K., and W. Szmielew. 1960. *Foundations of Geometry*, Amsterdam: North Holland.

Bottema, O., and B. Roth. 1979. *Theoretical Kinematics*, Amsterdam: North Holland.

Cartan, E. 1950. *Leçons sur la Géométrie Projective Complexe*, Paris: Gauthier-Villars.

Casson, A., and S. Bleiler. 1988. *Automorphisms of Surfaces after Nielsen and Thurston*, New York: Cambridge University Press.

Coxeter, H. S. M. 1968. *Non-Euclidean Geometry*, 5th ed., Toronto: University of Toronto Press.

Dembowski, P. 1968. *Finite Geometries*, New York: Springer-Verlag.

Do Carmo, M. 1976. *Differential Geometry of Curves and Surfaces*, Englewood Cliffs, N.J.: Prentice-Hall.

Einstein, A. 1921. *Relativity*, New York: Holt.

Ewald, G. 1971. *Geometry: An Introduction*, Belmont, Calif.: Wadsworth.

Fenchel, W. 1989. *Elementary Geometry in Hyperbolic Space*, New York: de Gruyter.

Gallo, D. M., and R. M. Porter, eds. 1987. *Analytical and Geometric Aspects of Hyperbolic Space*, New York: Cambridge University Press.

Greenberg, M. J. 1988. "Aristotle's Axiom in the Foundations of Hyperbolic Geometry," *Journal of Geometry* **33**:53–57.

———. 1979. "On J. Bolyai's Parallel Construction," *Journal of Geometry*, **12(1)**:45–64.

Helgason, S. 1962. *Differential Geometry and Symmetric Spaces*, New York: Academic Press.

Henkin, L., P. Suppes, and A. Tarski, eds. 1959. *The Axiomatic Method*, Amsterdam: North Holland.

Hilbert, D. 1987. *Foundations of Geometry*, 2nd English ed., La Salle, Ill.: Open Court.

Hughes, D. R., and F. C. Piper. 1973. *Projective Planes*, New York: Springer.

Klein, F. 1968. *Vorlesungen über Nicht-Euklidische Geometrie*, Berlin: Springer-Verlag.

Lenz, H. 1967. *Nichteuklidische Geometrie*, Mannheim: Bibliographisches Institut.

Milnor, J. 1982. "Hyperbolic Geometry—The First 150 Years," *Bulletin American Mathematical Society* **6**:9–24.

Moise, E. E. 1990. *Elementary Geometry from an Advanced Standpoint*, 3rd ed., Reading, Mass.: Addison-Wesley.

O'Neill, B. 1966. *Elementary Differential Geometry*, New York: Academic Press.

Redei, L. 1968. *Foundation of Euclidean and Non-Euclidean Geometries According to F. Klein*, Elmsford, N.Y.: Pergamon.

Schwerdtfeger, H. 1962. *Geometry of Complex Numbers*, Toronto: University of Toronto Press.

Sommerville, D. M. Y. 1958. *Elements of Non-Euclidean Geometry*, New York: Dover.

Spivak, M. 1970. *A Comprehensive Introduction to Differential Geometry*, 5 vols., Waltham, Mass.: Publish or Perish Press.

Stevenson, F. W. 1972. *Projective Planes*, New York, W. H. Freeman and Company.

Toth, L. F. 1964. *Regular Figures*, New York: Macmillan.

Wolfe, H. E. 1945. *Introduction to Non-Euclidean Geometry*, New York: Holt, Rinehart and Winston.

Yaglom, I. M. 1962. *Geometric Transformations*, 3 vols., Washington, D.C.: Mathematical Association of America.

AXIOMS

INCIDENCE AXIOMS

AXIOM I-1. For every point P and for every point Q not equal to P there exists a unique line *l* incident with P and Q.

AXIOM I-2. For every line *l* there exist at least two distinct points that are incident with *l*.

AXIOM I-3. There exist three distinct points with the property that no line is incident with all three of them.

BETWEENNESS AXIOMS

AXIOM B-1. If A * B * C, then A, B, and C are three distinct points all lying on the same line, and C * B * A.

AXIOM B-2. Given any two distinct points B and D, there exist points A, C, and E lying on $\overleftrightarrow{BD}$ such that A * B * D, B * C * D, and B * D * E.

AXIOM B-3. If A, B, and C are three distinct points lying on the same line, then one and only one of the points is between the other two.

AXIOM B-4. For every line l and for any three points A, B, and C not lying on l:

(i) If A and B are on the same side of l and B and C are on the same side of l, then A and C are on the same side of l.

(ii) If A and B are on opposite sides of l and B and C are on opposite sides of l, then A and C are on the same side of l.

CONGRUENCE AXIOMS

AXIOM C-1. If A and B are distinct points and if A′ is any point, then for each ray r emanating from A′ there is a *unique* point B′ on r such that B′ ≠ A′ and AB ≅ A′B′.

AXIOM C-2. If AB ≅ CD and AB ≅ EF, then CD ≅ EF. Moreover, every segment is congruent to itself.

AXIOM C-3. If A * B * C, A′ * B′ * C′, AB ≅ A′B′, and BC ≅ B′C′, then AC ≅ A′C′.

AXIOM C-4. Given any angle ⊀BAC (where by definition of "angle" $\overrightarrow{AB}$ is not opposite to $\overrightarrow{AC}$), and given any ray $\overrightarrow{A'B'}$ emanating from a point A′, then there is a *unique* ray $\overrightarrow{A'C'}$ on a given side of line $\overleftrightarrow{A'B'}$ such that ⊀B′A′C′ ≅ ⊀BAC.

AXIOM C-5. If ⊀A ≅ ⊀B and ⊀A ≅ ⊀C, then ⊀B ≅ ⊀C. Moreover, every angle is congruent to itself.

AXIOM C-6 (SAS). If two sides and the included angle of one triangle are congruent respectively to two sides and the included angle of another triangle, then the two triangles are congruent.

CONTINUITY AXIOMS

DEDEKIND'S AXIOM. Suppose that the set {l} of all points on a line l is the disjoint union $\Sigma_1 \cup \Sigma_2$ of two nonempty subsets such that no point of either subset is between two points of the other. Then there exists a unique point O on l such that one of the subsets is equal to a ray of l with vertex O and the other subset is equal to the complement.

ARCHIMEDES' AXIOM. If CD is any segment, A any point, and r any ray with vertex A, then for every point B $\neq$ A on r there is a number n such that when CD is laid off n times on r starting at A, a point E is reached such that $n \cdot CD \cong AE$ and either B = E or B is between A and E.

ARISTOTLE'S AXIOM. Given any side of an acute angle and any segment AB, there exists a point Y on the given side of the angle such that if X is the foot of the perpendicular from Y to the other side of the angle, XY > AB.

PARALLELISM AXIOMS

HILBERT'S PARALLEL AXIOM FOR EUCLIDEAN GEOMETRY. For every line l and every point P not lying on l there is at most one line m through P such that m is parallel to l.

EUCLID'S FIFTH POSTULATE. If two lines are intersected by a transversal in such a way that the sum of the degree measures of the two interior angles on one side of the transversal is less than 180°, then the two lines meet on that side of the transversal.

HYPERBOLIC PARALLEL AXIOM. There exist a line l and a point P not on l such that at least two distinct lines parallel to l pass through P.

SYMBOLS

P **ɪ** *l*   P is incident with *l* (p. 12)
{*l*}   the set of points lying on line *l* (p. 13)
AB   segment with endpoints A and B (p. 14)
$\overleftrightarrow{PQ}$ line through P and Q (p. 14)
≅   congruent (p. 15)
$\overrightarrow{AB}$ ray emanating from A through B (p. 16)
∢A   angle with vertex A (p. 17)
l ‖ *m*   line *l* parallel to line *m* (p. 19)
l ⊥ *m*   line *l* perpendicular to line *m* (p. 27)
△ABC   triangle with vertices A, B, C (p. 28)
□ABCD   quadrilateral with successive vertices A, B, C, D (p. 29)
S ∪ *T*   union of *S* and *T* (p. 30)
S ∩ *T*   intersection of *S* and *T* (p. 30)
H ⇒ *C*   statement *H* implies statement *C* (p. 40)
RAA   reductio ad absurdum method of proof (p. 42)
~*S*   negation of statement *S* (p. 44)
~   equivalence relation (p. 59)
$\mathscr{A}^*$ projective completion of $\mathscr{A}$ (p. 59).
A * B * C   point B is between points A and C (p. 73)
AB < CD   segment AB is smaller than segment CD (p. 88)

⊰ABC < ⊰DEF	angle ABC is smaller than angle DEF (p. 91)
$n \cdot CD$	segment CD laid off n times (p. 95)
$(⊰A)°$	number of degrees in angle A (p. 122)
$\overline{AB}$	length of segment AB (p. 122)
δABC	defect of triangle ABC (p. 130)
$\triangle DEF \sim \triangle ABC$	triangle ABC is similar to triangle DEF (p. 152)
$\Pi(PQ)°$	number of degrees in angle of parallelism associated to PQ (p. 197)
[ABCD	biangle ABCD (p. 209)
$r \mid s$	ray r is limiting parallel to ray s (p. 210)
$p(A)$	polar of the point A (p. 220)
A)(B	open chord with endpoints A and B (p. 228)
$P(l)$	pole of the chord l (p. 239)
(AB, CD)	cross-ratio of ordered tetrad ABCD (p. 248)
$d(AB)$	Poincaré length of Poincaré segment AB (p. 248)
R_m	reflection across line m (p. 111, 261)
$d'(AB)$	Klein length of segment AB (p. 268)
I	the identity transformation (p. 311)
H_A	half-turn about A (p. 327)
$\mathcal{P}^1(K)$	projective line over K (p. 352)
$PGL(2, K)$	projective group over K (p. 353)
$PSL(2, \mathbb{R})$	the real projective special linear group (p. 354)
C_n	cyclic group of order n (p. 364)
D_n	dihedral group of order $2n$ (p. 365)
sinh x	hyperbolic sine of x (p. 399)
cosh x	hyperbolic cosine of x (p. 399)
tanh x	hyperbolic tangent of x (p. 399)
$\Pi(x)$	angle of parallelism in radians (p. 402)
$(⊰A'AB')^r$	radian measure of ⊰A'AB' (p. 397)
$k = 1$	normalization of the plane curvature to be -1 (p. 401)
x^*	complementary length to x (p. 412)

NAME INDEX

SUBJECT INDEX

Parallel displacement 326, 337–
338
Parallel projection theorem 170,
174–175, 269
Parallelogram 29, 170, 174, 277
Pencil of lines 340, 455–456
Perpendicular bisectors 28, 31, 422
See also Concurrent lines
Perpendicular lines 27
in Klein model 238–240
Perspectivity 266, 336, 439
Pi (π) 7, 31, 176
Plane 12
absolute 455
affine 58
dyadic rational 107
elliptic 442, 452
Euclidean 102
generalized Euclidean 459
generalized hyperbolic 459
H 454–460
half 77
hyperbolic 452
incidence 382
inversive 284, 355
non-Legendrian 460
projective 58
rational 109
real 67, 114
semielliptic 460
semi-Euclidean 459
semihyperbolic 460
tangent 446
upper half 237
Plane separation axiom 76
Point 11
at infinity 59, 219, 238, 265, 269,
349–357, 376
ideal 212, 231, 238, 334–338,
455, 460
ultra-ideal 219–221, 240–241,
455

Polar 220, 259, 282, 328, 443,
456–457
Polarity 69, 220
Pole 90, 112, 219, 239, 442
Postulate 10
Power of a point 246
Principal curves 446
Projective completion
of an absolute plane, 455–457
of an affine plane, 59–62, 66
of the hyperbolic plane, 219–
221,227
Projective geometry xv, 58–62,
65–69, 219–221, 226, 240,
263–269, 289, 313–314, 336,
352–355, 383–384, 438–443,
455–462
Projectivity 336
Proof 41, 68, 298–301, 303, 306–
307
Pseudosphere 395–397
Pythagorean theorem 7, 26, 43,
114, 171–172, 207, 403
Pythagorean field 459

Quadrilateral 29
convex 127, 141, 170
diagonals of 29, 141
Lambert 160, 165–166, 170,
194, 198, 203, 235, 412–417
Saccheri 155, 160, 164, 170,
192–193, 203, 206, 235, 411
Quantifiers 45–48

Radians 256, 402
Radical axis 280–281, 284
Radius 15
Rays 16
coterminal 106
limiting parallel 195–198, 210,
231, 234